Advances in Physiological Research

Advances in Physiological Research

Edited by
H. McLennan
J. R. Ledsome
C. H. S. McIntosh
and
D. R. Jones

University of British Columbia
Vancouver, Canada

Plenum Press • New York and London

Library of Congress Cataloging in Publication Data

International Union of Physiological Sciences. Congress (30th: 1986: Vancouver, B.C.)
Advances in physiological research.

Proceedings of the 30th Congress of the International Union of Physiological Sciences held June 13–19, 1986 in Vancouver, Can.
Includes bibliographical references and index.
1. Physiology—Congresses. I. McLennan, Hugh, date. II. Title.
QP1.I5715 1986 599.01 87-14181

ISBN-13: 978-1-4615-9494-9 e-ISBN-13: 978-1-4615-9492-5
DOI: 10.1007/978-1-4615-9492-5

Proceedings of the 30th Congress of the International Union of Physiological Sciences, held June 13–19, 1986, in Vancouver, Canada

Softcover reprint of the hardcover 1st edition 1987
A Division of Plenum Publishing Corporation
233 Spring Street, New York, N.Y. 10013

PREFACE

The XXX Congress of the International Union of Physiological Sciences took place in Vancouver, Canada, in July 1986; and this Volume contains a selection of the Review Lectures which formed part of the Congress programme. They cover most of the areas of contemporary physiology and were presented by distinguished scientists from many parts of the world.

An innovation at this Congress was the inclusion in the programme of a number of lectures devoted to historical subjects. These lectures attracted large audiences at the meetings, and three of them also appear in this book.

Finally, the Plenary Lectures which formed part of the opening and closing ceremonies and which deal with some aspects of Canadian physiological history, find a place here as well.

The Editors are grateful to all of the authors who provided the manuscripts which go to make up this book, and to John Donald for his invaluable assistance in the preparation of the final text; as well of course to all of the contributors whose participation made the Congress the success which it was. It is hoped that this Volume will provide a useful memento of the event.

H. McLennan
J.R. Ledsome
C.H.S. McIntosh
D.R. Jones

CONTENTS

OPENING PLENARY LECTURE: PIONEERING IN PHYSIOLOGY IN BRITISH COLUMBIA

D. Harold Copp

Department of Physiology
University of British Columbia
Vancouver, B.C. V6T 1W5 CANADA

Mr. Chairman, honoured guests, fellow physiologists and friends. I appreciate this opportunity of addressing the opening session of the XXX Congress of Physiological Sciences, and I would like to add my welcome to those of my many old friends who are present on this platform. I have chosen as my title "Pioneering in Physiology in British Columbia" and will speak briefly of our City and University and then in more detail of my experience as a pioneer physiologist in the Department which I started in 1950 and in research in the field of calcium metabolism.

VANCOUVER

There were Indian villages in this region for centuries before the white man arrived in 1865 to log the tall straight Douglas Firs which were ideal for the masts and spars of sailing ships. Indeed, Stanley Park and part of Point Grey had been set apart as naval reserves for this purpose. In 1886, the sleepy sawmill town of Granville became the western terminus of the CPR, and with the arrival of the first transcontinental train, it was incorporated as the City of Vancouver. With typical exuberance, the City was burned down shortly thereafter and City Hall became a tent. The City Fathers had the wisdom, even amidst this disaster, to set aside the naval reserve as Stanley Park.

UNIVERSITY OF BRITISH COLUMBIA

From the time that the Province of British Columbia joined Canada in 1871, there was pressure to set up a Provincial University, and the University of British Columbia was incorporated by an Act of the Legislature in 1908. Land was cleared for the permanent campus on the magnificent site on Point Grey and a start was made on a science building when World War I intervened. Classes were begun in some old shacks near the Vancouver General Hospital which had previously been used by McGill College of B.C. After the war, there was increasing pressure to move to the permanent site at Point Grey, culminating in massive demonstrations and the Great Trek to the unfinished science building. The government responded promptly, the building was finished, and classes moved to the campus in 1925. However, the Great Depression hit B.C. hard and there was talk of closing the University. At that time, there were only two permanent buildings and a number of temporary huts. Following the Second World War, the University was faced with the challenge of the returning veterans who swelled the enrollment from 3,000 to 10,000 in a single year. This challenge was met by three great pioneers -- Norman McKenzie, the president, a man of vision who could see the future needs of the University and the Province; Gordon Shrum, a man of action who, without bothering to get permission from the Army or the Government in Ottawa, brought dozens of abandoned army huts to the campus to provide classrooms and living quarters for faculty and students; and Walter Gage, a man of heart who was devoted to the welfare of students and faculty.

DEPARTMENT OF PHYSIOLOGY

There was growing pressure to start a medical school and in 1949 Myron Weaver -- a physiologist and Associate Dean of Medicine at Minnesota was appointed Dean -- working out of an office at the end of a corridor in Shrum's Physics Department. The next year I was invited to head the Physiology Department in the new school. My reluctance to leave the University of California was overcome by my disgust with the loyalty oath which had just been imposed. I arrived on Labour Day, just before the medical class was to register. After depositing my family in a one bedroom hut in what was affectionately referred to as "Shrum's slum", I hastened over to Hut B6. This hut, which was shared with Anatomy and Pharmacology, was to house the Department for the next 12 years. I found to my dismay that none of the equipment which I had ordered was there -- not even a test tube. In those days, smoked drum kymographs were essential to a Physiology laboratory and I found that the Palmer company would not promise delivery for two years. I cabled them to deliver in six months or cancel. They delivered. I canvassed my friends in Canadian Departments of Physiology and finally obtained a loan of 30 retired Harvard kymographs from Reg Haist at the University of Toronto. Meantime, I had persuaded a sympathetic dean to postpone physiology classes for six weeks. I took advantage of this delay to visit a number of the leading North American Departments to learn how to teach Physiology and run a Department. I recall that I was allowed $5 a day for hotels and $5 for meals and taxis. It was an exciting experience, and I received help and encouragement from some of the great physiologists of the day -- Wallace Fenn in Rochester, Homer Smith in New York, John Fulton and C.N.H. Long at Yale, Hank McIntosh at McGill, Charles Best and Reg Haist at Toronto, Joe Doupe at Manitoba and Louis Jacques in Saskatoon. I returned inspired and ready to face the onslaught of 60 eager medical students with the aid of Edgar Black, a distinguished fish physiologist who had been recruited by Dean Weaver from the Biology Department. A dour Scot seven years my senior, I am sure he had some misgivings about my competence to head the department and was heard to mutter doubts about that "young whippersnapper from California". During that first year I gave most of the lectures, set up the laboratory and acted as secretary, technician and demonstrator. It was a busy year but we did get by on a budget of $18,000. Since many of our laboratories in those days depended on the use of pound dogs, I approached David Ricardo, executive secretary of the Vancouver SPCA, and he agreed that the Society would not oppose our approach to City Council to purchase pound dogs for the Department provided that they were treated humanely, were fully anaesthetized when experimented on and were disposed of painlessly afterwards. I in turn invited Dave to come to the Department at any time, unannounced, and to give a lecture to our students on the humane care of animals. The Department has continued its good relations with the SPCA and is currently involved in a research project with the Society and the University.

In 1952, on the advice of our Congress president Hank McIntosh, I recruited John Honour who had been Sir Thomas Lewis' right-hand man at University College London. John brought his great wealth of technical and laboratory experience to the Department. He also trained Kurt Henze, who so impressed me when I met him at a dog training class that I appointed him as departmental technician. Over the years, Kurt has been responsible for the smooth technical operation of the Department and student laboratories. I recall one evening when I stopped by the hut after a formal dinner party and found the parking lot jumping with frogs. Kurt came out immediately and we spent the rest of the night in a great frog roundup. In 1954 we received approval for a new faculty member and I appointed Carl Cramer whom I had known at the University of California. In 1957 we recruited Hugh McLennan from Dalhousie who has become a very distinguished neurophysiologist and is chairman of the Programme Committee for the Congress. In 1960 funds became available for construction of permanent buildings for the Basic Medical Sciences. It was ironic that the day we left our very inflammable hut and moved into our new building in 1962, someone set fire to the rubbish next to the air intake and the building had to be evacuated by the Fire Department. With our vastly improved facilities we were in a position to recruit new faculty and in 1965 I persuaded John Brown to join the Department. As the discoverer of two new hormones, GIP and motilin, he is a leader in the field of gastrointestinal hormones and organized the very successful Sixth International Symposium in this field which was held here last week. Three years later I recruited John Ledsome, an outstanding cardiovascular physiologist, who succeeded me as Head of the Department when I retired in 1980. He has been largely responsible for bringing this Congress to Vancouver and is chairman of the local Organizing Committee. As I look back on the last 36 years I take some pride in the distinction of those I appointed to this Department, including six Fellows of the Royal Society of Canada and one Fellow of the Royal Society of London. I should note that, with one exception, none have left. They must like it here. I also take some pride in the fact that we have one of the largest Honours programmes in the University and there is a marvellous esprit de corps among our graduate students.

CALCIUM REGULATION

I would now like to turn to another area of pioneering -- bone and calcium metabolism, including the hormones involved in its regulation. While I was at the University of California, I had worked on the problem of removing strontium-90 and plutonium from bone. I was not very successful in this endeavour, although I did show that strontium was removed from young rats fed a phosphate-free diet -- unfortunately, the calcium and bone disappeared too. This work on bone and calcium metabolism sharpened my interest in this neglected field and I was very pleased after coming to Vancouver when I was invited to participate in the Macy Conferences on

Metabolic Interrelations, with special reference to calcium. This select group of 25 represented most of those working in the field, and included such leaders as Baird Hastings, Joe Aub, Fuller Albright and Franklin McLean. When these conferences ended in 1953, I organized a small group meeting on Bones and Teeth at the 19th I.U.P.S. Congress in Montreal. After the dinner, a group of us "young turks" repaired to a pub on Peel Street to bemoan the end of the Macy Conferences and to plan something to take their place. Out of this grew the highly successful Gordon Conferences on Bones and Teeth. At the first meeting, organized by Felix Bronner, I had rashly agreed to give a keynote talk on the Homeostasis of Calcium and Phosphorus. I soon realized that I knew very little about the subject, and indeed, little was known. It was a virgin field -- always attractive to a young scientist. Everyone conceded that the level of extracellular calcium ion was critical to many physiological processes, including neuronal excitability, muscle contraction, membrane permeability, hormone release, enzyme activation and bone formation. The problem was that the standard Clark-Collip method for calcium determination had a variability of ±20-25%. In a laboratory in Hut B-6, we adapted the method of "Dynamite" Campbell of Toronto, which did measure plasma calcium precisely. It depended on titration of calcium with EDTA, using purpurate as indicator. With a grant of $4,000 from NRC, we could not afford expensive equipment. I borrowed a Klett colourimeter from the student laboratory and plotted the titration on graph paper. A critical part of the equipment was the eraser which made it possible to use the same piece of graph paper over and over. With this technique we were able to demonstrate the remarkable constancy of the plasma calcium in a series of male subjects at Shaughnessy hospital, even though there were wide fluctuations in plasma phosphate and calcium excretion.

CALCITONIN

Discovery

In normal dogs, we found that the plasma calcium level was rapidly restored after it had been raised by calcium infusion or lowered with EDTA. This precise control was lost after thyroparathyroidectomy. The failure to control hypocalcaemia was easily explained by the absence of the parathyroids, but there was no explanation for failure to control hypercalcemia. We now know that this was due to the removal of the calcitonin-producing C cells. We had actually obtained evidence for such control in 1958. A series of dogs were perfused intravenously with 1U/kg/h of parathyroid extract to simulate mild hyperparathyroidism (the dose was approximately 10 times the normal resting secretion rate). We observed a modest increase in calcium which persisted after the infusion was stopped. However, in a second group which were thyro-parathyroidectomized at the end of the perfusion, the plasma calcium shot up, and it was clear that the operation had removed a control of hypercalcaemia. I noted this in my lab book, but was not clever enough to appreciate its significance. In any case, I had just been appointed to a committee which met frequently in Ottawa to discuss government support of medical research and eventually led to the establishment of our Medical Research Council. Three years later, with the assistance of two medical students, we studied the effect of perfusing the thyroid-parathyroid apparatus of dogs with high and low calcium blood to determine whether the calcium level controlled the secretion of parathormone. We found, as expected, that a low level of calcium in the perfusate caused release of parathormone. However, we were surprised by the speed with which the plasma calcium fell during high calcium perfusion and realized that this could not have been due to suppression of the parathyroids since the plasma calcium rose when we removed the glands, just as it had in our experiments in 1958. When we found that the high calcium perfusates lowered calcium when they were injected into another dog, I realized that we had discovered a new hormone which I named "calcitonin", since it appeared to be involved in regulating the level or "tone" of calcium in the body fluids. I was naturally excited by this discovery, but felt that calcitonin could not be very important or it would have been discovered long ago. My colleagues obviously agreed, for they generally referred to it as "Copp's folly", and when Tony Care was actually convinced, we made him a charter member of the Calcitonin Club and a TRUE BELIEVER. In 1963 Iain MacIntyre at the Postgraduate Medical School in London actually confirmed our results, much to his surprise. In the same year Paul Munson who had read our paper while preparing an article on the parathyroids for Physiological Reviews, decided that calcitonin might explain some results which had been obtained in his laboratory a few years earlier. They had been parathyroidectomizing rats by hot wire cautery when a minor ether explosion prompted them to turn to surgical removal. To their surprise, the fall in plasma calcium was much less than that following cautery, and Paul speculated that this might have been due to release of calcitonin. Sure enough, they found that simple acid extracts of rat and hog thyroids had a dramatic hypocalcaemic effect when injected into young rats and they named their preparation "thyrocalcitonin" to indicate the gland of origin and its possible identity to calcitonin. For the next few years, this was the commonly accepted name for the hormone.

C cells

In 1966 the father of histochemistry, Tony Pearse (who is attending this Congress) showed that the cells which respond to hypercalcaemia were not the regular follicular cells of the thyroid but were in fact the parafollicular cells described by Nonidez which stain with silver, contain high levels of serotonin and most significantly, contain immunoreactive calcitonin. Tony called these "C" cells, to stand for calcitonin, or Canada or Copp. He also showed that they were derived from the ultimobranchial body of the embryo and ultimately from the cells of the neural crest as part of the neuroendocrine system. I was intrigued by these observations, since the ultimobranchial gland, derived from the last branchial pouch, persists as a separate gland in lower vertebrates. It has also been associated with calcium metabolism. As soon as I returned to Vancouver from my visit to Tony's lab, I rushed out to our Poultry Department to look for ultimobranchials. They are not hard to find if you first locate the brown oval thyroids in the neck, because below it lie the two parathyroids and finally the ultimobranchial. I tested the glands and found that they were loaded with calcitonin, while the chicken thyroid had none. It was clear that in lower vertebrates, calcitonin is an ultimobranchial hormone, and I was tempted to call it ultimobranchial calcitonin, which could be abbreviated as U.B.C. These observations brought a lot of new animals into the calcitonin race, and we proceeded to collect ultimobranchials from turkeys, dogfish and salmon. In all the species tested we found high levels of calcitonin.

Chemistry

I would now like to turn to the chemistry of the hormone and pay tribute to Paul Munson who prepared the first active biological extract and John Potts, whose laboratory has been responsible for the determination of the amino acid sequence of six calcitonins, including porcine calcitonin which was the first to be so characterized. The basic structure is similar to that of vasopressin, with a disulfide ring at the N terminus and an amide at the C terminus. We were naturally interested in ultimobranchial calcitonin, particularly since John Potts had said that it would never be possible to collect sufficient glands to isolate the hormone and determine its structure. This was a challenge to someone living in Vancouver, and I arranged to collect them from salmon, where they are located in the transverse septum just above the oesophagus. With the cooperation of Don Miller, who was Chancellor of the University and president of the Canadian Fishing Company, we organized an ultimobranchial plucking line and in the summer of 1967 we collected 100 kg of glands from approximately half a million salmon. In a remarkable example of scientific cooperation, the Armour company bulk-processed the glands, a graduate student in my laboratory, Ron O'Dor, isolated the pure hormone, John Pott's group in Boston determined the amino acid sequence, and the Sandoz company in Basel synthesized it -- all within 4 months. It was found that the salmon CT was 10-100 times as potent as the human hormone. This means that the cost of producing a unit of synthetic salmon calcitonin is 1/10 to 1/100 that of human or porcine CT. At present it is the form used world wide in the treatment of disease, with the exception of Japan, where an equally potent modification of eel calcitonin is used. The molecule has an ancient lineage and, using antibodies to human calcitonin, it has been found in the brains of primitive chordates, and even in unicellular organisms such as C.albicans and E.coli.

Gene Family

The other exciting development in the past 5 years has been the isolation by Rosenfeld and his colleagues of the calcitonin gene in rats and man, and the demonstration that it encodes for calcitonin and katacalcin in the thyroid C cells, and for calcitonin-gene-related-peptide (CGRP) in the nervous system. The latter resembles calcitonin in some respects, since there is a disulfide ring at the N terminus and an amide at the C terminus. It is also hypocalcaemic in young rats, although at much higher dose levels. An interesting observation is that CGRP lowers blood pressure and has a positive inotropic and chronotropic effect on the heart. It also appears to be released by cells in the posterior root ganglia with a distribution similar to that of substance P.

Action and Function

To return to calcitonin, its classical effect is to block bone resorption. Suppression of bone resorption explains the fall in plasma calcium and phosphate and the control of hypercalcaemia particularly in young animals and in patients with active osteolytic bone disease. In mammals, calcitonin appears to serve three functions: it controls hypercalcaemia, especially in young animals; it protects the skeleton during periods of calcium stress such as pregnancy and lactation; and it appears to prevent hypercalcaemia after ingestion of a calcium-rich meal -- especially important for cheese lovers. The secretion of both parathormone and calcitonin is controlled by the plasma calcium level through a very effective feedback mechanism, so that they serve as a kind of calciostat. A fall in plasma calcium turns on parathormone which increases bone resorption and raises the level; a rise releases calcitonin which has the opposite effect.

Extraosseous Effects

We now know that calcitonin also has many extraosseous effects especially at high "pharmacological" dose levels. In the kidney it has a calciuretic, natriuretic and diuretic effect at a dose 1/300 of that of furosimide. In the stomach it suppresses gastric secretion and experimental ulcer formation and it is also anorexic through an inhibitory effect on the appetite centre in the hypothalamus. However, its most intriguing effects are in the central nervous system where calcitonin receptors have been demonstrated in cells of the limbic system and in the periaquaductal grey matter which is involved in pain pathways. This explains the powerful analgesic effect of the hormone first observed in patients treated for Paget's disease. When injected intracerebrally it increases the levels of beta-endorphin, and also has an analgesic effect which is not blocked by naloxone and is independent of endorphins. On a weight basis, it is 30-50 times as effective as morphine without the problems of developing tolerance or addiction (except possibly in my own case). Currently it is undergoing clinical trials for potential use in relieving the intractable pain of terminal bone cancer, with some encouraging results.

Clinical Effects

I would now like to turn to the clinical uses of this hormone, for which I originally thought there would be no practical applications. There is no doubt that the radioimmunoassay for calcitonin provides an ideal marker for medullary carcinoma of the thyroid -- a cancer of the C cells associated with very high blood levels of CT. Since this is an inherited disease, it is particularly useful for screening family members to detect the cancer in its early stages. It has also been shown to have considerable therapeutic application. I was intrigued by an advertisement of the Greyhound Corporation telling how a school of salmon had taught them how to treat bone disease. (Greyhound had taken over calcitonin along with the Armour Company, and has in turn been taken over by Revlon). Calcitonin is indeed the treatment of choice in Paget's disease -- a bone disorder occurring in perhaps 5% of the population over 50. It acts to restore the bone to normal and to relieve the bone pain. As might be expected from its physiological role, it is also useful in reducing hypercalcaemia resulting from hyperparathyroidism or bone cancer. There is also increasing evidence that it may be involved in postmenopausal osteoporosis -- that crippling and I believe preventable disorder resulting from the rapid bone loss which occurs after the cessation of ovarian function. This results in very thin bones and fractures which occur spontaneously in the spine, wrist and hips. It is estimated that this disease costs the North American health care system up to $3 billion a year, and there is now general agreement that it can be largely prevented by cyclical low dose replacement of estrogen and progestin to mimic the normal menstrual cycle, along with exercise and adequate calcium intake. There is a sharp drop in calcitonin levels after the menopause and it has been suggested that estrogen replacement, by increasing calcitonin, may protect the skeleton as it does during pregnancy and lactation. It is significant that black women have high levels of calcitonin and do not get osteoporosis. Calcitonin has now been approved for use in treatment of this condition and early results of clinical trials are encouraging. Baylinck at the University of Washington has shown a positive calcium balance and a small but significant increase in total bone mass in patients treated with the hormone. In studies in Italy, treatment resulted in a significant reduction in the incidence of fractures in postmenopausal women.

CALCIUM REGULATION IN FISH

Since my retirement six years ago, I have become involved in another pioneer field -- that of calcium regulation in fish. It is a new field, with perhaps a dozen participants, and I have been involved in the organization of four workshops, including one which will be held in southern France next October. In contrast to land vertebrates, in which calcium homeostasis depends primarily on bone and kidney, in fish -- even in fresh water -- there is an ample supply of calcium in the water flowing past the gills. Regulation depends primarily on control of the calcium uptake by the gills, and involves prolactin from the pituitary and a hormone from the corpuscles of Stannius, endocrine glands which are unique to fish. We have named it TELEOCALCIN because of its involvement in calcium metabolism in teleost fish and have shown that it inhibits gill uptake of calcium. We have isolated it in what we believe to be a pure form and have obtained a partial amino acid sequence which appears to be unique. Now thanks to a new grant which I recently received from the Medical Research Council of Canada, we are prepared to press ahead.

CONCLUSION

As I look back on my career, I realize how lucky I was to be involved as a pioneer in the development of the Department of Physiology here, and in the discovery and development of calcitonin. I am very grateful for the opportunities I have had and am proud of what has been achieved.

CLOSING PLENARY LECTURE: MICHEL SARRAZIN, 1659-1734

Hugh McLennan

Department of Physiology
University of British Columbia
Vancouver, B.C. V6T 1W5 Canada

At this the close of the XXX International Physiological Congress and the second to be held in Canada, the Organizing Committee felt that it might be appropriate to look back in history to the beginnings of biological research in this country, to the man whom the Canadian Physiological Society regards as the father of Canadian physiology. Indeed the Society sponsors an annual lecture at its winter meetings which is named for him, and in 1985 I had the honour to present the lecture from which my remarks today are culled.

The remarkable person of whom I speak was named Michel Sarrazin (Fig. 1), who was born in 1659 in the Côte d'Or region of Burgundy, the son of the bailiff of the Abbaye des Citeaux. We know nothing further about him until in 1685, at the age of 26, he arrived in Québec which was the capital of New France with the title of "surgeon". In those days this of course did not imply either a lengthy or a rigourous training, but there is little doubt that during and following the crossing a person with any medical experience would be badly needed.

The fleet carried the new Governor of the colony, the Marquis de Denonville, together with a contingent of 6-700 new troops. Parenthetically another passenger was the new Bishop of Québec, the Abbé de St.-Vallier who had been described by a contemporary as "too young, too much of a perfectionist, and too austere" but who nevertheless had been named Bishop because of his aristocratic origins. He indeed turned out not to be too popular: his predecessor Bishop Laval who remained in Québec,

Fig. 1. The only known portrait of Michel Sarrazin, from the archives of the Hôtel-Dieu de Québec.

came to refer to him as "a scourge sent by God to punish everyone for their sins". M. de St.-Vallier will briefly appear again in due course.

To come back to the convoy however, as was expected in those days soldiers had a far better chance of surviving a war against the Indians, the English, the Spaniards or anyone else than of making it alive across the Atlantic. True to form, on one of the ships 60 of 140 men had died of typhus or scurvy during the voyage; and by the time they had all landed there were 300 desperately sick men to be cared for. There was only one hospital in Québec at the time, the Hôtel-Dieu du Précieux Sang, which had been founded in 1639 and possessed only 50 beds divided into wards for men, women, officers and the "mentally afflicted". It is perhaps not surprising that Sarrazin's first official job was to be impressed into service as Surgeon-Major to the troops of the colony, an appointment which was confirmed five years later by royal warrant and which he retained to the end of his life.

How he and his colleagues, with the nuns who staffed the hospital coped with this emergency is not recorded; but Sarrazin pops up in the historical record every now and then over the next few years. In 1693 he was appointed to the staff of the hospital in recognition of his services not only to the military but to the inhabitants of the entire colony which included the settlements of Trois Rivières and Montréal which were situated respectively 140 and 270 km. up river from Québec. Medical forays into the hinterland may have been acceptable in summer when travel by water was possible; but can only have been formidable to face in winter. Nevertheless he was reported as a visiting surgeon in Montréal in December of 1692 (he was a patient in the Hôtel-Dieu de Montréal, apparently suffering from exhaustion), so that one must presume such journeys were not uncommon.

Thus far his career probably resembled that of many other barber-surgeons throughout the French possessions; but at about this time (1692-93), Sarrazin made a courageous decision. I have mentioned that his training as a surgeon would not have been extensive; by contrast the education of a physician was much more rigourous. Although it was still rooted in the teachings of Galen and the philosophical concepts of the Middle Ages, medical training in the late 17th century was in fact becoming more scientific as a consequence of the work of Harvey, Sydenham, Malpighi and others; and thanks to the patronage and support of the King, the Ecole de Médecine in Paris was at the forefront of modern thinking. There was even a regularly published medical research journal entitled "Le Progrès de la Médecine". There Sarrazin enrolled himself in 1693 for what was normally a seven-year course, which he successfully completed in four.

There was however another particularly significant occurrence during Sarrazin's sojourn in Paris which, one suspects, was seminal for his development into a true biologist. One of his fellow students was a man named Joseph Pitton de Tournefort who was a botanist but who also completed his medical studies in 1698 and shortly thereafter was appointed Professor of Botany at the Jardin Royale des Plantes and also became a member of the Académie Royale des Sciences. Both of these organizations, the Garden and the Academy, received much encouragement from Louis XIV; and there was a close connection too between the Garden and the Faculty of Medicine. Indeed three professors of the Faculty, including presumably de Tournefort, were seconded to the Garden to teach botanical pharmacology. To Sarrazin's relationship with de Tournefort I shall return shortly.

On September 8, 1697, Sarrazin returned to New France as its first physician, arriving as usual with shiploads of desperately ill passengers. One among them on this crossing was Bishop de St.-Vallier, and the disease from which they were suffering was named "fièvre pourprée" which was either typhus or a form of plague: that is was contagious is evidenced by the fact that eight of the nursing sisters also contracted it. It is a testament to Sarrazin's skill that to a condition which in the past had always had a heavy mortality he lost not one patient, not even the unpopular Bishop, which inspired the Mother Superior of the Hôtel-Dieu to writer to the Governor: "Que Dieu bénisse un si sage, si vigilant et habile médecin!", to which she added the fervent wish that he might be granted a royal salary so that he could be persuaded to remain in the colony. In fact that wish was shortly after granted.

I do not intend to dwell too long on Sarrazin's clinical and other responsibilities -- among other things about this time he became a member of the Supreme Council which advised the Governor of the colony -- for this talk after all is intended to extol him as a biologist. On March 4, 1699 he was named a corresponding member of the Académie des Sciences, the same day that Sir Isaac Newton was made a foreign member, interestingly enough; and de Tournefort whom he had known in Paris was the recipient of the correspondence. In 1704 he sent to de Tournefort at the Jardin Royale a collection of 200 living specimens of unique Canadian plant species with a complete catalogue of their characteristics and possible usefulness, and with the requirements for their cultivation and propagation. Ten years later all of his plants were alive and flourishing in Paris. One of the novel species which he sent was the North American pitcher plant, now the floral emblem of our Province of Newfoundland, which de Tournefort named "Sarracena canadensis" in his honour. (It has since been renamed "Sarracena purpurea"). De Tournefort wrote: "Sarracena is named after the distinguished Dr. Sarrazin, medical doctor, anatomist and botanist of the King, who examined this plant, described it and graciously sent it to me from Canada".

Sarrazin's interest in plants continued throughout his life. In 1730 he sent a communication to the Academy on the four varieties of sugar maple and on the conditions under which maximum yields of sugar were produced, which was published in the Transactions; but perhaps his most remarkable if at the time seemingly unappreciated study was a practical and experimental one. The Supreme Council was concerned with agriculture in the colony, and in 1715 asked Sarrazin to look into the whole question of the growth and yield of wheat and other grains. One of the problems of course was the short growing season, and Sarrazin's solution was to obtain samples from another northern country, Sweden, of so-called winter wheat and rye which are sowed in the autumn. His experiment produced the next year "a greater amount of fine flour than summer wheat", but apparently the approach was too revolutionary and it did not catch on -- only with the opening up of the Prairie wheatfields a century and a half later was Sarrazin's solution shown to be an appropriate one.

His scientific interests however were not exclusively botanical. In 1704 he provided the Academy through de Tournefort with a detailed anatomical description of the beaver with special emphasis on its unique features as a diving animal. This was followed after de Tournefort's death in 1708 with communications to the Abbé Bignon, who was President of the Academy when de Tournefort and Sarrazin became associated with it, and to M. de Réaumur, which dealt with such other Canadian animals as the seal, the wolverine and the "loup-marin". He observed the seven different kinds of skin possessed by the porcupine, and speculated whether this animal really could throw its quills when assailed (he concluded that it probably could not). However his zoological masterpiece was presented to the Academy in 1725 through de Réaumur, and is a description of the muskrat illustrated by 16 drawings of excellent clarity which he made himself, and in which he comments in detail on the changes in the stomach which take place during digestion, and of the different effects elicited by summer versus winter diets. He complained in a letter of the difficulty of obtaining and preserving fresh specimens for examination -- they were often half rotten, he says -- so although he did try it is perhaps not surprising that the skunk defeated him, because "it had a dreadful smell, capable of making a whole parish desert".

In 1712 Sarrazin finally found the time to marry and raise a family of seven children, four of whom survived him. Thanks to his royal salary, by that time considerably augmented, and the property brought to him by his wife after her father's death he became quite a rich man; but his forays into the world of property development were not successful and although he retained the respect of the community and his place on the Council, towards the end of his life he became practically a pauper. It was perhaps fated that when the ships arrived in the autumn of 1734 he would contract "une fièvre maligne" from his patients. He was admitted to the Hôtel-Dieu on September 6 and died two days later in his 75th year, to be buried with scant ceremony in the pauper's cemetery.

We will leave Michel Sarrazin with these words which the Governor wrote after his death:

> L'Académie des Sciences avec laquelle il a été en correspondence pendant de longues années pour des recherches de botanique et d'anatomie, lui a donné souvent des preuves de son estime. Il a servi le Roi dans les hôpitaux avec un zèle et une application peu ordinaire. Ses bonnes qualités, ses moeurs irréprochables, l'ont fait aimer pendant qu'il a vécu en ce pays et regretter après sa mort plus que nous ne pouvons affirmer.

WALTER BRADFORD CANNON: PHYSIOLOGIST AND CITIZEN OF THE WORLD

A. Clifford Barger

Department of Physiology and Biophysics
Harvard Medical School
25 Shattuck Street
Boston, MA 02115 U.S.A.

Forty years after his death, Walter Bradford Cannon (Fig. 1), an internationally renowned physiologist, organizer of the 1929 International Physiological Congress in Boston and a longtime officer of the American Physiological Society, is still making headlines. The lead article on the front page of the Bozeman Montana Daily Chronicle of September 9, 1985 heralded the discovery by two mountaineers of a note apparently left on the top of Mount Cannon by Walter and his bride, Cornelia (Fig. 2), 85 years ago. The story of the discovery of the note was originally greeted with some skepticism, but Dr. James McMillan, a physiologist in the Department of Biology at Montana State University in Bozeman, suggested that a facsimile of the note be sent to me for verification of the handwriting. There is little doubt concerning the authorship of the note -- it is Cannon's.

On their honeymoon in 1901 Cornelia and Walter Cannon explored the United States forest preserve now known as Glacier National Park. As they crossed Lake McDonald they saw the impressive peaks of Goat Mountain (Fig. 3) which, they were told, had not yet been climbed. They decided that they would try to reach the summit. The next day, with a local guide, they began their ascent, and after some terrifying experiences, reached the top. In his autobiographical memoir, The Way of an Investigator[1], Cannon wrote:

1. W.B. Cannon, The Way of an Investigator. Norton: New York, 1932, pp 22-27.

Fig. 1. Portrait of Walter Bradford Cannon painted by Marie D. Page, 1930.

Fig. 2. Cornelia James shortly before her marriage to W.B. Cannon in 1901.

Fig. 3. Photograph of Mt. Cannon (Courtesy of Dr. James McMillan).

> The glory of the view disclosed to us was hardly to be imagined. The mountain we had scaled, situated slightly west of the main range of the Rockies, allowed us to look away north and south for perhaps a hundred visible miles in either direction. The snow capped peaks, the glaciers, the lakes, the water falls, all gleaming in the sunshine . . .
>
> That we might leave evidence of our climb we wrote on a scrap of paper a brief account of it and the date, and put the record [Fig. 4] in a small bottle. Over it we built a cairn of flat stones. So far as we are aware, no one has climbed the mountain since that time. It is quite possible that through the intervening decades the paper has disintegrated. The cairn should be there, however, and the bottle under it.

When the Cannons descended to the base of the mountain they met a United States government geological survey team. One of the members of the team took notes as they described their climb to the top of the mountain. Several years later the peak was renamed Mount Cannon in honour of the first couple to scale it. This event was only the first of Cannon's many pioneering feats. Cannon would prove to be a many-faceted pioneer: in research on gastrointestinal motility, on the role of emotional factors in the disturbances of physiological processes, the emergency function of the autonomic nervous system, homeostasis and chemical transmission of nerve impulses. He was also a pioneer in medical education with the development of the "case method" of teaching (now continued as the weekly "Cabot" cases

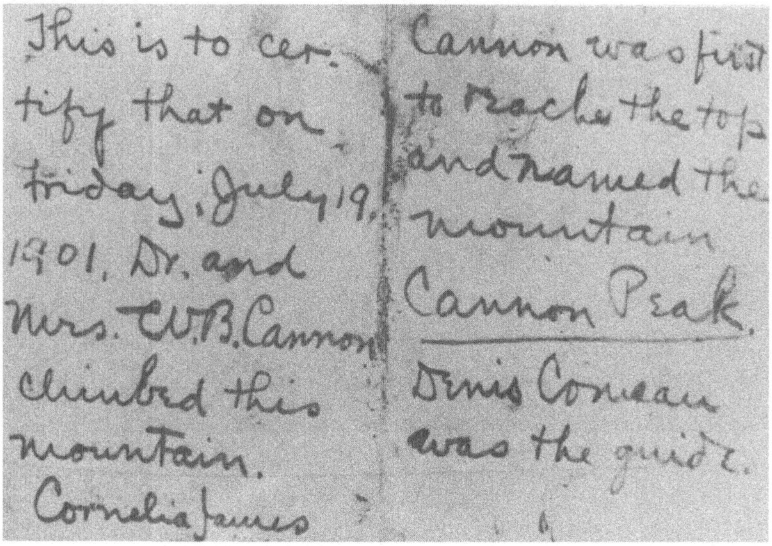

This is to cer-
tify that on
friday, July 19,
1901, Dr. and
Mrs. W.B. Cannon
climbed this
mountain.
Cornelia James
Cannon was first
to reach the top
and named the
mountain
Cannon Peak.
Denis Comeau
was the guide.

Fig. 4. Note left by Cornelia and Walter Cannon on top of Goat Mountain, July 19, 1901.

Fig. 5. William Beaumont (1785-1853).

Fig. 6. Walter Cannon (1871-1945) (left) and Albert Moser (1870-1903) (right) as students at the Harvard Medical School.

Fig. 7. Henry Pickering Bowditch (1840-1911).

in the New England Journal of Medicine), and in defense of freedom for scientific inquiry, both at home and abroad.

Cannon was born on October 19, 1871 at Prairie-du-Chien, Wisconsin, a small town on the upper Mississippi River. It would appear that Cannon's studies on gastrointestinal physiology were predestined, for the town of Prairie-du-Chien contains the site of old Fort Crawford, where, in the 1820s, William Beaumont (Fig. 5) made many of his classic observations on gastric function in his patient with a permanent gastric fistula. Cannon entered Harvard College in 1892 and did his first research with C.B. Davenport on the determination of the direction and rate of movement of unicellular organisms induced by light (Davenport and Cannon, 1897). He graduated in 1896, summa cum laude, with both the A.B. and M.A. degrees. In his first year at the Harvard Medical School he found the many hours of lectures so tedious that he began to search for a more productive way to use his time. Certain that he could master some of the courses without attending all of the lectures, Cannon, with a fellow student of the second year class, Albert Moser (Fig. 6), asked Professor Henry Pickering Bowditch (Fig. 7), the Chairman of the Department of Physiology, for advice regarding a research project in which they might engage. They could not have come at a more propitious time. Bowditch, impressed by the opportunities presented by the recently discovered Röntgen rays, suggested that they use the X-rays, and an opaque contrast medium, to test the concept of deglutition proposed by Hugo Kronecker and Samuel Meltzer -- namely, that the pressure developed by the muscles of the mouth during the act of swallowing was sufficient to force liquids and solids into the stomach without oesophygeal peristalsis. Thus it was Bowditch who had the vision to guide Cannon to the use of X-rays for the study of gastrointestinal motility -- a field in which Cannon's contributions later placed him alongside such giants as Beaumount and Pavlov.

Earlier, however, when Cannon first heard of the discovery of X-rays, he greeted the announcement with some scorn. In her unpublished memoir, Life with a Scientist[2], Cornelia James, later to become Mrs. Cannon, recorded Cannon's introduction to the miracle of X-rays in December 1895:

----

2. C.J. Cannon Life with a Scientist. Unpublished memoir in the Cannon Archives, Countway Library, Harvard Medical School.

> I well remember the first time WBC heard of the X-rays which were to play so important a role in his early investigation and, in the end, to be the cause of his death. It happened when I was in Radcliffe College. We were on a street car together on our way to a Lowell Lecture. In front of us stood a man and a short-haired woman of masculine cast. She was talking to her escort about the "new rays". "They say that you can see right through the body and make out separate bones as if the flesh was transparent!"
>
> We looked at each other with understanding smiles. Some more Boston psychics we thought. But the next day the paper was full of the news of the Röntgen rays, and we recalled our scornful attitudes with chagrin.

On December 9, 1896, a year after Röntgen's discovery, Cannon first used X-rays for the study of gastrointestinal function. Some years later he recalled his introduction to the use of Röntgen rays in a letter to Dr. John Fulton at Yale[3]:

> It was thought best to try first a small dog as a subject and I was commissioned to get a card of globular pearl buttons for the dog to swallow. Dr. Dwight, Professor of Anatomy, and Dr. Bowditch, Dr. Codman and I were the only witnesses. We placed a fluorescent screen over the dog's oesophagus, and with a greenish light on the tube shining below we watched the glow of the fluorescent surface. Everyone was keyed up with intense excitement. It was my function to place the button as far back as possible in the dog's throat so that he would swallow it. Nothing was seen and as the intensity of our interest increased someone exploded: "Button, button, who's got the button?" We all broke out in a sort of hysterical laughter.

Nevertheless, on December 29, 1896, only 3 weeks later, at a meeting of the American Physiological Society in Boston, the phenomenon of deglutition, as exhibited by the goose when swallowing radio-opaque material, was informally demonstrated to the members by the means of the Röntgen rays. This was the first public demonstration of the movements of the alimentary tract in the conscious animal by the use of the new technology. Cannon's first full-length paper on gastrointestinal motility (Cannon, 1898) was entitled "The movements of the stomach studied by the means of the Röntgen rays." This classic paper reported on the movements

---

3. W.B. Cannon to J.F. Fulton April 16, 1942, in Cannon Papers, Countway Library, Harvard Medical School

of the various parts of the stomach in the unanaesthetized animal and included studies on the role of the pyloric sphincter. The last part of the paper, "The inhibition of stomach movements during emotion", is evidence of the kindling of Cannon's interest in neural control of the gastrointestinal tract and in psychosomatic medicine. Cannon noted that:

> ... early in the research an unlikeness was noted in the action of male and female cats. The peristalsis seen with only few exceptions in female cats failed to appear in most of the males, although both had received exactly the same treatment. Along with this difference was a very striking difference in behavior when bound to the holder; the females would lie quiet, mewing occasionally, but purring as soon as they were gently stroked. The males, on the contrary, would fly into a violent rage .... On account of this difference only female cats were used ....
>
> A few days later an observation on a female with kittens explained the absence of gastric movements in the male. While the peristaltic undulations were coursing regularly over the cat's stomach, she suddenly changed her peaceful sleepiness, began to breathe quickly and struggle to get loose. As soon as the change took place, the movements of the stomach entirely disappeared.

Cannon began to pet the cat and observed renewed peristalsis when the animal began to purr. He noted that any distress was accompanied by a total suspension of the motor activities of the stomach, and concluded: "It has long been common knowledge that violent emotions interfere with the digestive process, but that the gastric motor activity should manifest such extreme sensitiveness to nervous conditions is surprising".

The role of the emotions in the digestive process repeatedly came to the fore while Cannon continued his work with X-rays. He became more and more intrigued by the effects of emotional factors on gastrointestinal motility. On October 1, 1910, Cannon wrote in his diary: "In morning . . suggested research on adrenalin during emotion to Hoskins (by letter)".[4] This entry, and a letter to Roy G. Hoskins (Fig. 8) who had just received his Ph.D. under Cannon, is the first written statement of Cannon's shifting interest to the sympathetic nervous system.

----

4. W.B. Cannon Entry for October 1, 1910 in diary in Cannon Papers, Countway Library, Harvard Medical School.

Fig. 8. Roy G. Hoskins (1880-1964).

Fig. 9. Daniel de la Paz.

Fig. 10. Joseph E. Uridil (1891-1954).

Fig. 11. Sir Henry Hallett Dale (1875-1968).

Cannon, assisted by Daniel de la Paz (Fig. 9), a native of the Philippines, and Cannon's first foreign fellow, began a series of studies on adrenaline released from the adrenal glands into the vena cava during stress. In 1911 they published their paper on "The emotional stimulation of adrenal secretion" (Cannon and de la Paz, 1911). This was the first of a series of publications that led to Cannon's elaboration of the emergency function of the sympathetic nervous system and the development of his concepts on chemical transmission. In 1921, at the same time that Otto Loewi was demonstrating the transmission of the peripheral effects of the vagus nerve by a chemical mediator (Loewi, 1921), Cannon and Joseph Uridil (Fig. 10) reported the acceleration of the denervated and sensitized heart when the hepatic nerves were stimulated (Cannon and Uridil, 1921). As Sir Henry Dale (Fig. 11) noted some years hence:

> Similar effects at a distance, transmitted by the circulation, were later recognized by Cannon and his colleagues as the result of stimulating other sympathetic nerves; and it seems clear that he had not been far from the discovery which gained Loewi the Nobel Prize.[5]

Indeed, we have Cannon's own reaction to the announcement of the 1936 Nobel Prize in a letter to Uridil dated November 21, 1935[6]

> If only we had taken the hint that came from the observations which we made in 1920, we should have been credited with a quite novel demonstration in the history of physiology. As it was, Loewi made his observations about the same time, and by following them up has the credit of establishing the fact of chemical mediation of nerve impulses.

Cannon's extraordinary manual dexterity and surgical skill enabled him to remove the entire sympathetic nervous system. The observations on the sympathectomized animals gradually led him to the development of his ideas concerning the role of the sympathetic nervous system in the maintenance of the steady-state of the internal environment. In a series of

---

5. H.H. Dale, Walter Bradford Cannon, 1871-1945. Obituary Notices Fellows Roy. Soc. 1947; 5: 407-23.

6. W.B. Cannon to J.E. Uridil November 2, 1935, in Cannon Papers.

Fig. 12. George N. Stewart (1860-1930).

Fig. 13. Julius M. Rogoff (1883-1966).

investigations continuing for almost a decade, Cannon elaborated on Claude Bernard's concepts of the constancy of the milieu intérieur, and coined a word, homeostasis (from the Greek homeo [like or similar] and stasis [conditional]), to describe his own concept (Cannon, 1926).

Throughout Cannon's early career, most of his research was promptly accepted by physiologists as well as by clinicians, and generally with a great deal of praise for his ingenuity. To be sure, he had debates from time to time with some of his peers, such as that with Henry Dale and Anton Carlson, but such skirmishes were by and large fleeting. During the summer of 1916, however, George N. Stewart (Fig. 12) at Western Reserve Medical School, with the aid of his assistant Dr. Julius M. Rogoff (Fig. 13), launched an attack on the validity of Cannon's theory of the emergency function of the adrenals. It was the first time that the entire corpus of one of Cannon's scientific investigations was challenged. Stewart, a physiologist from the University of Edinburgh who was extraordinarily well-trained in physics and mathematics, proved to be a formidable opponent.

In The Way of an Investigator, Cannon wrote:[7]

> During my long experience in physiological research I have had my share of controversy. For about a decade there was a sharp difference of testimony between the group working at the Harvard Physiological Laboratory, on the one side, and two physiologists at Cleveland, Ohio, in the opposition .... The Harvard group sustained the view that secretion into the blood stream from the medulla of the adrenal glands is much increased whenever there is a wide spread activity of the sympathetic division of the involuntary or autonomic nervous system -- for example, in conditions of asphyxia, pain, and great emotional excitement. Our critics, on the other hand, contended that the secretion from the adrenals is constant and unvarying and they found fault with the method employed by the Harvard Laboratory ....
>
> One hears relatively little of this conflict now. Through the decades and from many quarters of the world, investigators have confirmed the evidence and conclusions reported by the Harvard group. In spite of confirmation, however, for years an impression persisted that because conditional control of adrenal secretion was once in question, it continued to be in question.

----

7. W.B. Cannon The Way of an Investigator, pp 100-101.

Fig. 14. Göran Liljestrand (1886-1968).

In fact, the misguided and acrimonious criticisms of Stewart and Rogoff, and the ensuing public controversy, may have been a major factor in the negative decisions of the Nobel Prize Committee regarding Cannon's nominations in the years 1921, 1928 and 1932. The criticisms voiced by Stewart and Rogoff were emphasized in each of the Nobel reviewer's critiques for those years. By 1934, however, Göran Liljestrand (Fig. 14), Professor of Pharmacology at the Karolinska and a member of the Nobel Committee, changed his opinion dramatically. Although he mentioned the studies of Stewart and Rogoff in his critique, it was merely to dismiss them out of hand. Cannon's investigations were described in the most laudatory of terms and Liljestrand concluded that Cannon's thesis regarding the emergency theory and the role of the adrenals had been amply verified both in Cannon's laboratory and by several groups in Japan. Liljestrand noted that:

> Cannon has presented a theory according to which changes arising during pain and emotional conditions can be attributed as a series of appropriate reflexes, causes which lead to an increased defense preparedness in the body. Mobilization of sugar counters fatigue and promotes muscle action; the increased heart rate, dilatation of the bronchi also lead to an adaptation to stress, the increased blood supply serves the same purpose, while the increased coagulability of the blood serves to preserve the integrity of the body ....

Liljestrand then went on to state that:

> if any specific discovery stands above all the others it would be the discovery that emotions induce an increased adrenalin secretion .... But to this has to be added principles of great importance in our understanding of the sympathetic-adrenal system's function and to the knowledge of emotions. The discovery was made some time ago but it is only through recent articles that it has really been verified. In my opinion Walter Bradford Cannon's discovery of the association of the emotions and the sympathetic-adrenal nervous system is of enough originality and importance to be considered for the Nobel Prize in Physiology.[8]

However, ironically, the prize in 1934 went to two other Harvardians, George Minot and William Murphy who received the award, along with George Whipple of Rochester, for their successful treatment of pernicious anaemia.

Cannon was nominated again for the Nobel Prize in 1935 and once more had the strong backing of Liljestrand. Liljestrand wrote: "I have come to exactly the same conclusion as I did last year, namely that Walter Bradford Cannon's studies on the sympathetic-adrenal system, and especially his discovery of the relationship between emotions and the sympathetic-adrenal system" were worthy of the Nobel Prize.[9] But again Cannon lost out and the prize was awarded in 1935 to Hans Spemann for his discovery of the organizer effects in embryonic development. In 1936 the prize went to Otto Loewi and Sir Henry Dale for their studies on chemical transmission of nerve impulses.

Despite his deep involvement in science, Cannon believed strongly that the scientist was also a citizen and, as a citizen, had an obligation to defend freedom. In his view, freedom was an essential element for productive scholarship. For 17 years he was chairman of the American Medical Association's Council on Defense of Medical Research and the major architect and spokesman in the successful battle against the anti-vivisectionists in the early nineteen hundreds. His belief in the

---

8. G. Liljestrand <u>Nobel Committee Archives</u>, 1934, Stockholm, Sweden.

9. *Ibid*, 1935.

Fig. 15. W.B. Cannon and colleagues at the Peking Union Medical School, 1935.

Fig. 16. Ivan P. Pavlov (1849-1936) and Walter Cannon.

brotherhood and universality of science took Cannon to the Peking Union Medical College in 1935 (Fig. 15) and involved him in the formation of the Medical Bureau to Aid Spanish Democracy as well as the organization of the American-Soviet Medical Society, the American Bureau for Medical Aid to China and the United China Relief. Since these activities engendered much political controversy, it is well to examine the roots of some of these affairs.

Cannon's work on the gastrointestinal tract was responsible, in part, for his long and continuing interest in Russia, especially in the school of physiology sired by Pavlov. Initially Cannon and Pavlov came to know each other because of their common interest in the problems of digestion. At the beginning of their friendship, they exchanged letters and papers, but did not meet. In 1923, before the International Physiological Congress in Scotland, Pavlov, in response to Cannon's written entreaties, visited the United States. In 1929 Pavlov (Fig. 16) returned to the United States to attend the International Physiological Congress in Boston at which Cannon was the Chairman of the Organizing Committee. In 1935 the two physiologists cemented their friendship further when Cannon visited Russia to attend the International Physiological Congress in Leningrad. On this occasion, Cannon dominated the Congress, not because of the scientific paper he presented but because of his prefatory statement of the relation of freedom to scientific research:

> During the last few years, how profoundly and unexpectedly the world has changed. Nationalism has become violently intensified until it is tinted with bitter feeling. Governments whose strength seemed deeply rooted in fixed traditions have vanished like phantoms only to be replaced by strange new forms and agents. The world-wide economic depression has greatly reduced the material support for scholarly efforts. In consequence lameness is already at hand and paralysis is threatening. Creative investigators of high international repute have been degraded and subjected to privations. Some universities have been closed. Others have been deprived of their ideal social function of providing a sanctuary for scholars where the search for truth is free and untrammeled, and where novel ideas are welcome and evaluated. As scientific investigators commonly associate with universities, these conditions have serious meaning for all of us.

The prescience and wisdom of Cannon's remarks have been forgotten, and he is often portrayed as naive in political matters. Cannon well understood the Soviet Union as a social experiment; and while he applauded

Fig. 17. Juan Negrín (1892-1956).

Soviet governmental support of science, he did not overlook the harsher aspects of Soviet policy.

Through his friendship and correspondence with scientific leaders such as Juan Negrín of Spain and Otto Loewi of Austria, Cannon very early saw the rising danger of fascism too. In 1930, while Cannon was serving as an exchange professor in France, he motored with the members of his family to Spain to visit some of his colleagues and former students. At Madrid, the professor of physiology and acting dean of the medical school was Juan Negrín (Fig. 17). Negrín was deeply involved in the planning and the superintending of the erection of the impressive structures which were to be part of the University City. He hoped that this would be a centre of culture for Spain. Despite his planning for the future, Negrín voiced his concerns to Cannon about the political conditions in the country. However, neither Cannon nor Negrín sensed how rapidly conditions would deteriorate in Spain. On May 19, 1933, Cannon wrote to Negrín as follows:

> I have often thought of you and the magnificent university city which you were planning and eagerly fostering when I went to Madrid three years ago. The recent events in Germany have led me to ponder over a remark which you made at that time. You said that you could put up splendid buildings and possibly provide funds for university activities, but the difficulty

would be that of securing outstanding scholars who might be suitable occupants of these new structures.

You will pardon me, I am sure, if I seem to be intruding into a situation where I should not intrude. Nevertheless, since I have heard you have been able to get your construction in spite of the turbulence of the revolution, I am venturing to suggest that the driving of Jewish scholars out of Germany offers an opportunity to obtain the services of outstanding investigators who would instantly make Madrid a center of scholarly activity.[10]

Cannon had already begun his attempts to rescue scholars from the Nazi menace. Cannon and Negrín met again in 1935 at the International Physiological Congress shortly after Negrín had accepted a position in the Spanish government. Cannon wrote in The Way of an Investigator[11] that:

early in 1936, only a few months after our meeting, the election occurred which brought into power the liberal element of the Spanish population and aroused in the conservative element deep apprehension. The story since that time is well known. Franco, with the open support of troops of Hitler and Mussolini, with plentiful equipment and tanks, heavy guns and armed airplanes supplied by these aggressive dictators, attacked the unprepared, ill equipped, and much depleted army of the republic ....

Cannon went on:

In a letter from Dr. Negrín, written to me after he became premier of the republic, he declared that he could not understand the misleading charge of communist control of Spain except as a conspiracy on the part of Hitler and Mussolini to prejudice world opinion against the legitimate Spanish government.

Cannon was deeply frustrated that he could not take a more active role in the support of the Negrín government. He noted:

The Neutrality Committee in London was, in effect, supporting Franco and his ruthless fascist collaborators. In the circumstances the only action the friends of the republic could take was that of providing medical and surgical supplies, clothing, and food. For

----

10. W.B. Cannon to J. Negrín, May 19, 1933 in Cannon Papers.

11. Cannon, op. cit., pp 160-162.

more than three years I served as National Chairman of the Medical Bureau to Aid Spanish Democracy. During that time we sent to Spain medical personnel and surgical instruments, hospital equipment and ambulances, amounting in value to more than a million dollars. Naturally enough I was charged of being a Bolshevik, a supporter of communists, an enemy of the Roman Church, and in general a red....

It was some years before the United States government was ready to admit the mistakes they had made in their Spanish-American policy. Cannon went on:

After being subjected to much hostile criticism and spending much time in what I could for the Spanish Loyalists, I confessed to having a sense of satisfaction as I read the testimony of Sumner Welles [Under Secretary of State] in his Time for Decision. In 1944, eight years after the start of the war in Spain, he admitted that many Americans were beginning to realize more accurately the real question raised by fascist and Nazi policy towards that country was whether the government, elected by the people and representing democratic ideals and aspirations, should be overthrown with impunity, not by revolution inside Spain but by armed forces of Hitler and Mussolini. "In the long history of the foreign policy of the Roosevelt Administration," he concluded, "there has been, I think no more cardinal error in the policy adopted during the civil war in Spain".[12]

Cannon was also deeply disturbed by the treatment accorded Jewish scientists in Germany, and was fearful that the same fate would befall his Austrian colleagues, particularly Otto Loewi (Fig. 18) of Graz. As noted above, Loewi and Cannon were both involved in studies of chemical transmission of nerve impulses as early as 1921. Their friendship, however, began in 1929 at the International Physiological Congress in Boston. In 1930 Cannon wrote to Loewi, and enclosed the summaries of two papers which he had recently submitted for publication and which supported Loewi's thesis on chemical transmission. Cannon wrote:

You will be gratified to see this brought to your views; it must be a keen satisfaction to you after seeing your work denied validity, to find at last that it is being regarded as thoroughly sound and of fundamental significance. You have my hearty congratulations.[13]

---

12. Ibid, p. 162.

13. W.B. Cannon to O. Loewi December 19, 1930, Cannon Papers.

Fig. 18. Otto Loewi (1873-1961).

Cannon was so very impressed with the investigations of Loewi that he invited him to give the prestigious Dunham Lectures at Harvard in 1933. Many of the Dunham lecturers have gone on to win the Nobel Prize. The lectures by Loewi were well received and he was delighted by his warm reception in Boston. On June 21, 1933, Cannon sent a letter of thanks to Loewi for the lectures and for his note of appreciation of Boston's hospitality and added: "It is a great satisfaction to all of us who are watching events in Germany and Austria that so far Austria has fought effectively Nazi influence".[14] Nine months later Cannon wrote once more:

> It was very good indeed to see your hand writing again and to learn that in spite of the unsettled state of affairs in Austria you are able to keep on with your scientific work. After seeing how the university condition has been profoundly upset in Germany, I have been watching with the greatest interest the course of events in Austria - anxious, of course - least there should be a spread of the dark clouds from Germany into your country. I most certainly hope that you will be

---

14. Ibid, June 21, 1933.

> spared the degradation of humane spirit which has manifested itself in your native land".[15]

Loewi apparently remained optimistic about the state of affairs in Austria and tried to reassure Cannon who replied on January 11, 1935:

> It was a real pleasure for me to know that by the cooperation of France and Italy, Austria is to be protected from a reunion with Germany. I suppose there might have been a time in the past that such a reunion would have been welcome and quite possible. I cannot conceive of anything just now which would be worse than bringing Austria under the domination of the Nazi regime. I only hope that ignorance and fanaticism may not threaten you in Austria as a consequence of the proximity of German influence. I should think if anything the example of Germany would be an abhorrence rather than an invitation.[16]

When Sir Henry Dale and Otto Loewi were awarded the Nobel Prize in 1936 for their discoveries relating to the chemical transmission of nerve impulses, Cannon wrote letters of congratulations to both Nobel laureates. In his reply Loewi stated, "Many thanks for your dear congratulations. The Dunham Lectures, which I owe to you, was the first real acknowledgement for the work. I shall always be indebted to you." Again and again, we were to see that the Nobel Prize offered no protection to Jewish scholars in Germany and in Austria. On March 29, 1938, Cannon received an urgent message from Dr. Hans J. Loewi, Otto's son, informing him that his father had been arrested and jailed with two of his sons on March 12. Cannon immediately alerted the American Physiological Society to the plight of Loewi and also solicited the help of Professor Felix Frankfurter of the Harvard Law School, a man of considerable influence in the Roosevelt administration. After Loewi's release from prison two months later, Cannon worked unceasingly to obtain a position for him in the United States. Finally with the help of his colleague, Dr. Homer W. Smith, Loewi was appointed Research Professor of Pharmacology at New York University Medical School. The Loewi affair was but one example from the long chronicle of Cannon's efforts to save refugees from oppression.

----

15. _Ibid_, March 3, 1934.

16. _Ibid_, Jan 11, 1935.

Fig. 19. Bernardo A. Houssay (1887-1971).

On her death bed, when Cannon was only ten years old, his mother called him to her bed side and said tenderly, "Walter, be good to the world".[17] Sixty-four years later at the memorial exercises for Cannon in 1945, Bernardo A. Houssay of Argentina (Fig. 19) and his associates, in their letter of condolence serendipitously concluded with the same theme:

> Many tears have been shed at his departure but they are not bitter ones; he has left only sweet and tender memories because he has been good to the world.[18]

ACKNOWLEDGEMENTS

Some of the material in this account is taken from the book, Walter Bradford Cannon: The Life and Times of a Young Scientist which is in press at the Harvard University Press. I thank my co-authors, Saul Benison and Elin Wolfe for permission to use excerpts. I also thank Birthe Creutz for

----

17. W.B. Cannon The Way of an Investigator, p. 15.

18. Walter Bradford Cannon, 1871-1945: A Memorial Exercise held at the Harvard Medical School, November 5, 1945, p. 66.

her help in preparing the manuscript. Financial assistance has been received from the National Library of Medicine, the Commonwealth Fund, the Josiah Macy, Jr. Foundation and the Eleanor Naylor Dana Charitable Trust. We appreciate the permission granted by the Nobel Assembly to quote from their archives.

REFERENCES

Cannon, W.B. (1898) The movements of the stomach studied by means of the Röntgen rays. Am. J. Physiol. 1, 359-375

Cannon, W.B. (1926) Physiological regulation of normal states: some tentative postulates concerning biological homeostatics. In: Jubilee Volume to Charles Richet. Editions Mèdicales: Paris, pp 91-93.

Cannon, W.B. and de la Paz, D. (1911) Emotional stimulation of adrenal secretion. Am. J. Physiol. 28, 64-70.

Cannon, W.B. and Uridil, J.E. (1921) Studies on the conditions of activity in endocrine glands. VIII Some effects on the denervated heart of stimulating the nerves of the liver. Am. J. Physiol. 58, 353-364.

Davenport, C.B. and Cannon, W.B. (1897) On the determination of the direction and rate of movement of organisms by light. J. Physiol. (Lond.) 21, 22-32.

Loewi, O. (1921) Über humorale Übertragbarkeit der Herznervenwirkung. Pflügers Arch. 189, 239-242.

SYNAPSES OF THE CENTRAL NERVOUS SYSTEM FROM SHERRINGTON TO THE PRESENT

J.C. Eccles

Max-Planck-Institut für biophysikalische Chemie
Göttingen, West Germany

The neurones of the central nervous system had long been recognized and in the latter part of the last century Sherrington had accepted the neurone theory of Râmon y Cajal and many other neuro-anatomists, that the neurones were structurally independent. He rejected the alternative reticular theory of Gerlach and Golgi according to which in the central nervous system the neurones were in continuity in a net-like structure with connectivities that could be vaguely seen in the inadequate histological preparations of that time. Sherrington recognized that the neurone theory involved functional communication between neurones at sites of contiguity that probably were made by structures called baskets or knobs or boutons. In writing on the functional activity of the central nervous system in 1897 Sherrington felt the need for some specific term for these zones of functional connection between neurones. On the advice of a Greek scholar of Cambridge he coined the word synapse for the surface of functional interaction between neurone and neurone.

The technical facilities of that time were far too inadequate for direct investigations on the mode of action of the postulated synapses. Sherrington's great contribution was to utilize the simplest spinal reflexes to gain insight into the special properties exhibited by transmission of nerve impulses across synapses in contrast to the well known transmission along nerve trunks. In the first chapter of his great book, The Integrative Action of the Nervous System, he listed eleven special properties of reflexes that presumably were attributed to the synapses incorporated in their pathways. Properties such as delay,

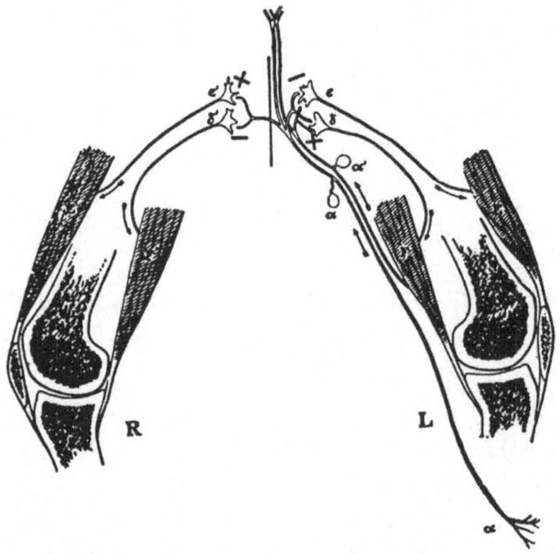

Fig. 1. Drawing of the simplest synaptic connections of the cat spinal cord showing pathways from a cutaneous receptor that give an ipsilateral flexor reflex (+) with extensor inhibition (-) and reciprocally on the other side (-) and (+) respectively (Sherrington, 1906).

irreversibility, fatigue, summation, longer duration of action, sensitivity to anoxia and drugs would be accepted today as synaptic properties but in a more refined form. Already Sherrington had recognized inhibitory synaptic action as being opposite to excitatory synaptic action as illustrated in his diagram (Fig. 1).

It was, however, not until 1925 that Sherrington adventured on a refined theoretical treatment of synapses. He recognized that, when an impulse impinges on a synapse, it is transformed into more enduring states that had opposed actions on the neurone, and these he named central excitatory state, c.e.s. and central inhibitory state, c.i.s. These concepts forestalled in a remarkable manner the eventual discovery by intracellular recording 26 years later of the excitatory postsynaptic potential, EPSP, and the inhibitory postsynaptic potential, IPSP. In 1925 I arrived at Oxford to work with Sherrington (Fig. 2) and I can vividly remember the impact of these clear theoretical developments.

Electrical recording was still in its infancy and Sherrington had perfected optical isometric recording from reflexly contracting muscles with the concept of the motor unit for interpreting the responses of motoneurones in reflexes that were being studied in a progressively more refined way. Sherrington's Ferrier Lecture of 1929 admirably describes the

Fig. 2. Sir Charles Sherrington at Oxford in 1929.

beliefs we had then of excitatory and inhibitory synapses. I quote from his summary:

> Though trains of impulses are the sole reactions which enter and leave the central nervous system, nervous impulses are not the sole reactions functioning within that system. States of excitement which can sum together, and states of inhibition which can sum together and states which represent the algebraic summation of these two, are among the central reactions. The motoneurone lies at a focus of interplay of these reactions and its motor unit gives their net upshot, always expressed in terms of motor impulses and contraction. The central reactions can be much longer lasting than the nerve impulses of nerve trunks.

The time course of the c.e.s. was investigated at that time by utilizing the summation of two convergent inputs to motoneurones with recording of the motor twitch response. In the simplest responses two subliminal inputs evoked reflexes that were largest with simultaneity and progressively smaller with lengthening of stimulus interval, to be zero at about 15 msec., much as would be expected for the EPSP today (Fig. 3).

Sherrington and I demonstrated this at the Oxford meeting of the Physiological Society in 1930. Unfortunately the spinal cat developed a continuous small twitching of the tibialis anterior muscle that would have

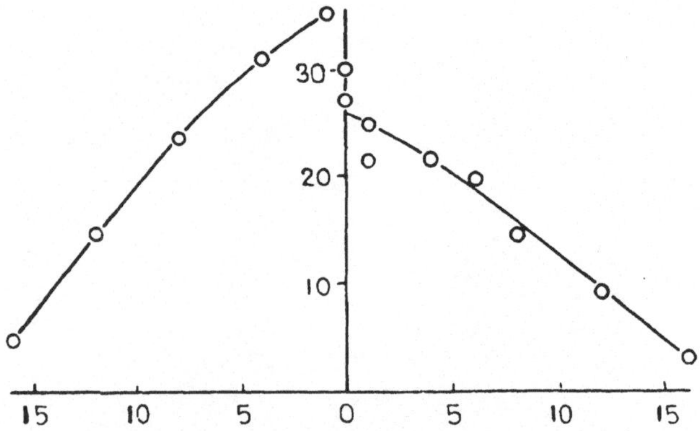

Fig. 3. Time courses (msec.) of facilitation of flexor reflexes by inputs that alone were subliminal (Eccles and Sherrington, 1930).

ruined the demonstration. Sherrington got the idea that the twitching would be inhibited by squeezing the cat's tail. It worked, and we had a most successful demonstration of reflex facilitation as revealed by the optical myograph. As each group came to see the demonstration, Sherrington sitting close to the cat with his hand on it could squeeze the tail unobserved, while I gave the nerve stimuli at different intervals to demonstrate for the first time the time course of c.e.s. We kept our secret, laughing together afterwards at our success.

Adrian and Bronk at Cambridge and Denny-Brown at Oxford had been most successfully studying the synaptic mechanisms generating the rhythmic discharges of motoneurones. Later, Hoff and I used antidromic impulses to analyse the mechanism of rhythmic impulse discharge from a single motoneurone subjected to a steady synaptic activation.

Meanwhile acetylcholine transmission at ganglionic and neuromuscular synapses had been demonstrated by Dale, Feldberg, Gaddum, Brown and associates. The controversy arose because it seemed that chemical transmission would be too slow for the initial fast synaptic action of about 1 msec. latency. There was much study of electrical models, called ephapses, displaying fast transmission from one excitable unit to another by electrical currents across the close apposition. The chemical versus electrical hypotheses were published in an extensive review in 1936 with much help from G.L. Brown. Though fighting hard scientifically there were the most cordial relations between Sir Henry Dale and his associates and me, so much so that some onlookers thought that it was like shadow boxing! Ranged with me on the electrical side were Erlanger, Gasser, Lorente de Nó, Fessard and Arvanitaki.

Fig. 4. Sir Charles Sherrington at Ipswich in 1936.

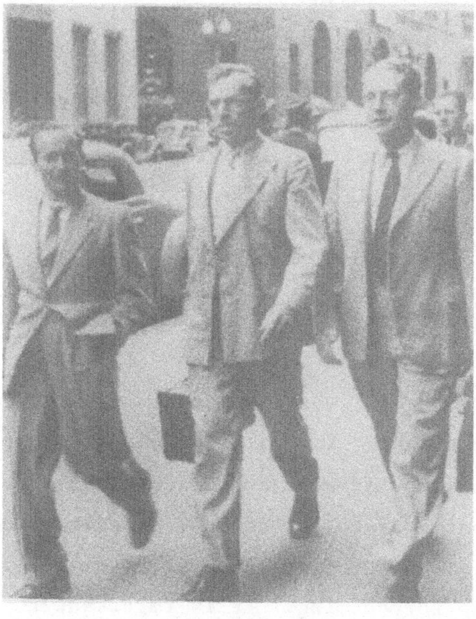

Fig. 5. Kuffler, Eccles and Katz in Sydney, 1942.

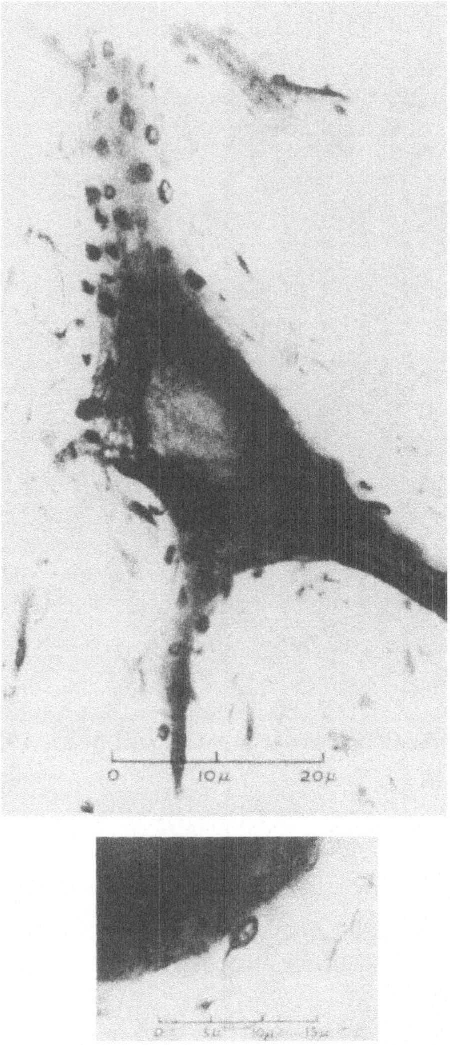

Fig. 6. A neurone of the cat spinal cord showing synaptic boutons on its surface with below a single bouton and its presynaptic fibre.

In 1936 I visited Sherrington in his retirement at Ipswich and talked of synapses (Fig. 4) and in 1937 I left Oxford for Sydney to be later joined by Katz and Kuffler (Fig. 5) where we continued studying the endplate potential (EPP) concluding in 1942 from pharmacological studies of EPP that even the fast transmission was due to acetylcholine. However, there was no such clear pharmacology for central synapses, where acetylcholine was certainly not the fast synaptic transmitter, and no alternative transmitter was forthcoming. It is important to realize that we were still in the age of light microscopy. One of the best pictures

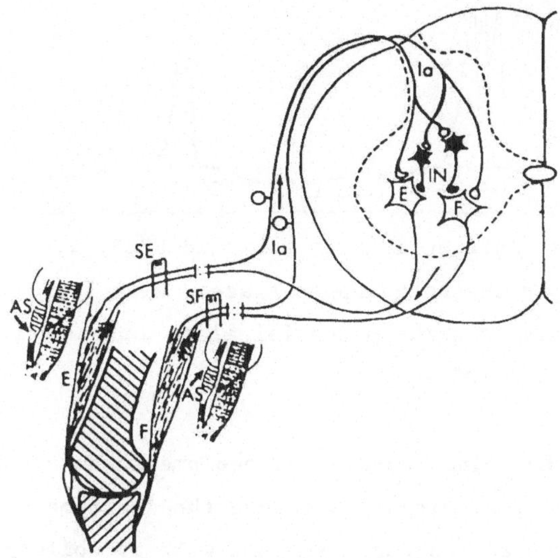

Fig. 7. Drawing of extensor and flexor muscles of a cat knee joint showing pathways of Ia afferent fibres from the muscle spindles (AS) to give monosynaptic excitation to homonymous motoneurones and via an interneurone (IN) inhibition to the antagonist motoneurone in a symmetrical arrangement. These inhibitory interneurones were not discovered until much later (Eccles, Fatt and Landgrén, 1954.

(Fig. 6) shows the synapses on the surface of a motoneurone with below a single loop synapse in profile with a gap of about 1 μm.

Meanwhile during the war years Lloyd and Renshaw at the Rockefeller Institute produced the best studies of authentic monosynaptic reflexes and their interaction in facilitation and inhibition. Fig. 7 shows the inputs into the spinal cord of Ia afferent impulses from antagonistic muscles at the knee joint. What we may call unitary synaptic responses had a time course of decay up to about 15 msec. intervals for both excitatory and inhibitory synapses, much as in our simplest c.e.s. responses at Oxford (Fig. 3). After the war I was struggling to catch up with the Rockefeller investigators, and in particular I was concerned to test Lloyd's idea that enduring presynaptic activity was responsible for the excitatory synaptic facilitation of up to 15 msec.

In 1951 I decided to try the hazardous procedure of intracellular recording by a technique being developed at that time by Fatt and Katz in London for isolated neuromuscular junctions. The motoneurones were rather large targets for shooting at, but were several millimeters below the

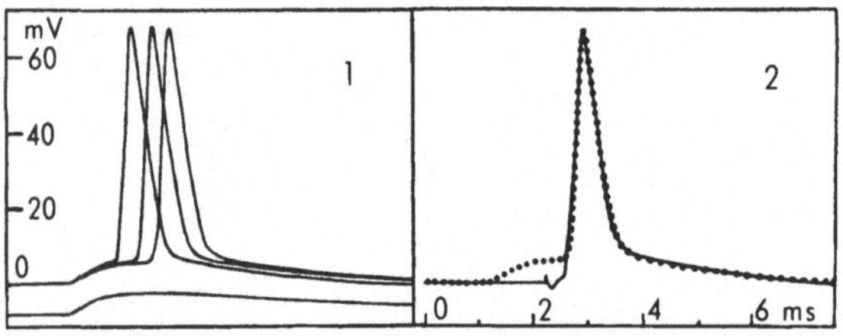

Fig. 8. 1. The first intracellularly recorded EPSPs with superimposed spike potentials in the three upper traces, with EPSP alone below.
2. An antidromic spike potential alone and superimposed on an EPSP (Brock et al., 1951)

dorsal surface of the spinal cord. The equipment was homemade, as was everything in those pioneering days before the instrument companies got into the act. It was an arduous task, but with my colleagues Coombs and Brock in 1951 in Dunedin we had almost immediate success (Fig. 8) with EPSPs and neuronal spike potentials which were observed to be like the EPPs and muscle impulses of Fatt and Katz.

There still was the enigma of the mode of action of inhibitory synapses that we had been investigating intensively by extracellular recording in the late 1940's. So intense was my involvement that in 1947 I had a dream of how inhibitory inputs could act on motoneurones by the anelectrotonic action of Golgi interneurones (Fig. 9). The dream was in accord with all experimental data. It was soon published in Nature by Chandler Brooks and myself and stood up to all our extracellular testing. After the initial intracellular studies in 1951 we had sufficient experience to test inhibitory synaptic action intracellularly. We could get pure synaptic inhibition by the Ia afferents from antagonistic muscles (Fig. 7). The crucial test was on the fateful night of August 20, 1951. Fig. 10A shows the actual first recordings. We had predicted that on the electrical model the trace would go up, whereas it would be down for chemically mediated synaptic transmission. It went down! Immediately I rejected the electrical model for inhibition (Fig. 9) and embraced the chemical for both excitatory and inhibitory central synapses. Fig. 10B shows later intracellular recordings with graded IPSPs and the approximately mirror relationships of EPSPs and IPSPs as was envisaged in the Sherringtonian days of c.e.s. and c.i.s. and the later Lloyd interaction experiments. It is of interest that as discovered by Korn and Faber the electrical model for inhibition (Fig. 9) actually holds for

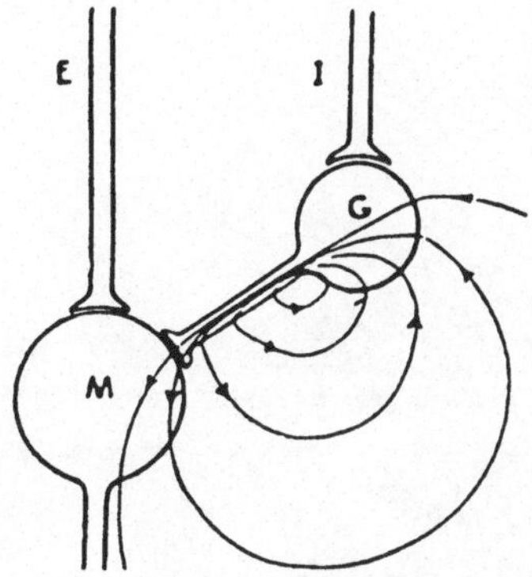

Fig. 9. The proposed circuitry for electrical inhibition of an I input via a Golgi cell (G) to give an anelectrotonic action onto the excitatory focus of an E input on a motoneurone (Brooks and Eccles, 1947).

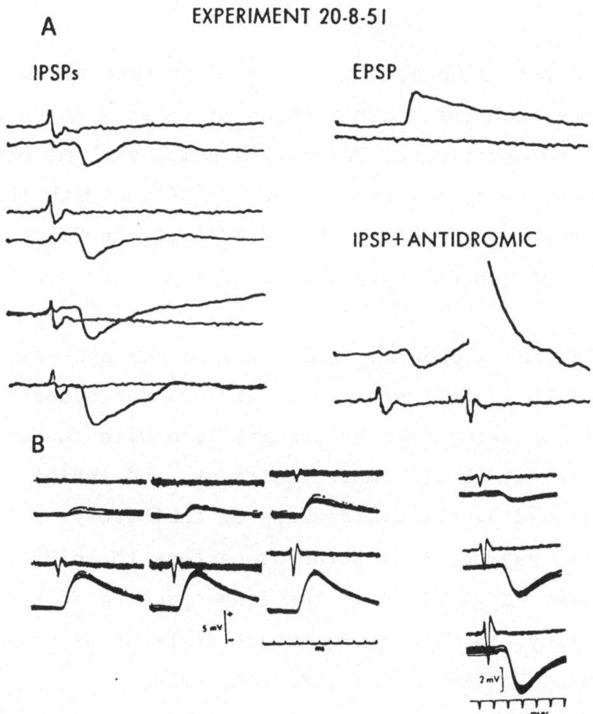

Fig. 10 A. First intracellular recordings of Ia IPSPs (a brief hyperpolarization recorded downwards) of a cat biceps-semitendinosus neurone. In right column EPSP and IPSP + antidromic response of same neurone. (Brock, Coombs and Eccles, unpublished). B. Graded EPSPs to left and graded IPSPs to the right: recorded in Canberra by Coombs, Curtis and Eccles.

Fig. 11. Celebration party in Feldberg's house, London, in February 1952.

inhibition of fish Mauthner cells and for the initial brief component of basket cell inhibition of the cerebellar Purkinje cell.

In 1952 I arrived in London for a series of lectures and conferences on central synapses with the enthusiasm of my belated conversion. Fig. 11 shows a convivial celebration at Feldberg's home, with my cross-hand shake of chemical Feldberg to electrical Lorente de Nő and with other notable participants, Marthe Vogt, Fessard, Richard Jung, the Ritchie's and others. So began a new era of central synapses, starring EPSPs and IPSPs.

Meanwhile electron microscopy had revealed the extremely close synaptic contact with a cleft of only about 200Å and synaptic vesicles that eventually came to be recognized as being filled with the as yet unknown transmitter substances. Furthermore excitatory and inhibitory synapses could be distinguished by the ovoid shape of inhibitory vesicles discovered by Uchizono and the asymmetrical structure across the cleft of the excitatory synapses. Fig. 12 shows the dense packing of boutons on a motoneurone (Poritsky, 1969) and the unique differences of excitatory and inhibitory synapses (Uchizomo, 1965; Bodian, 1966).

As the next stage of my Synaptic Saga I choose to present the story of the inhibitory synaptic pathway. In 1952 one still drew the Ia pathway as making excitatory synapses on its own motoneurones and by collateral branches inhibitory synapses on the antagonistic motoneurones. It was assumed that the same transmitter had opposite synaptic actions. However, there was a longer delay of about 0.8 msec. for IPSPs as compared to EPSPs (Fig. 10B), and Paul Fatt proposed that this time could be occupied by an interneurone interposed on the inhibitory pathway. As shown in Fig. 7 such interneurones were soon discovered. A further discovery at that time was the inhibitory pathway from axon collaterals of motoneurones via a unique kind of interneurone first discovered by Renshaw, and that we investigated in the early Canberra days (Fig. 13.). It was remarkable in that it was powerfully excited by the same nicotinic acetylcholine action that operated for the terminals at the neuromuscular junction. We immediately thought of Sir Henry Dale (Fig. 14) who in 1935 had proposed that the same chemical transmitter mechanism would be operating at all of the synapses made by a neurone in accord with the metabolic unity of a neurone. I immediately wrote to Dale of our discoveries and there was a very happy exchange of letters. I coined the term Dale's Principle. He was a little coy about the term, but never forbade it, and we continued in an even more friendly relationship, as can be recognized from his letters. Of course, the phrase chemical transmitter mechanism covers the recent discovery of multiple transmitter substances. Fig. 15 displays a simple diagram by Swanson that shows Dale's Principle with two different transmitters.

As illustrated in Fig. 16 there are two great classes of neurones in the central nervous system, those working by excitatory synapses and those by inhibitory synapses, and there are no ambivalent neurones. Intensive experimental work was involved in establishing this basic simplicity, which is now generally accepted, as can be seen in all diagrams of central pathways.

From the 1950's onwards there has been the immense enterprise of identifying the central synaptic transmitter substances. At first Dale and ourselves had thought that one or two excitatory and one or two inhibitory transmitters would be enough, as seemed to be the case for peripheral synapses. The cholinergic excitation of Renshaw cells (Fig. 13) was almost unique, and today we can accept that glutamate and aspartate are the principal central excitatory substances with quick ionotropic actions. For central inhibitory synapses GABA is undoubtedly the principal transmitter, with glycine as a subsidiary, particularly for inhibitory synapses of the spinal cord. This investigation of chemical transmitters has provided the principal challenge for neuropharmacologists and neurochemists. A whole host of transmitter substances has been discovered in addition to these five original transmitters. There are the amines such as serotonin,

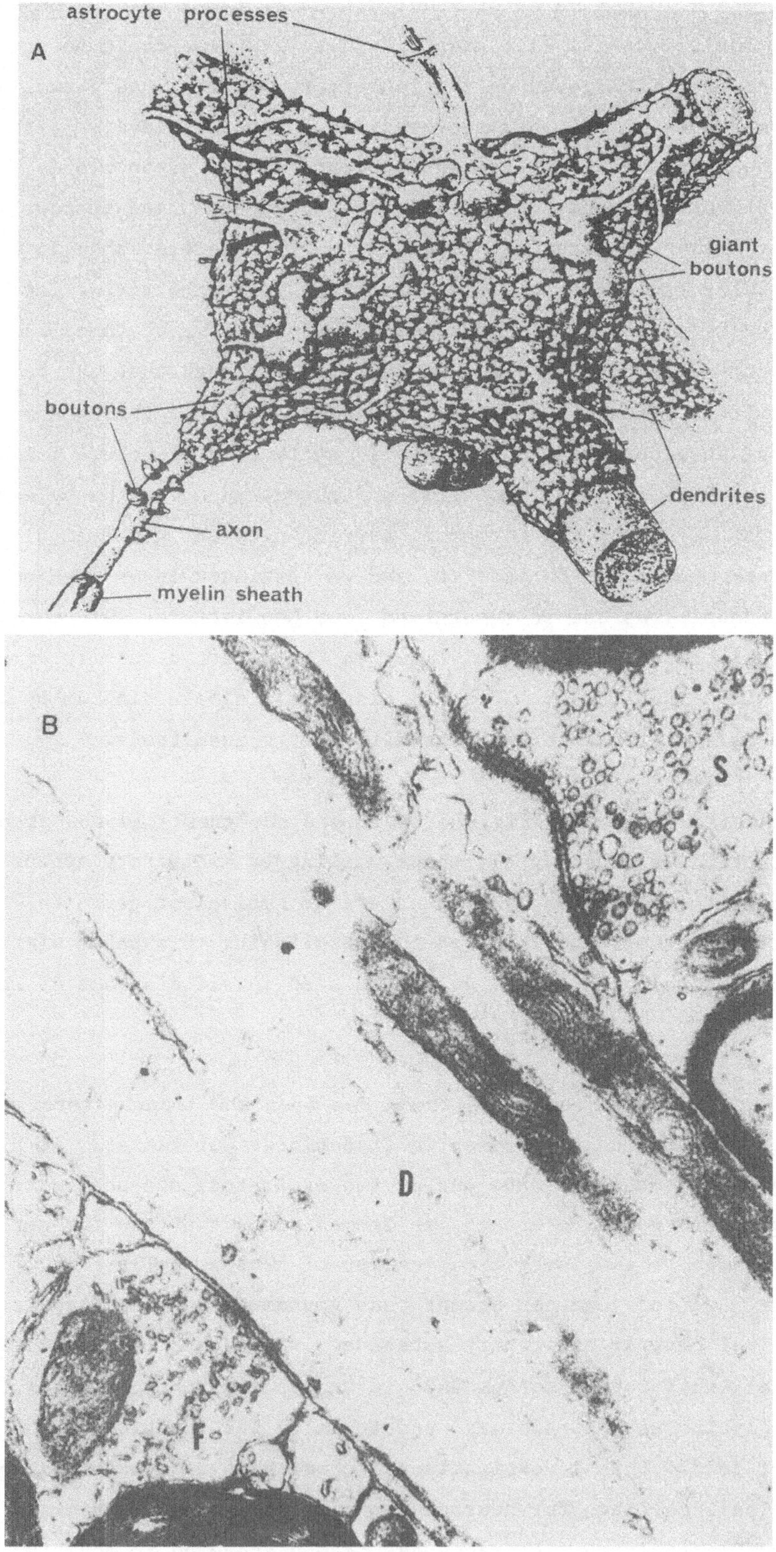
A
astrocyte processes
giant boutons
boutons
dendrites
axon
myelin sheath
B
S
D
F

Fig. 12. (previous page) **A**. Synapses on the surface of a motoneurone. The ← tightly packed surface scale of synaptic knobs (boutons), some being large (Poritsky, 1969). **B**. Electron micrograph of two synaptic knobs on the apical dendrite (D) of a cortical pyramidal neurone. That to the right is excitatory with spherical vesicles (S) and dense staining on each side of the synaptic cleft, and that to the left is inhibitory with ellipsoid vesicles (F) and a much lighter staining of the membranes on each side of the synaptic cleft (Bodian, 1966).

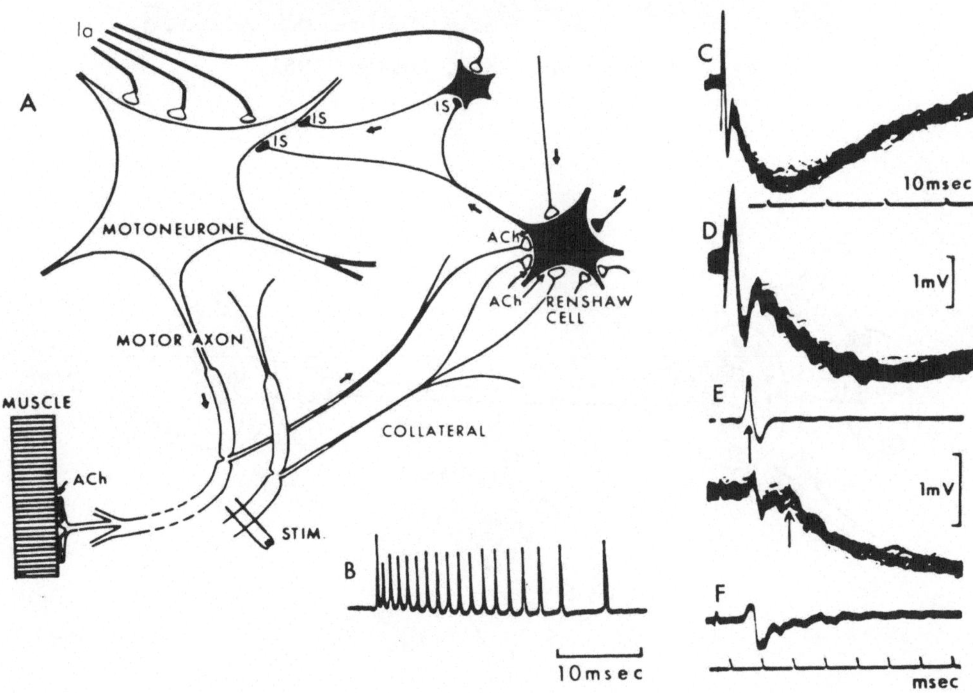

Fig. 13. A. Renshaw cell excited by axon collaterals of motoneurones to give an intense repetitive discharge, B, that exerts powerful inhibitory action on the motoneurone, a negative feedback. The IPSPs in C, D show the rhythmic discharge of the Renshaw cell and E, F give the latency of the inhibition between the two arrows, F being extracellular (Eccles, Fatt and Koketsu, 1954).

Fig. 14. Sir Henry Dale, London, 1952.

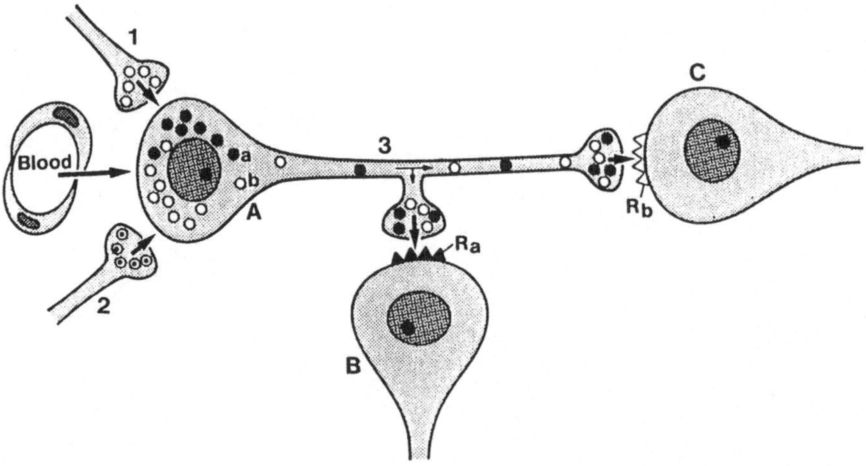

Fig. 15. A modern illustration of Dale's Principle showing neurone A generating two types of synaptic vesicles a and b that both travel down the axon (3) to both boutons of that axon on B and C neurones (Swanson, 1983).

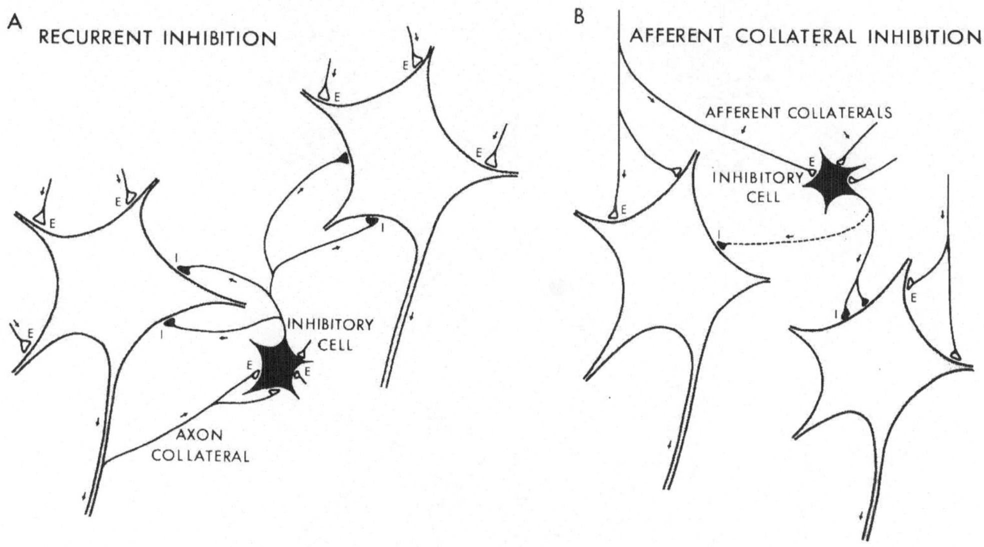

Fig. 16. Drawing of inhibitory neurones in circuits giving recurrent inhibition (A) and feed-forward inhibition (B).

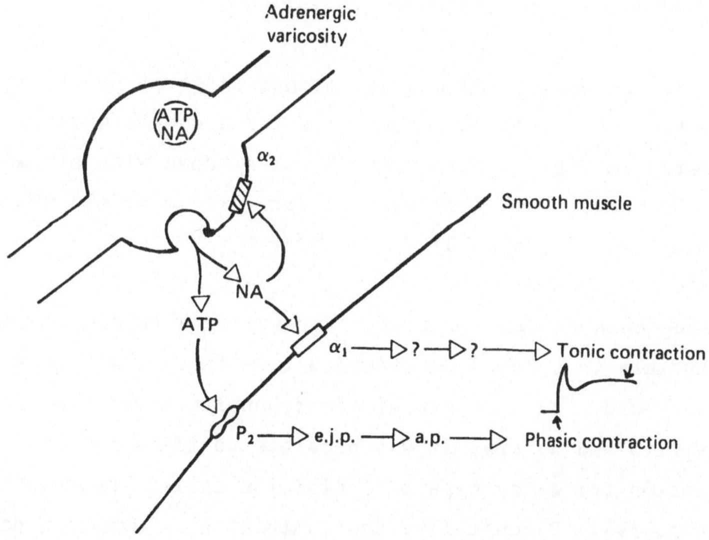

Fig. 17. Drawing of an adrenergic varicosity making a primitive synapse on a smooth muscle cell, showing a vesicle liberating two transmitters, ATP and noradrenaline that give respectively a phasic and a tonic contraction of the smooth muscle (Sneddon and Westfall, 1984).

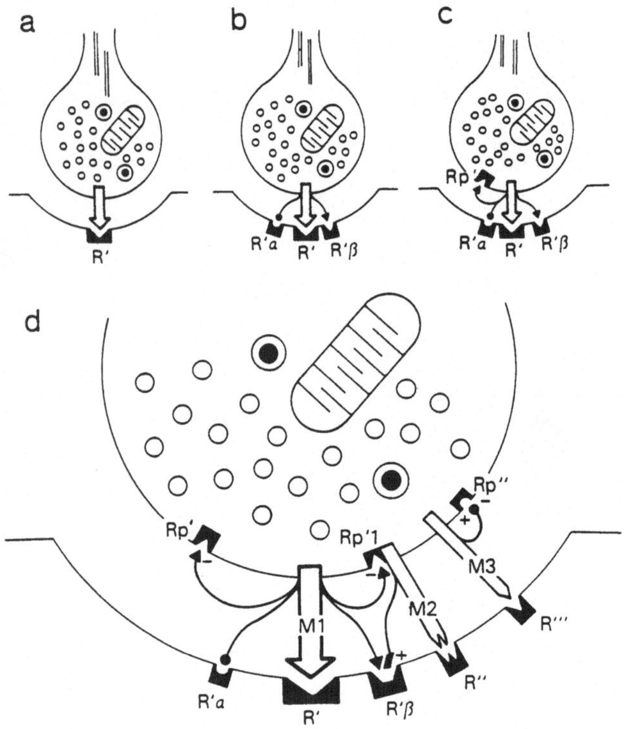

Fig. 18. Drawings showing progressive stages in synaptic complexity with the release of multiple transmitters that act on several types of postsynaptic receptor sites and also feed back to presynaptic receptor sites (Lundberg and Hökfelt, 1983).

noradrenaline and dopamine, and now an immense range of peptides is being discovered, often many different substances for a single synaptic bouton, as is illustrated in Fig. 17 for a peripheral synapse with ATP and noradrenaline by Sneddon and Westfall and for various levels of complexity by Lundberg and Hökfelt (Fig. 18).

It had long been recognized that afferent nerve inputs evoked slow negative potentials that could be recorded from the dorsal roots or the adjacent spinal cord. As a result of comprehensive investigation by Schmidt and Willis and others, it was possible to illustrate the manner of synaptic operation for a new type of inhibition called presynaptic inhibition (Fig. 19). Essentially, the inhibition is due to a diminution of transmitter output by a depolarization of the presynaptic terminals by unique synaptic boutons attached to them. Presynaptic inhibition diminishes in importance as one ascends the central nervous system and it is virtually absent in the cerebral and cerebellar cortices.

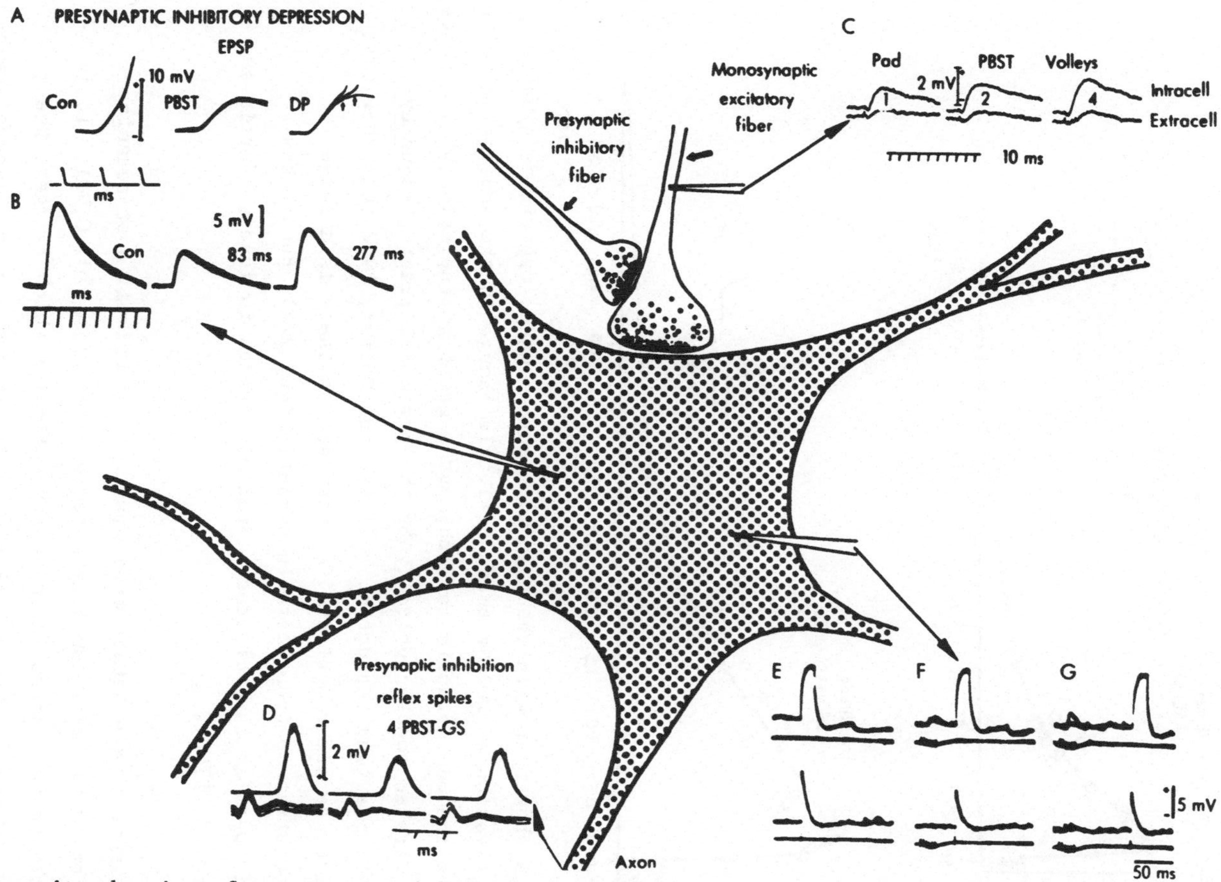

Fig. 19. Composite drawing of presynaptic inhibition showing location of presynaptic inhibitory synapse and the various testing procedures. A and B show the presynaptic inhibitory reduction of the EPSPs. In C there is the presynaptic depolarizing action of the excitatory fibre, that reduces the EPSPs (A,B) and the impulse discharge (D), but does not change the electrical properties of the postsynaptic neurone, upper traces E to F and G, but there is a reduced EPSP E to F and G (lower traces).

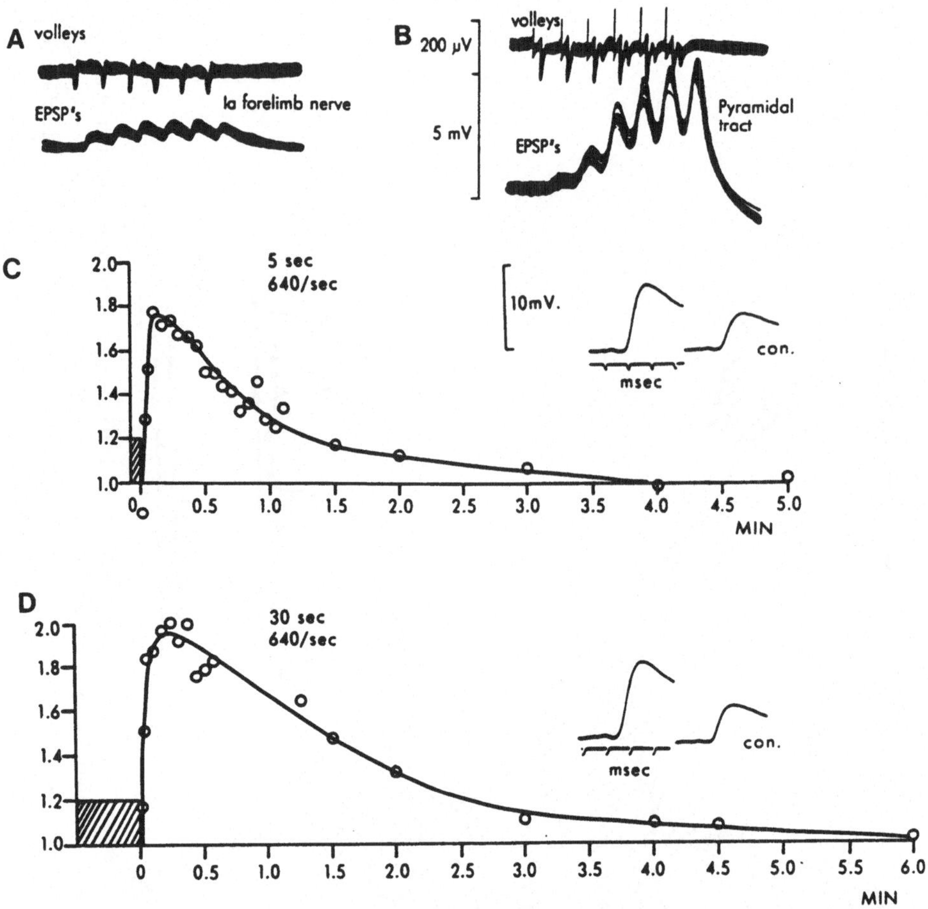

Fig. 20. A,B show that with the same motoneurone under repetitive synaptic activation at 200 Hz there is little change in the successive EPSPs for the monosynaptic Ia path, but a great potentiation of the EPSPs generated monosynaptically by the pyramidal tract (frequency potentiation) (Landgrén, Phillips and Porter, 1962) C,D show that after severe repetitive Ia tetanization there is a large potentiation (post-tetanic potentiation, PTP), but it has disappeared in 6 minutes. Specimen traces in insets (Curtis and Eccles, 1960).

Even after severe tetani with thousands of stimuli post-tetanic potentiation (PTP) lasts for only a few minutes (Fig. 20, C,D,). It is attributable to Ca^{2+} accumulation presynaptically which triggers the increased transmitter output. In the 1960's it was a great disappointment that PTP could not be proposed as a model for memory. We were looking at the wrong synapses, as can be seen by comparing Fig. 20A with B.

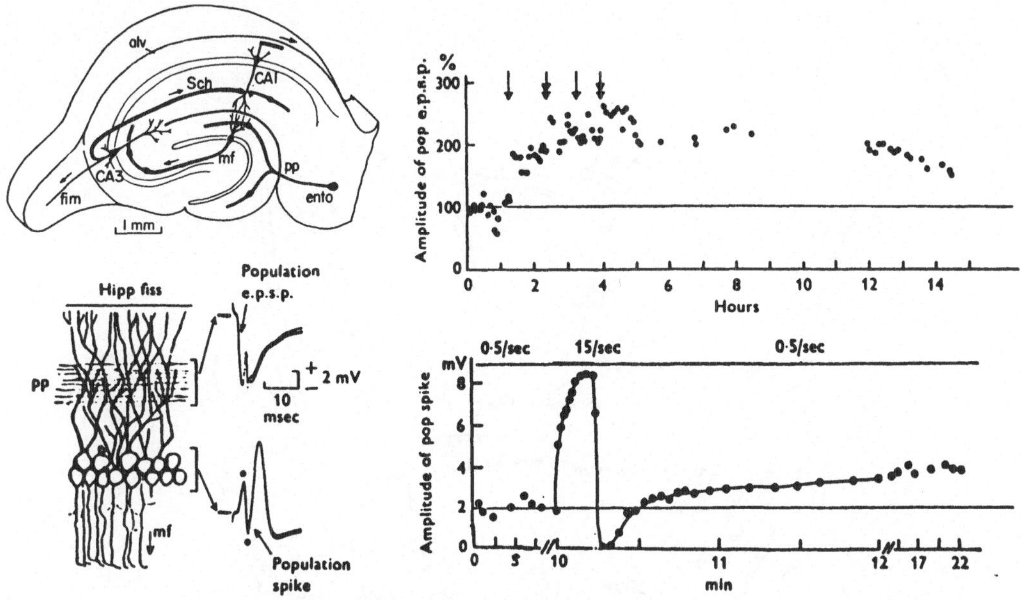

Fig. 21. Slice of hippocampus showing the principal pathways. Input from entorhinal cortex (ENTO) via the perforant pathway (pp) excites the dendrites of the granule cell with its mossy fibre (mf) axon. Below, the pp makes synapses with the dendrites of the granule cells exciting a population EPSP recorded extracellularly as a negative wave. In the upper frame four bursts of pp stimulation (15 Hz for 15 sec., arrows) result in an increase in the population EPSP that is still large 10 hours after the last stimulation. A similar long-term potentiation (LTP) is exhibited by the population spike recorded extracellularly at the level of the granule cell bodies (Bliss and Lømo, 1973).

In 1973 Bliss and Lømo published their most convincing demonstration of a model for memory. After moderate stimulation of a few hundred stimuli, EPSPs extracellularly recorded were potentiated for hours (Fig. 21), days or weeks, which is about 10,000 times longer than PTP. So long-term potentiation (LTP) was recognized immediately as the basis for a model of memory. There is much evidence that this prolonged potentiation is due to Ca^{2+} influx postsynaptically. Fig. 22 illustrates a transformation of our concepts on the nature of the LTP (Gustafsson and Wigström, 1986; Eccles, 1986b). The hypothesis is that the postsynaptic receptor of the spine synapse is a double receptor structure. There is much recent evidence that a single synaptic bouton of a spine synapse has two distinct receptor sites for the transmitter, glutamate. One is the conventional

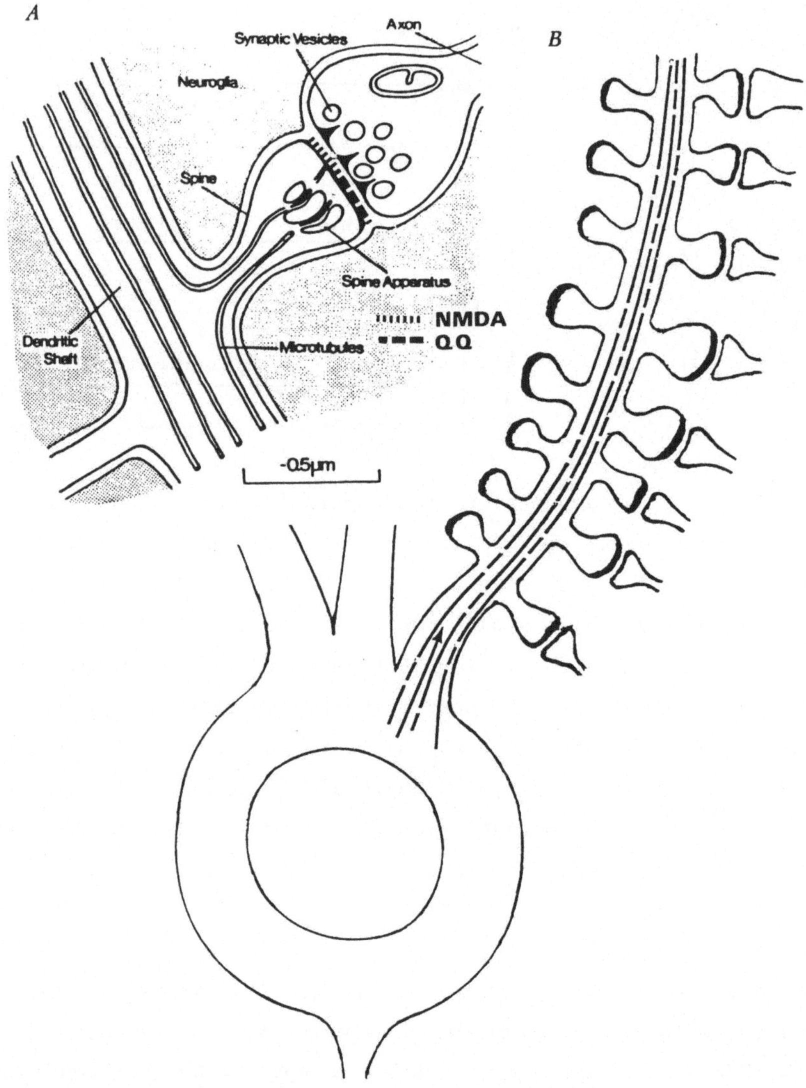

Fig. 22. A shows a spine synapse (of Fig. 12A) on the dendrite of a hippocampal neurone. The postsynaptic glutamate receptor is shown with a double composition. One part is the normal depolarizing synapse that is depolarized by the transmitter glutamate, acting on receptors that are probably those specially receptive to quisqualate, hence labelled QQ, and the other part is specially receptive to N-methyl-D-aspartate (NMDA) (Eccles, 1986b).

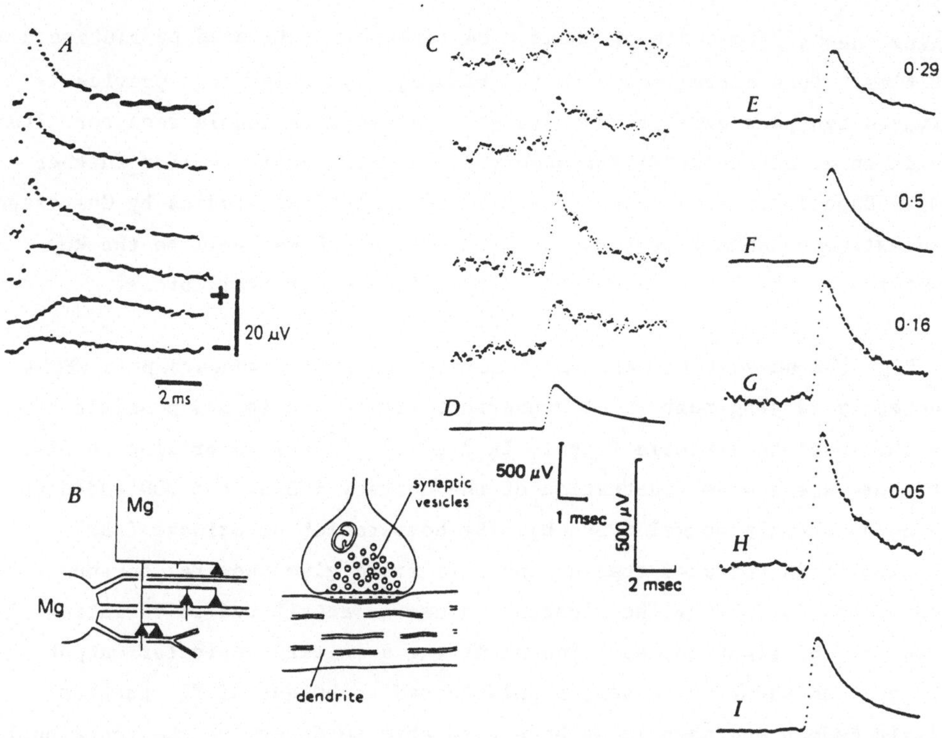

Fig. 23. Intracellular recording of a motoneurone activated by input from a single Ia afferent fibre. Full description in text (Jack et al. 1981).

quisqualate (QQ) receptor that when activated opens channels to Na^+ and K^+ ions, thus producing the large depolarization of the EPSP. The other is specifically excited by N-methyl-D-aspartate (NMDA) and blocked by 2-amino-5-phosphonovalerate (APV). Activation of this NMDA receptor by glutamate normally gives a negligible depolarization. However, when the extracellular Mg^{2+} concentration is greatly reduced or when the receptor is subjected to a steady large depolarization, glutamate-activated NMDA receptors give a large slow depolarization of the spine lasting for about 60 msec. with a large influx of Ca^{2+} ions through the opened Ca^{2+} channels (Gustafsson and Wigström, 1986). APV prevents this depolarization of NMDA receptors and the opening of the Ca^{2+} channels. As would be expected, LTP is also blocked by APV.

In the normal operation of a cortical neurone such as that partly shown in Fig. 22B there would be activation of hundreds of synapses with the current passing through the QQ receptors to give a large EPSP. Furthermore this cooperativity would produce sufficient depolarization for opening the Ca^{2+} channels of those NMDA receptors activated by glutamate at that time. This correlates with the finding that only those previously activated synapses exhibit LTP. These synapses with double receptor sites provide an effective model for memory and a great challenge for further study. Especially notable are the precise analytical studies by Gustafsson and Wigström on NMDA receptors for glutamate which may well be the key structures in the laying down of memories in the cerebral cortex.

Fig. 10B shows intracellular recording of graded monosynaptic EPSPs produced by varying numbers of convergent fibres. It is now possible to cut down the input to a single fibre. In Fig. 23C it was surprising to find that there was a wide fluctuation of the unitary EPSPs, but 800 added up gave an apparently normal EPSP (D). By horseradish peroxidase (HRP) injection it is revealed that a single Ia fibre gives several boutons to a motoneurone (B). Hence the fluctuation can be attributed to variations in the bouton contribution, each bouton giving a unitary vesicular output every now and then. By a very sophisticated technique of fluctuation analysis Redman and associates have been able to determine the contribution of individual boutons, there being in the illustrated case four such quantal emitters, with probabilities shown in Fig. 23E to H, together with the averaged EPSP of each. It turns out that the probability of vesicular exocytosis is less than one for a bouton, with usual values of about 0.3. A similar range of probabilities has been found in other species of synapses. Here we have an important new synaptic property.

In attempting to understand the synaptic mechanism underlying this probability, we turn to the idealized bouton drawn by Akert as a result of electronmicroscopic and freeze fracture techniques (Fig. 24). Each bouton confronts the synaptic cleft by a paracrystalline structure composed of presynaptic dense projections and synaptic vesicles in hexagonal array in close contact with the presynaptic membrane fronting the synaptic cleft. Evidently the presynaptic vesicular grid must work in a global manner conserving vesicles so that an impulse causes a single exocytosis with a low probability in accord with the fluctuation analysis of the unitary EPSPs in Fig. 23.

In his book the quantum physicist Margenau (1984) suggested that a non-material mental event such as an intention or thought could influence neural events at microsites without violating the conservation laws of physics. In a recent article (Eccles, 1986a) I have proposed that because

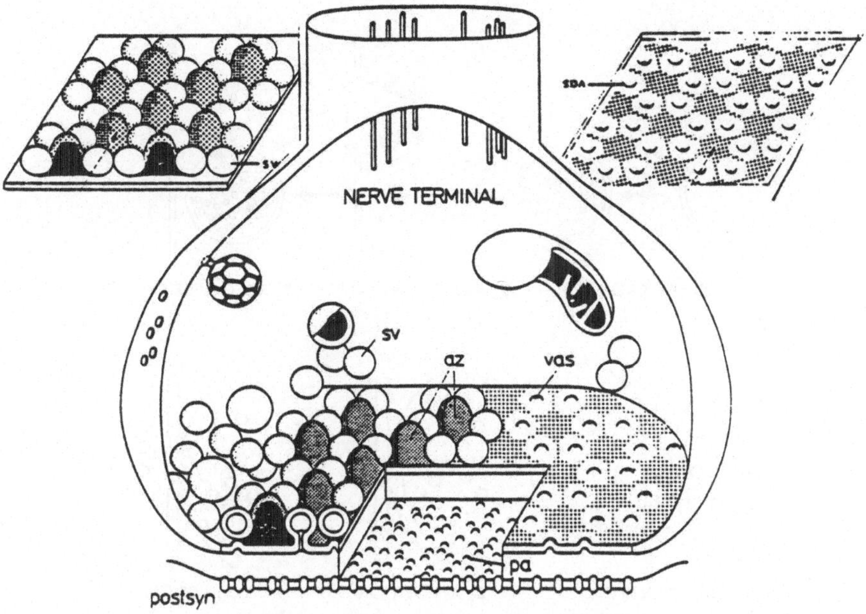

Fig. 24. Idealized synaptic bouton showing in partial perspective the presynaptic vesicular grid confronting the synaptic cleft with synaptic vesicles (sv) in hexagonal array between the presynaptic dense projections (az). The bouton is partly cut away to show the subsynaptic membrane, and to the right only the base plate (vas) of the presynaptic vesicular grid is shown (Akert et al., 1975).

of the probability of vesicular emission, the presynaptic vesicular grids are ideally fitted to be the targets for non-material mental events such as an intention for some movement. There need only be a change in probability of vesicular emission. Fig. 25 illustrates the movement of a synaptic vesicle up to the presynaptic membrane and its emission of transmitter (exocytosis). The structural change in opening the gate requires a gap of about one millionth of a centimeter in the double membrane about one millionth of a cm. thick, so that the mass involved in exocytosis is about 10^{-18} g., which is of the order of magnitude for the effect of a quantal probability field as suggested by Margenau for the operation of a non-material mental event without breaking the conservation laws of physics.

It is of the greatest importance that the synaptic design allows this microsite operation for the explanation of the hitherto intractable mind-brain problem. The effect of a probability variance of one vesicular emission is many orders of magnitude too small for an explanation of

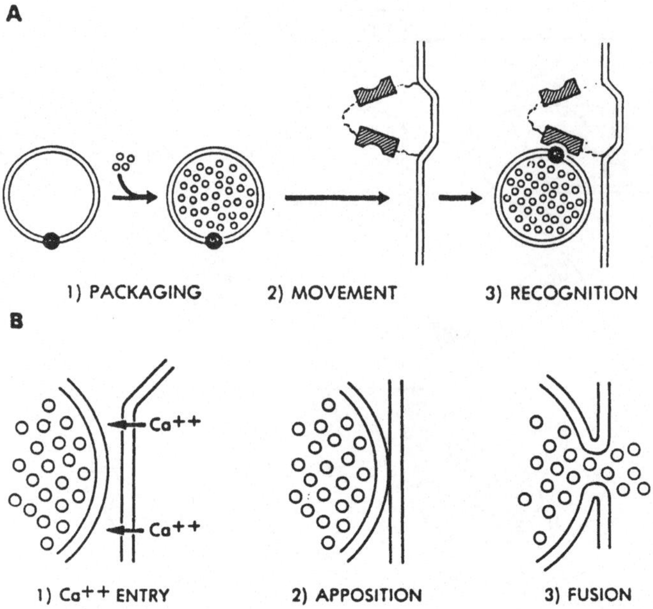

Fig. 25. Stages of synaptic vesicle development, movement and exocytosis. (A) The three steps involved in filling a vesicle with transmitter and bringing it to attachment to a presynaptic dense projection of triangular shape. (B) Stages of exocytosis with release of transmitter into the synaptic cleft, depicting the essential role of Ca^{2+} input from the synaptic cleft (Kelly et al. 1979).

neuronal discharges observed, for example, in the supplementary motor area in carrying out an intention to move. However, there are about 10,000 spine synapses on one pyramidal cell and many pyramidal cells are assembled in bundles, so that there is great amplification.

This hypothesis that mental events act on probabilistic synaptic events in a manner analogous to the probability fields of quantum mechanics seems to open up an immense field of scientific investigation both in quantum physics and in neuroscience. Sherrington would have been fascinated by the tremendous developments that have come from his simple concept of the synapse.

REFERENCES

Akert, K., Peper, K. and Sandri, C. (1975) Structural organization of motor end plate and central nervous synapses. In: Cholinergic Mechanisms. Waser, P.G., ed. Raven: New York, pp 43-57.

Bliss, T.V.P. and Lømo, T. (1973) Long-lasting potentiation of synaptic transmission in the dentate area of the anaesthetized rabbit following stimulation of the perforant path. J. Physiol. (Lond.) 232, 331-356.

Bodian, D. (1966) Synaptic types on sprinal motoneurons: an electron microscopic study. Bull. Johns Hopk. Hosp., 119, 16-45.

Brock, L.G., Coombs, J.S. and Eccles, J.C. (1951) Action potentials of motoneurones with intracellular electrode. Proc. Univ. Otago Med. Sch., 29, 14-15.

Brooks, C. McC. and Eccles, J.C. (1947) An electrical Hypothesis of Central Inhibition. Nature, 159, 760-764.

Curtis, D.R. and Eccles, J.C. (1960) Synaptic action during and after repetitive stimulation. J. Physiol. (Lond.), 150, 374-398.

Eccles, J.C. (1986a) Do mental events cause neural events analogously to the probability fields of quantum mechanics? Proc. Roy. Soc. B:227, 411-428.

Eccles, J.C. (1986b) Mammalian systems for storing and retrieving information. In: Cellular Mechanisms of Conditioning and Behavioural Plasticity. Woody, C.D., ed. Plenum: New York, in press.

Eccles, J.C., Fatt, P. and Koketsu, K. (1954) Cholinergic and inhibitory synapses in a pathway from motor-axon collaterals to motoneurone. J. Physiol. (Lond.) 126, 524-562.

Eccles, J.C., Fatt, P. and Landgrén, S. (1954) The "direct" inhibitory pathway in the spinal cord. Aust. J. Sci. 16, 130-134.

Eccles, J.C., Katz, B. and Kuffler, S.W. (1942) Effect of eserine on neuromuscular transmission. J. Neurophysiol. 5, 211-230.

Eccles, J.C. and Sherrington, C.S. (1930) Reflex summation in the ipsilateral spinal flexion reflex. J. Physiol. (Lond.) 69, 1-28.

Gray, E.G. (1959) Electron microscopy of synaptic contacts on dendritic spines of the cerebral cortex. Nature, 183, 1592-1593

Gustafsson, B. and Wigström, H. (1986) Hippocampal long-lasting potentiation produced by pairing single volleys and brief conditioning tetani evoked in separate afferents. J. Neurosci. 6, 1575-1582.

Jack, J.J.B., Redman, S.J. and Wong, K. (1981a) The components of synaptic potentials evoked in cat spinal motoneurones by impulses in single group Ia afferents. J. Physiol. (Lond.) 321, 65-96.

Kelly, R.B., Deutsch, J.W., Carlson, S.S. and Wagner, J.A. (1979) Biochemistry of Neurotransmitter Release. Ann. Rev. Neurosci. 2, 399-446.

Landgrén, S., Phillips, C.G. and Porter, R. (1962) Minimal synaptic actions of pyramidal impulses on some alpha motoneurones of the baboon's hand and forearm. J. Physiol. (Lond.) 161, 91-111.

Lundberg, J.M. and Hökfelt, T. (1983) Coexistence of peptides and classical neurotransmitters. Trends in Neurosci. 6, 325-333.

Margenau, J. (1984) The Miracle of Existence. Ox Bow Press: Woodbridge

Poritsky, R. (1969) Two and three dimensional ultrastructure of boutons and glial cells on the motoneural surface in the cat spinal cord. J. Comp. Neurol. 135, 423-452.

Sherrington, C.S. (1906) The Integrative Action of the Nervous System. Yale University Press: New Haven

Sherrington, C.S. (1925) Remarks on some aspects of reflex inhibition. Proc. Roy. Soc. B:97, 519-545

Sherrington, C.S. (1929) Some functional problems attaching to convergence (Ferrier Lecture). Proc. Roy. Soc. B:105, 332-362.

Sneddon, P. and Westfall, D.P. (1984) Pharmacological evidence that adenosine triphosphate and noradrenaline are co-transmitters in the guinea-pig vas deference. J. Physiol. (Lond.) 347, 561-580.

Swanson, L.W. (1983) Neuropeptides - new vistas on synaptic transmitters. Trends in Neurosci. 6, 294-295.

Uchizono, K. (1965) Characteristics of excitatory and inhibitory synapses in the central nervous system of the cat. Nature 207, 642-643.

THE GASKELL EFFECT AND A HUNDRED YEARS ON

O.F. Hutter

Institute of Physiology
University of Glasgow
Glasgow G12 8QQ, United Kingdom

In choosing W.H. Gaskell as fountainhead for this lecture, I court the danger of over-familiarity, if not with the man, then with his work. For every student of physiology knows about the myogenic origin of the heartbeat and about the layout of the autonomic nervous system, which facts owe their establishment to Gaskell in large measure. However, not all of Gaskell's work gained equally ready acceptance. His discovery of the hyperpolarizing effect of vagus stimulation attracted controversy from the beginning, and the chequered history of this phenomenon is my topic today.

Walter Gaskell went up to Cambridge in 1865. As a student he took Mathematics before beginning to study for a medical career. He attended classes in Physiology given by Michael Foster, and at Foster's suggestion he cut short his clinical training at University College London and entered upon physiological research (Langley, 1915).

Gaskell's work on the nature of cardiac rhythmicity and its nervous control appeared in the 1880's. To set the stage, allow me to take you back to the middle of the 19th century, when Remak, Ludwig and Bidder and Rosenberger discovered the ganglion cells of the frog heart. At about the same time also, the existence of rhythmically active neurones in the respiratory centre was first recognized. Thus it became natural to assume that the intracardiac ganglion cells likewise discharge rhythmically and thereby cause the heart to beat. As a corollary, the inhibitory action of the vagus on the rhythm of the heart was attributed to some interference with the presumed intracardiac motor centres. It all amounted to a coherent theory according to which nerve cells initiated all forms of muscular activity; and for several decades this theory was little questioned, perhaps because it was so consistent with the hierarchical social order of the age.

The first challenge to this supposedly total hegemony of nerve cells over muscle cells, came from Theodore Engelmann of Utrecht, who found peristaltic activity in sections of the ureter devoid of ganglion cells. The apex of the ventricle, which is likewise free of ganglion cells, was also found to be capable of self-sustained rhythmic activity, especially if it was stretched a little. But this finding was open to the interpretation that stretch acts as a substitute stimulus for the stimulus normally emanating from cardiac ganglion cells. It remained for Gaskell (1883) to brush away such subterfuges and to discard the neurogenic theory of the heart-beat in its entirety. In experiments done mostly on the tortoise heart, Gaskell found no correlation between the rhythmical power of different parts of the heart and the presence of ganglion cells. Today, it would be difficult to repeat these experiments exactly. They were done concurrently with Sidney Ringer's work, when physiologists at large still used saline mixed with blood to keep tissues moist. It could well be that in some instances the potassium concentration was lower than would now be considered normal, and that this promoted spontaneous activity, particularly of ventricle strips[1]. However that may be, Gaskell found that, after a period of standstill of variable duration, all parts of the tortoise heart are capable of developing a rhythm. Hence he concluded as follows:

> Since the purely myogenic rhythm of the apex is closely related to that of the ventricle and therefore...to that of the auricle, and since no line of demarcation can be drawn between the rhythm of the auricle and that of the sinus, the logical conclusion is that the rhythm of the sinus and therefore of the whole heart depends on the rhythmical properties of the muscular tissue of the sinus, and not upon any special rhythmical nervous apparatus (Gaskell, 1883)

In so rendering the neurogenic theory redundant, Gaskell took a bolder step than it now seems: for at first he had no alternative role to propose for the ganglion cells in the heart. That had to await completion of his

----

1. In later years, Gaskell (1900, p. 176) failed to obtain spontaneously beating preparations from tortoise ventricle.

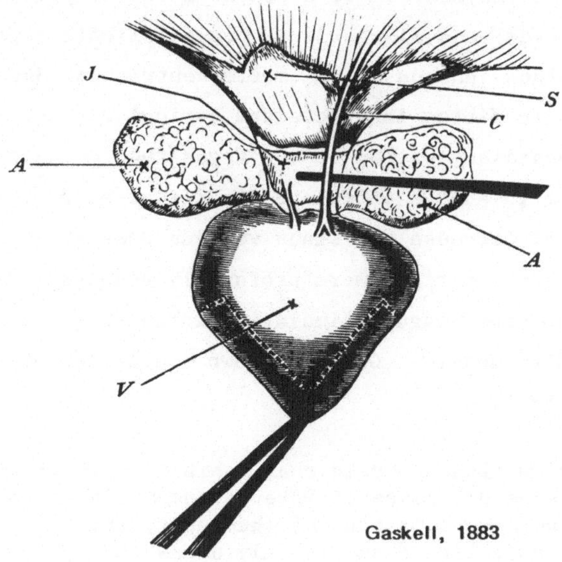

Fig. 1.

subsequent studies on the myelination of visceral nerve fibres. Only then was he able to declare the ganglion cells of the heart to be:

> part of the great group of ganglion cells which are situated in the course of the small-fibred efferent nerve supply to the viscera.... They are cells connected only with the inhibitory fibres of the vagus and as such are simply part and parcel of the mechanisms of inhibition.

This new role assigned by Gaskell to the intracardiac ganglion cells implied that the seat of vagus inhibition must be the heart muscle itself. In point of fact, the Brothers Weber had already noted, when they first discovered the action of the vagus, that in frog the vagus both slowed the heart and weakened the beat. But sight of the reduction in contractility was lost subsequently, in controversies which originated from the admixture of sympathetic fibres in the vagus of the frog. That this anatomical complexity may be overcome by stimulating the intracranial vagal roots was first shown by Haidenhain. Adopting this method, Gaskell re-established that vagus stimulation diminishes the contractions of the frog heart; and he went on to show that this effect is not simply a consequence of the concomitant slowing in rate. To make this point, Gaskell availed himself of an anatomical peculiarity of the tortoise heart. Fig. 1 is a woodcut from Gaskell's paper of 1883. It illustrates a tortoise heart from the posterior aspect. The special feature brought out by the seeker is a bundle lying outside the heart. It contains a coronary vein, which drains into the sinus venosus; and alongside it pre-ganglionic vagal fibres running from the sinus venosus to the atrio-ventricular junction. From there post-ganglionic fibres innervate the auricular musculature. Gaskell called this extra-cardiac portion of the vagus the "coronary nerve". He contrived to ligate it and to stimulate its distal end. This left the heart rate unchanged (because the sinus venosus remained uninfluenced) but the contractions of the auricle were profoundly reduced. Here then, was unequivocal evidence that vagus stimulation influences also the contractile properties of cardiac muscle. Gaskell's own evaluation of this and related findings reads as follows:

> the general view hitherto held that inhibition is to be explained by some processes of interference which take place in the nervous apparatus of the heart, and therefore the muscular tissue is only a passive instrument in the hands of such a mechanism. Such a mechanism is obviously untenable in the face of evidence which now exists to show that all the properties of the muscular tissue are deeply affected by the stimulation of the inhibitory fibres.... (Gaskell, 1886a)

Gaskell then takes a giant conceptual leap forward in the following inimitable words:

> There is to my mind no greater mystery involved in the conception of a nerve of inhibition than in the conception of a nerve of contraction.... Strong evidence in favour of this would be afforded if it could be proved that stimulation of an inhibitory nerve brings about an electrical change in a tissue of an opposite sign to that which occurs when the motor nerve is stimulated. (Gaskell, 1886b)

Such proof Gaskell was soon able to furnish.

The preparation which Gaskell used is shown in Fig. 2. The heart of the tortoise is cut through between the sinus and auricles, which causes the auricles and the ventricle to stand still for some time. The severed heart is held fixed by a clamp on the truncus arteriosus. A lever is attached to the apex of the right auricle. The apex is injured by heating and the resulting demarcation current is led off with non-polarizable electrodes to a galvanometer. The coronary bundle is left intact and with it the vagal nerve supply to the auricles. In the neck, the vagus nerve is isolated and lifted clear for stimulation.

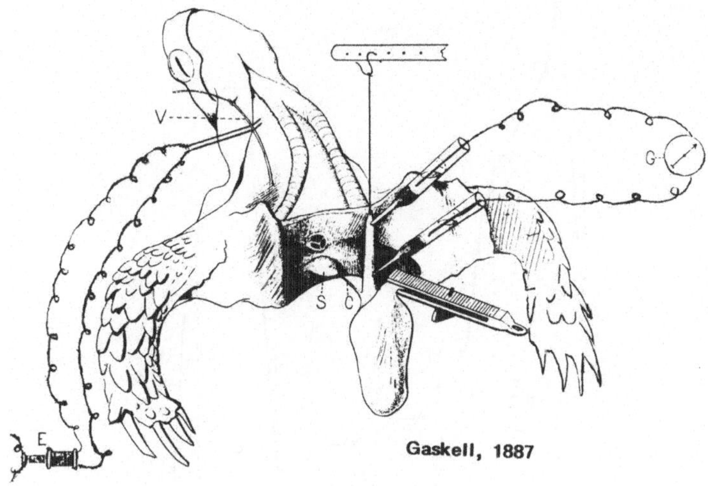

Fig. 2

We may pause to marvel at this masterly choice of experimental conditions. With the heart quiescent and no action currents, the full sensitivity of the galvanometer was available to detect any change in the demarcation current. With the auricle suspended and drained of blood, the shunting of demarcation current would be minimal. And with the vagus lifted free for stimulation and the coronary nerve the only connection between the body and the heart, stimulating current would not reach the galvanometer.

Gaskell's arrangements for data processing - as it is now called - were equally rigorous, if primitive. The demarcation current was registered every five seconds by an observer stationed at the galvanometer, who was not necessarily aware when the vagus was stimulated. The plot of Fig. 3 is the result of an experiment done by Gaskell on 6th July 1886, a hundred years ago, almost to the day. The ordinate represents divisions of the galvanometer scale and the fine divisions of the abscissa intervals of five seconds. During the course of the experiment the base line demarcation current fell progressively. Nowadays, we can recognize this as due to the gradual "healing over" of the injured apex. The periods of stimulation are indicated by the arrows. Each time the right vagus was stimulated the demarcation current increased, that is to say, a positive variation was observed. In today's language, this represents an hyperpolarization of the intact tissue at the base of the auricle. The phenomenon was abolished by atropine and by ligation of the coronary nerve,

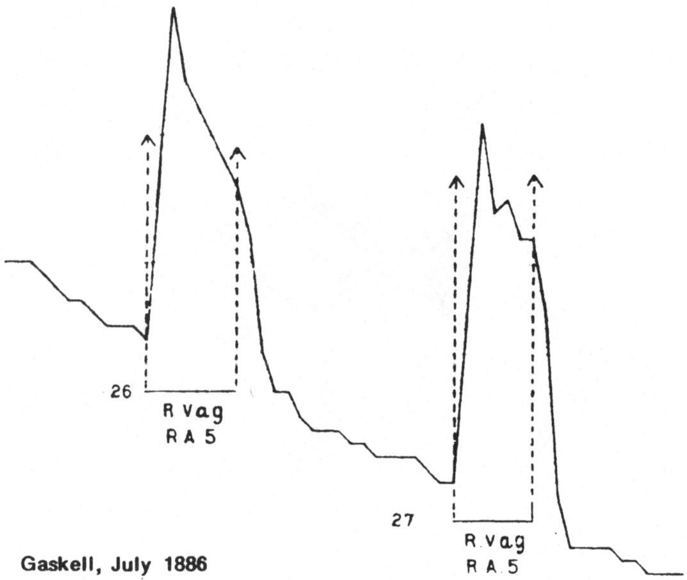

Fig. 3.

proof enough that it was not an artefact. As regards the mechanical state of the auricle, this remained uninfluenced by vagus stimulation.

Gaskell (1887a) published his findings in a Festschrift for Carl Ludwig's seventieth birthday. Gaskell had spent his first year after giving up clinical studies with Ludwig in Leipzig; so it was fitting for him to offer a tribute to that master of experimental physiology. To keep his English speaking colleagues informed, an abbreviated account appeared in the Journal of Physiology (Gaskell, 1887b).

The scene now shifts from Cambridge to Oxford where the Physiological Society held a meeting in July, 1887. In the Chair was Sir John Scott Burdon Sanderson. While still at University College London, Burdon Sanderson and Page had laid the foundation of cardiac electrophysiology with the capillary electrometer and photographic recording. They even reached the conclusion that the peak amplitude of the negative variation, that is the monophasic action potential, exceeds the resting demarcation potential (Burdon Sanderson and Page, 1879). On moving to Oxford in 1883, Burdon Sanderson took with him the then Sharpey Scholar at University College, Francis Gotch. Gotch had been brought up on the capillary electrometer, and with it he gave many demonstrations on electrophysiological topics to the Physiological Society. His demonstration at the 1887 Oxford meeting is recorded by the following entry in the Proceedings.

> Mr. Gotch showed an experiment in which a sensitive capillary electrometer was placed in connection with the beating auricle of the Tortoise heart. The heart was prepared according to Dr. Gaskell's plan.... The electrometer showed the existence of a demarcation current and the usual rhythmical monophasic excitatory electromotive change. The trunk of the right vagus nerve was excited and the heart inhibited. It was observed that during inhibition the top of the mercurial column remained steady at the level which it assumed during diastole. (Gotch, 1887)

Thus in the beating preparation, Gotch failed to find the hyperpolarization Gaskell had observed in the quiescent auricle. However, the credibility of Gaskell's effect was rescued by Burdon Sanderson (1887). He reminded the Meeting that an increase in demarcation current such as had been found by Gaskell galvanometrically might be driven by a potential difference so small as to be undetectable by the capillary electrometer.

That evening the twenty members of the Society attending the Meeting, including Gaskell, dined at Magdalene College. We shall never know how the conversation went between Burdon Sanderson, Gaskell and Gotch. Certainly, their outlook differed. Gaskell considered the quiescent, but inhibited auricle preparation to be in a state of perfect rest. Accordingly, he interpreted the positive variation he had observed on vagus stimulation as the electrical sign of processes opposite in direction to those occurring during excitation. And since excitation was known to cause breakdown of metabolites, that is catabolism, Gaskell proposed that inhibition promotes the build up of metabolic reserves, that is anabolism. Of course, it was not really necessary for Gaskell to commit himself to such a specific theory. But, according to Langley (1915), he was a man given to generalization and he was seeking to assign a survival value to inhibitory nerves. Burdon Sanderson (1887), on his part, leaned to the more conservative view of the vagus as a quelling nerve "of which the function is to quiet down an existing excitation". This implied that in the quiescent auricle used by Gaskell there would need to be present a remnant of excitation. In 1887, neither party was able to carry the argument much further. Here it is salutary to recall that Arrhenius had then only just published his classic paper on the "Dissociation of Electrolytes"; and the

Fig. 4.

Twentieth Century had arrived before Bernstein introduced the concept of membrane potential into physiology.

Fig. 4 is a portrait of Gaskell in later life. His interests had by then turned to morphological problems; however he continued to teach physiology and to inspire his students. Schafer's Text Book of Physiology published in 1900 carries a chapter by Gaskell on cardiac muscle. It is a masterful account, and includes an historical overview (which may be consulted for references, here omitted, to work done in the mid-19th Century), but in essence it is a restatement of the position in 1886/87; for little of immediate bearing on the nature of vagus inhibition had been added in the meantime.

The subject came to life again in 1908. By then, Willem Einthoven had invented the string galvanometer. In experiments on dogs, Einthoven (1908) was impressed with the fact that during vagal stimulation no changes appeared in the electrocardiogram which could be interpreted as increased positivity of the basal regions of the heart; and this caused him to voice doubts about Gaskell's original observation. Across the Atlantic, Meek and Eyster (1912) similarly found no change in the baseline of the electrocardiogram during vagus inhibition; but before drawing conclusions,

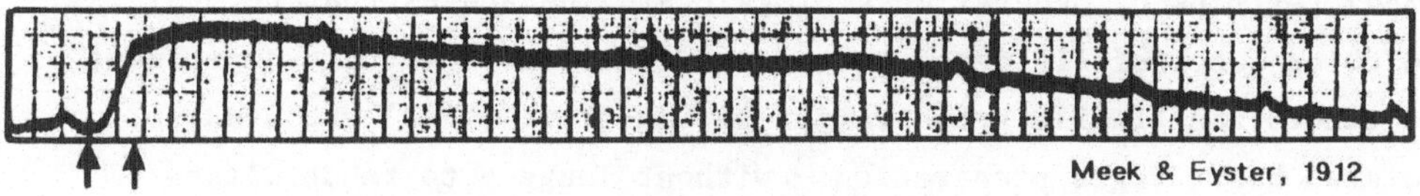

Fig. 5.

they wisely determined to use the string galvanometer under conditions more closely comparable to those of Gaskell's original experiments.

Fig. 5 is a record from Meek and Eyster's exemplary paper. Their preparation differed slightly from that of Gaskell in that only the right auricle, from which the demarcation current was recorded, was rendered quiescent. The left auricle remained connected to the sinus venosus, and spread of action currents from it accounts for the small superimposed rhythmic deflections of the trace. Between the arrows, the vagus was stimulated. This slowed the heart and duly caused a positive variation in the right auricle, as in Gaskell's experiment. In Meek and Eyster's work the positive variation amounted to about 0.3 mV, that is to say about 1/100th of the magnitude of the monophasic action potential recordable under similar conditions.

Now one would have thought that an effect so small would not have drawn Einthoven's renewed attention. But he did return to the attack because he held a theory which could not accommodate a positive variation without a mechanical relaxation. This led him eventually to claim that Gaskell's original observation, and Meek and Eyster's repetition thereof, were both due to a mechanical artefact. Not to beg a question, Einthoven attributed the Gaskell effect to stretch of the auricles secondary to a vagally induced contraction of smooth muscle in the lungs (Einthoven and Rademaker, 1917).

This episode helps to explain why Gaskell's effect was under a heavy cloud in the early decades of this century; and with it the very idea that inhibitory and excitatory nerves produce opposing effects on the membrane potential of the cells upon which they act. This allowed the resurrection of the theory that inhibition is a secondary effect of excitation (Lucas, 1917). However, Sherrington (1906) recognized in Gaskell's positive variation a mechanism consonant with the central inhibitory state he had postulated. William Bayliss also retained faith in Gaskell, and in his famous textbook of General Physiology (1915) he states that he, Bayliss, had twice successfully demonstrated the effect as a lecture experiment. The record was finally set straight by Samojloff (1923) who used an isolated heart vagus preparation - without lungs - to rehabilitate Gaskell's finding beyond reasonable doubt.

Arriving now in modern times, with the chemical theory of synaptic transmission established and the ionic nature of the nerve action potential elucidated, the ionic mechanism of the Gaskell effect was ripe for investigation. First on the scene were Burgen and Terroux (1953). Working at McGill University, they applied the intracellular microelectrode to measure the resting potential of cat auricle. In the presence of acetylcholine, or carbachol, they found the membrane potential to obey more closely the behaviour of a potassium electrode. Hence they concluded that an increase in potassium permeability underlies the inhibitory hyperpolarization, that is to say, the positive variation of old.

At this point, it becomes possible to put a modern gloss on the position taken up by Burdon Sanderson in 1887. Arguably, the potassium equilibrium potential may be regarded as the perfect state of rest and any divergence from it as a remnant of excitation. If so, then indeed the vagus can be regarded as a "quelling nerve, the function of which is to quiet down an existing excitation". As regards Gaskell's anabolic theory, its modern equivalent would be that a metabolically driven ion pump produces the inhibitory hyperpolarization. Interestingly enough, this was the interpretation first put on the inhibitory hyperpolarization of spinal motoneurones (Brock, Coombs and Eccles, 1952).

Now if the vagus quells excitation by driving the membrane potential to the potassium equilibrium potential, then the largest effects should be seen in the pace-maker fibres of the heart, where the tendency to spontaneous depolarization is most pronounced. Fortunately, the arrival of the intracellular microelectrode made the pace-maker fibres amenable to direct study, and there soon accumulated information on the effects of vagus stimulation both in the cold-blooded and in the mammalian heart.

On the left of Fig. 6 are shown examples of the effect of vagus stimulation on the sinus venosus of the frog (del Castillo and Katz, 1955; Hutter and Trautwein, 1955, 1956) and on the right the effect of vagus stimulation on the sino-atrial node of the rabbit (Bouman, 1965; Toda and

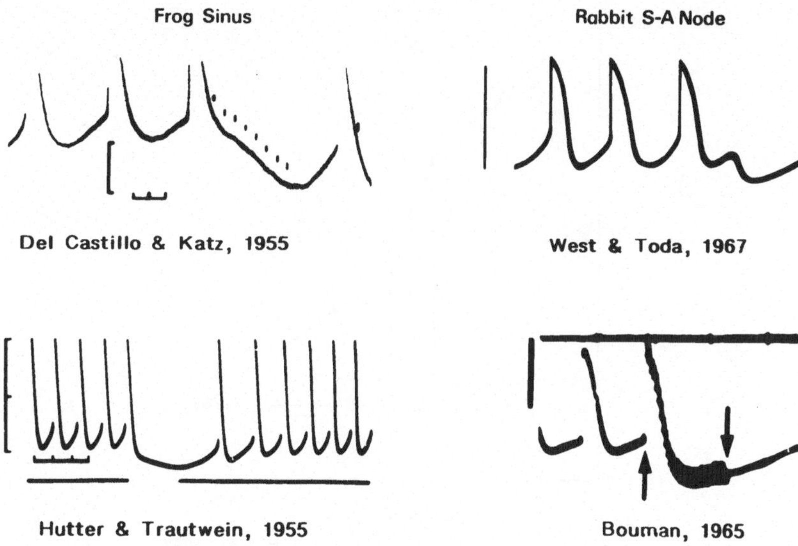

Fig. 6.

West, 1967). In every instance, the membrane potential undergoes hyperpolarization. That this is an approach towards the potassium equilibrium potential was shown by Trautwein and Dudel (1958), who measured the reversal potential of the effect. In addition, experiments with radioactive isotopes provided direct evidence for an increase in potassium permeability. The result of an experiment on an isolated sinus venosus loaded with ^{42}K is illustrated in Fig. 7. It shows that upon stimulation of the vagus at 10/sec the rate of potassium efflux increases more than two-fold (Hutter, 1957). Applied ACh produces even larger effects (Harris and Hutter, 1956; Hutter, 1961).

Fig. 8 shows yet another experimental approach. On top is an ordinary intracellular record from a rabbit SA-node. Upon vagus stimulation, the membrane undergoes a large hyperpolarization because the extracellular potassium concentration - at 2.5mM - is in this case relatively low. The lower trace is from a potassium sensitive microelectrode and it shows a 0.2mM increase in extracellular potassium concentration (Kronhaus, Spear, Moore and Kline, 1978). Net loss of potassium on vagus stimulation is in fact an old observation, first made by Howell and Duke (1908). They considered the possibility that arrest of the heart by the vagus and arrest of the heart by high potassium might have a common origin. But of course, on that hypothesis vagal arrest should be accompanied by depolarization, for which there was never any evidence.

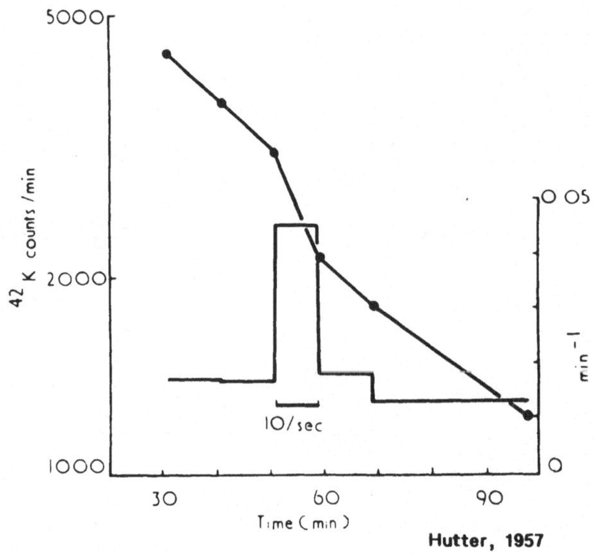

Fig. 7.

Returning to modern times, much new detail about the acetylcholine activated potassium current has been discovered by voltage clamping. The first preparations so studied were atrial trabeculae with the sucrose gap technique (Garnier, Goupil, Nargeot and Ojeda, 1976; Giles and Noble, 1976; ten Eick, Newrath, McDonald and Trautwein, 1976). Next we witnessed the successful application of the double microelectrode technique to isolated preparations of sino-atrial node (Noma and Trautwein, 1978). This method also yielded the first information on single channel properties by current noise analysis (Noma, Peper and Trautwein, 1979). More recently still, the patch clamp technique has been applied to isolated atrial cells (Sackman, Noma and Trautwein, 1983).

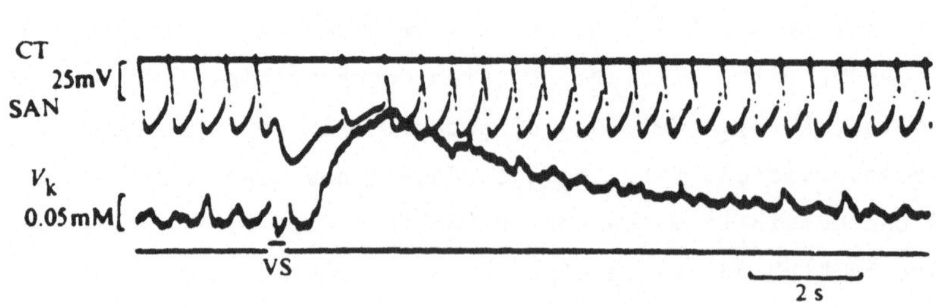

Fig. 8.

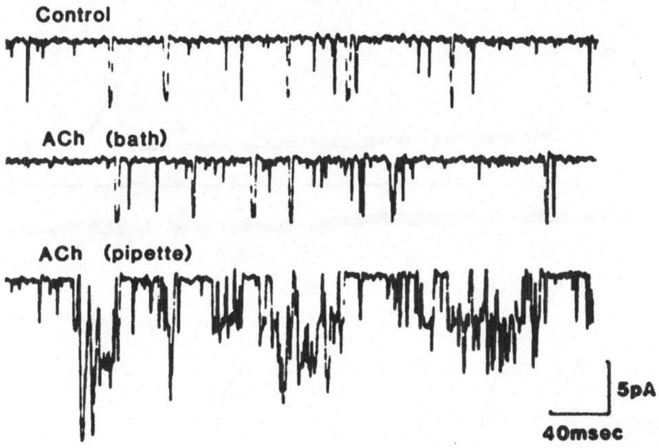

Fig. 9.

Fig. 10 shows potassium channel currents recorded by Soejima and Noma (1984) from a patch left attached to an isolated rabbit atrial cell. The top trace illustrates channel openings that occur even in the absence of acetylcholine. The middle trace shows that addition of acetylcholine to the bath, and hence to the general cell surface, does not increase the basal activity of the channels within the patch. Only when acetylcholine is introduced within the patch clamp electrode, as in the bottom trace, does channel activity greatly increase. Now, although such experiments show that diffusible intracellular messengers play <u>no</u> part in potassium channel activation, a multistep process <u>is</u> involved. This was first suggested by the sigmoidal timecourse of the hyperpolarizing effect of acetylcholine (Hartzell, Kuffler, Stickgold and Yoshikami, 1977; Pott, 1979), and has been borne out by the demonstration, simultaneously in several laboratories this last year, that a guanyl nucleotide binding protein couples the occupation of the muscarinic receptor by acetylcholine to the facilitation of potassium channel opening (Pfaffinger, Martin, Hunter, Nathanson and Hille, 1985; Breitwieser and Szabo, 1985; Sorota, Tsuji, Tajima and Pappano, 1985).

At this point I would like to broach the question: what is the mechanism of the negative effect of vagus stimulation or acetylcholine on the contractility especially of auricular musculature? As will be recalled it was that phenomenon which convinced Gaskell "that all properties of the muscular tissue are deeply affected by stimulation of inhibitory fibres".

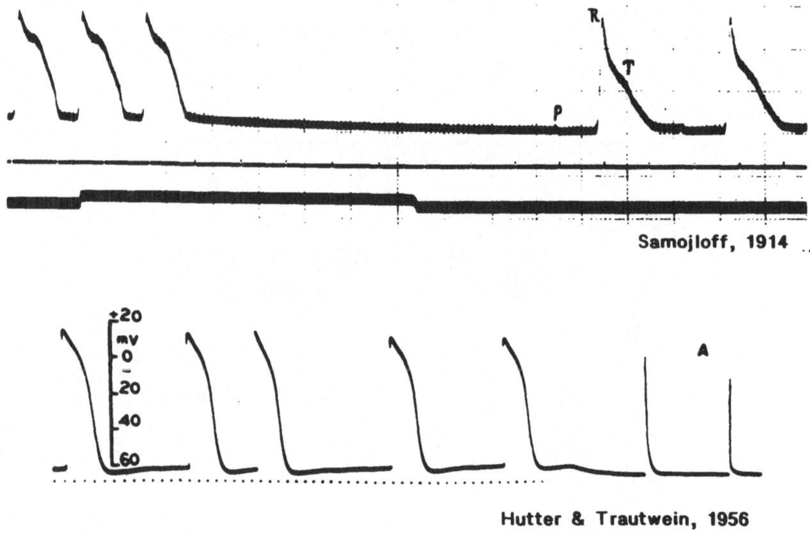

Fig. 10.

The splendid string galvanometer record from a frog ventricle, reproduced as Fig. 10 (top) was published in 1914 by Samojloff. It shows, for the first time I think, an effect of vagus stimulation on the plateau phase of the monophasic action potential: as the heart comes out of vagal arrest, so the plateau phase is foreshortened. The same effect can be seen in Fig. 10 (bottom) in more extreme form. In this tortoise preparation external stimuli were delivered at the depth of vagus inhibition. The action potentials so elicited are reduced to a needle-like shape (Hutter and Trautwein, 1956).

Now in the 1950's the role of outward potassium current in producing repolarization had just been established. It was therefore all too easy to advance the increase in potassium permeability by acetylcholine, as the explanation for the foreshortening of the action potential and in turn for the diminution of the contractions. To be sure, there were some awkward facts. For instance, in driven atrial fibres it was common to find action potentials shortened by vagus stimulation, or acetylcholine, without detectable hyperpolarization (Hoffman and Suckling, 1953; Hutter and Trautwein, 1956). But at first, this could be reconciled by supposing that the resting membrane potential was already close to the potassium equilibrium potential and therefore insensitive to an increase in potassium permeability.

An opportunity to assess how effective is an increase in potassium permeability in modifying the shape of the cardiac action potential, came along in 1961 when Denis Noble worked on his first mathematical model. I then prompted him to add a potassium conductance comparable to that produced by vagus stimulation. On the basis of the Takeuchis' (1960) work on the current-voltage relation of the acetylcholine induced motor-end plate current - then the only available model - we chose to make it an additional linear K-conductance. The computer duly produced a needle shaped action potential (Noble, 1962), and at the time that seemed to wrap up everything tidily.

Some years later, Garnier, Nargeot, Ojeda and Rougier (1978) showed that in heart muscle the acetylcholine induced K conductance in fact rectifies inwardly, so that it passes little outward current during the plateau phase of the action potential. But by then the plot had thickened again anyway, owing to the arrival on the scene of the slow inward current as a major new factor in shaping the cardiac action potential and in excitation-contraction coupling (Niedegerke and Orkand, 1966; Reuter, 1966; Hagiwara and Nakajima, 1966). As a result of this development, the question "How does sympathetic nerve stimulation augment the contractions of heart muscle?" soon received an answer in terms of an increase in slow inward calcium current, later shown to be mediated by cAMP (Reuter, 1967; Vassort, Rougier, Garnier, Sauviet, Coraboeuf and Gargouil, 1969; Tsien, Giles and Greengard, 1972). But for the best part of a decade no-one envisaged that modulation of the slow inward current might also play a part in the negative effect of acetylcholine on contractility; because the well documented and beguilingly simple potassium permeability hypothesis seemed capable of accounting for all effects of acetylcholine.

As so often happens, it required workers away from the main centres of activity to escape from a prevailing theory. In this case the new ground was broken by Prokopezuk, Lewartowski and Czarnecka (1973). They found that ACh, given after a calcium channel blocker, produces little additional effect on either the configuration of the action potential or the strength of the contraction, and on this basis they proposed that acetylcholine itself may act by inhibiting the slow inward current. Voltage clamp experiments, done in several laboratories, soon confirmed this suggestion (Giles and Tsien, 1975; Ikemoto and Goto, 1975; Giles and Noble, 1976; ten Eick, et al., 1976). But there emerged quantitative differences between the various preparations used as regards the concentration of ACh required

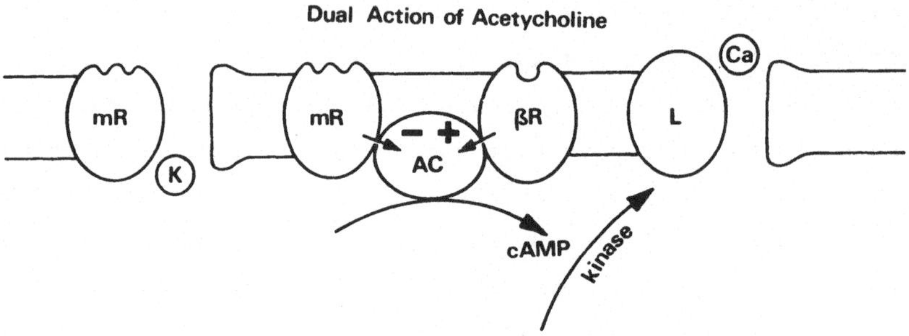

Fig. 11.

and the magnitude of the effect. One point, however, is now generally agreed: ACh is particularly effective in suppressing the slow inward current after it has been augmented by isoprenaline (Carmeliet and Mubagwa, 1986). This had led to the following picture of the dual mode of action of acetylcholine.

On the left of Fig. 11 is an inwardly rectifying potassium channel and the controlling muscarinic receptor complex. Shown in the centre is adenyl cyclase and its controlling mechanism. Beta-adrenoceptor receptor occupation stimulates and muscarinic receptor occupation inhibits the enzyme. Such cAMP as is formed as a result of this tug of war, promotes opening of calcium channels through the cascade process involving protein kinase. This hypothesis has been put on a solid footing by Hescheler, Kameyama and Trautwein (1986) who showed that ACh fails to defacilitate inward calcium current if cAMP or the catalytic sub-unit of protein kinase is introduced into the cell artificially. In this diagram (Fig. 12) I have glossed over the pivotally important GTP binding proteins. But even so, it will be obvious that here is a system prone to give quantitatively different results in metabolically different circumstances. Here also is the kind of system by which the augmentor and inhibitory nerve supply to the heart can bring about, through an intracellular messenger, those "deep

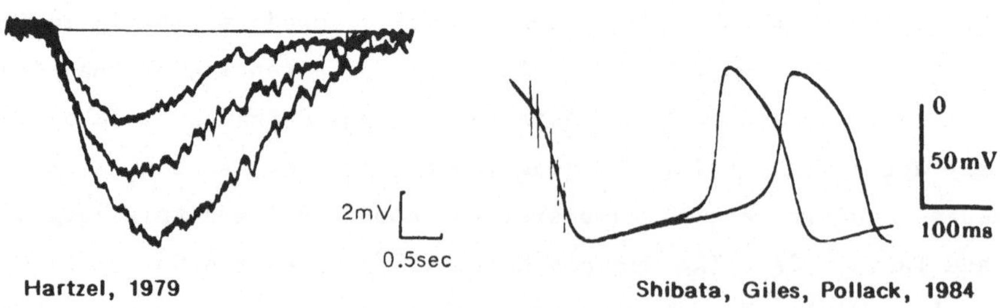

Fig. 12.

alterations in all the properties of the muscular tissue" which Gaskell suspected to exist.

Now there is one final twist to the story. In 1969, Paes de Carvalho, Hoffman and de Paula Cervalho suggested that the slow upstroke of the action potential in sino-atrial node and atrio-ventricular node fibres is of different ionic origin than the faster upstroke in the rest of the heart. About the same time, it was found that pacemaker activity is insensitive to TTX but easily suppressed by Mn ions (Yamagishi and Sano, 1966; Babskii, Berdyaev and Khorunzhii, 1973). Taken together, these findings indicated that the depolarizing current in pacemaker tissue is the inward current flowing through calcium channels, and this was confirmed in voltage clamp experiments by Noma and Irisawa (1976) and Brown, Giles and Noble (1977). The question therefore arises, "how far does suppression of the inward current by ACh account also for the slowing of the heart on vagus stimulation?" Or, may we retain the earlier-established view that an increase in potassium permeability is responsible? Here controversy still reigns.

On the left of Fig. 12 is a recent demonstration by Hartzell (1979) of the hyperpolarizing effect of vagus stimulation. A frog sinus venosus was rendered quiescent by treatment with a calcium entry blocker. The graded effects were produced respectively by 1, 2 or 3 stimuli to the vagus. Since even a single stimulus sufficed to produce an hyperpolarization, it is hard to escape the conclusion that at least a small increase in potassium permeability will occur also in the beating sinus venosus whenever the vagus is active.

On the right of Fig. 12 is a recent result by Schibata, Giles and Pollack (1985) which questions the significance of any increase in potassium permeability. Action potentials from a rabbit primary pacemaker are so superimposed as to bring out a change in frequency. The slowing shown was produced by a train of six field stimuli, which released acetylcholine from the vagal nerve endings. No hyperpolarization was detected in this experiment. Rather, the slowing was due to a reduction in the slope of the last portion of the pacemaker depolarization, which is attributable to a reduction of inward calcium current.

More than a simple species difference between frog and rabbit is here involved. For in voltage clamp experiments on isolated rabbit atrial cells

Iijima, Irisawa and Kameyama (1985) have recently found acetylcholine to be more effective in increasing potassium permeability than in suppressing the calcium current. It may well be that there is no single answer to the problem; that as regards pacemaking also, the relative importance of the two stabilizing effects of acetylcholine depends on the general condition of the cell, in particular the background activity of adenyl cyclase. From the historical point of view, at all events, we may look at the two contrasting results reproduced in Fig. 12 as well nigh exact modern counterparts of the contrasting finding which faced the members of the Physiological Society at that Oxford Meeting a century ago.

REFERENCES

Babskii, E.B., Berdyaev, S.Yu. and Khorunzhii, V.A. (1972) (Action of manganese ions on the automatic activity of frog heart pacemakers.) Dokl. Acad. nauk, 209, 171-174.

Bayliss, W.M. (1915) Principles of General Physiology. Longmans: London.

Bouman, L.N. (1965) De Weking van de Nervus Vagus op de Prikkelvorming in de Sino-Auriculaire Knoop. S.O.L. Offsetdruk: Amsterdam

Breitwieser, G.E. and Szabo, G. (1985) Uncoupling of cardiac muscarinic and β-adrenergic receptors from ion channels by a guanine nucleotide analogue. Nature 317, 538-540.

Brock, L.G., Coombs, J.S. and Eccles, J.C. (1952) The recording of potentials from motoneurones with an intracellular electrode. J. Physiol. (Lond.) 117, 431-460.

Brown, H.F., Giles, W. and Noble, S.J. (1977) Membrane currents underlying activity in frog sinus venosus. J. Physiol. (Lond.), 271, 783-816.

Burdon Sanderson, J. (1887) Proceedings of the Physiological Society. J. Physiol. (Lond.), 8, XXVII-XXIX.

Burdon Sanderson, J. and Page, F.J.M. (1879) On the time relations of the excitatory process in the ventricle of the heart of the frog. J. Physiol. (Lond.), 2, 384-442.

Burgen, A.S.V. and Terroux, K.G. (1953) On the negative inotropic effect in the cat's auricle. J. Physiol. (Lond.), 120, 449-464.

Carmeliet, E. and Mubagwa, K. (1986) Changes by acetylcholine of membrane currents in rabbit cardiac Purkinje fibres. J. Physiol. (Lond.), 371, 201-218.

Del Castillo, J. and Katz, B. (1955) Production of membrane potential changes in the frog's heart by inhibitory nerve impulses. Nature, 175, 1035.

Einthoven, W. (1908) Weiteres über das Elektrokardiogramm. Pflügers Arch., 122, 517-584.

Einthoven, W. and Rademaker, A.C.A. (1917) Uber die angebliche positive Stromschwankung in der Schildkrotenvorkammer bei Vagusreizung. Pflügers Arch., 166, 109-143.

Garnier, D., Goupil, N., Nargeot, J. and Ojeda, C. (1976) Comptes rendus de l'Association des Physiologistes. J. Physiol. (Paris), 72, 8A.

Garnier, D., Nargeot, J., Ojeda, C. and Rougier, O. (1978) The action of acetylcholine on background conductance in frog atrial trabeculae. J. Physiol. (Lond.), 274, 381-396.

Gaskell, W.H. (1883) On the innervation of the heart, with especial reference to the heart of the tortoise. J. Physiol. (Lond.), 4, 43-127.

Gaskell, W.H. (1886a) On the structure, distribution and function of the nerves which innervate the visceral and vascular systems. J. Physiol. (Lond.), 7, 1-80.

Gaskell, W.H. (1886b) The electrical changes in the quiescent cardiac muscle which accompany stimulation of the vagus nerve. J. Physiol. (Lond.), 7, 451-452.

Gaskell, W.H. (1887a) Ueber die elektrischen Veranderungen, welche in die ruhende Herzmuskel die Reizung der Nervus vagus begleiten. In: Beitrage zur Physiologie, Ludwig, C., ed., Vogel: Leipzig, pp 114-131.

Gaskell, W.H. (1887b) On the action of muscarin upon the heart and on the electrical changes in the non-beating cardiac muscle brought about by stimulation of the inhibitory and augmentor nerves. J. Physiol. (Lond.), 8, 404-414.

Gaskell, W.H. (1900) The contraction of cardiac muscle. In: Textbook of Physiology Schafer, E.A., ed., Pentland: Edinburgh, Vol II, pp 169-227.

Giles, W. and Noble, S.J. (1976) Changes in membrane currents in bullfrog atrium produced by acetylcholine. J. Physiol. (Lond.), 261, 103-123.

Giles, W. and Tsien, R.W. (1975) Effects of acetylcholine on membrane currents in frog atrial muscle. J. Physiol. (Lond.), 246, 64-66P.

Gotch, F. (1887) Proceedings of the Physiological Society. J. Physiol. (Lond.), 8, XXVI-XXVII.

Hagiwara, S. and Nakajima, S. (1966) Differences in Na and Ca spikes as examined by application of tetrodotoxin, procaine and manganese ions. J. gen. Physiol., 49, 793-806.

Harris, E.J. and Hutter, O.F. (1956) The action of acetylcholine on the movements of potassium ions in the sinus venosus of the heart. J. Physiol. (Lond.), 133, 58P.

Hartzell, H.C. (1979) Adenosine receptors in frog sinus venosus: slow inhibitory potentials produced by adenine compounds and acetylcholine. J. Physiol. (Lond) 293, 23-49.

Hartzell, H.C., Kuffler, S.W., Stickgold, R. and Yoshikami, D. (1977) Synaptic excitation and inhibition resulting from direct action of acetylcholine on two types of chemoreceptors on individual amphibian parasympathetic neurones J. Physiol. (Lond.), 271, 817-846.

Hescheler, J., Kameyama, M. and Trautwein, W. (1986) On the mechanism of muscarinic inhibition of the cardiac Ca current. Pflügers Arch., 407, 182-189.

Hoffman, B.F. and Suckling, E.E. (1953) Cardiac cellular potentials: effect of vagal stimulation and acetylcholine. Am. J. Physiol., 173, 312-320.

Howell, W.H. and Duke, W.W. (1908) The effect of vagus inhibition on the output of potassium from the heart. Am. J. Physiol., 21, 51-63.

Hutter, O.F. (1957) Mode of action of autonomic transmitters on the heart. Brit. Med. Bull., 13, 176-180.

Hutter, O.F. (1961) Ion movements during vagus inhibition of the heart. In: Nervous Inhibition, Florey, E., ed., Pergamon: New York, pp 114-123.

Hutter, O.F. and Trautwein, W. (1955) Effect of vagal stimulation on the sinus venosus of the frog's heart. Nature, 176, 512.

Hutter, O.F. and Trautwein, W. (1956) Vagal and sympathetic effects on the pacemaker fibers in the sinus venosus of the heart. J. gen. Physiol., 39, 715-733.

Iijima, T., Irisawa, I. and Kameyama, M. (1985) Membrane currents and their modification by acetylcholine in isolated single atrial cells of the guinea-pig. J. Physiol. (Lond.), 359, 485-501.

Ikemoto, Y. and Goto, M. (1975) Nature of the negative inotropic effect of acetylcholine on the myocardium. Elucidation on the bullfrog atrium. Proc. Jap. Acad., 51, 501-515.

Kronhaus, K.D., Spear, J.F., Moore, E.M. and Kline, R.P.T. (1978) Sinus node extracellular potassium transients following vagal stimulation. Nature, 275, 322-324.

Langley, J.N. (1915) Walter Holbrook Gaskell, 1847-1914: Obituary Notice. Proc. Roy. Soc. B:88, XXVII-XXXVI.

Lucas, K. (1917) The Conduction of the Nerve Impulse. Longmans: London

Meek, W.J. and Eyster, J.A.E. (1912) Electrical changes in the heart during vagus stimulation. Am. J. Physiol., 30, 271-277.

Niedegerke, R. and Orkand, R.K. (1966) The dual effect of calcium on the action potential of the frog's heart. J. Physiol. (Lond.), 184, 291-311.

Noble, D. (1962) A modification of the Hodgkin-Huxley equations applicable to Purkinje fibre action and pace-maker potentials. J. Physiol. (Lond.), 160, 317-352.

Noma, A. and Trautwein, W. (1978) Relaxation of the ACh-induced potassium current fluctuations in the rabbit sino-atrial node. Pflügers Arch., 377, 193-200.

Noma, A., Peper, K. and Trautwein, W. (1979) Acetylcholine-induced potassium current in the rabbit sinoatrial node cell. Pflügers Arch., 381, 225-262.

Paes de Carvalho, A., Hoffman, B.F. and de Paula Cervalho, M. (1969) Two components of the cardiac action potential. I. Voltage-time course and the effect of acetylcholine on atrial and nodal cells of the rabbit heart. J. gen. Physiol. 54, 607-635.

Pfaffinger, P.J., Martin, J.M., Hunter, D.D., Nathanson, N.M. and Hille, B. (1985) GTP-binding proteins couple cardiac muscarinic receptors to a K channel. Nature, 317, 536-538.

Pott, L. (1979) On the time course of the acetylcholine-induced hyperpolarization in quiescent guinea-pig atria. Pflügers Arch., 380, 71-77.

Prokopezuk, A., Lewartowski, B. and Czarnecka, M. (1973) Cellular mechanism of the inotropic action of acetylcholine on isolated rabbit and dog atria. Pflügers Arch., 339, 305-316.

Reuter, H. (1966) Strom-Spannungsbeziehungen von Purkinje extracellularen Calcium-Konzentrationen und unter Adrenalineinwirkung. Pflügers Arch., 287, 357-367.

Reuter, H. (1967) The dependence of slow inward current in Purkinje fibres on the extracellular calcium-concentration. J. Physiol. (Lond.), 192, 479-492.

Sackmann, B., Noma, A. and Trautwein, W. (1983) Acetylcholine activation of single muscarinic K^+ channels in isolated pacemaker cells of the mammalian heart. Nature, 303, 250-253.

Samojloff, A. (1914) Die Vagus and Muscarinwirkung. Pflügers Arch., 155, 471-522.

Samojloff, A. (1923) Die positive Schwankung des Ruhestromes an Vorhofe des Schildkrotenherzens bei Vagusreizung (Gaskells Phänomen) Pflügers Arch., 199, 579-594.

Sherrington, C.S. (1906) Integrative Action of the Nervous System (reprinted 1947) University Press: Cambridge

Shibata, E.F., Giles, W. and Pollack, G.H. (1985) Threshold effects of acetylcholine on primary pacemaker cells of the rabbit sino-atrial node. Proc. Roy. Soc. B.: 233, 355-378.

Soejima, M. and Noma, A. (1984) Mode of regulation of the ACh-sensitive K-channel by the muscarinic receptors in rabbit atrial cells. Pflügers Arch., 400, 424-431.

Sorota, S., Tsuji, X., Tajima, T. and Pappano, A.J. (1985) Pertussis toxin treatment blocks hyperpolarization by muscarinic agonists in chick atrium. Circ. Res., 57, 748-758.

Takeuchi, A. and Takeuchi, N. (1960) On the permeability of end-plate membrane during the action of transmitter. J. Physiol. (Lond.), 154, 52-67.

ten Eick, R., Newrath, H., McDonald, T.F. and Trautwein, W. (1976) On the mechanism of the negative inotropic effect of acetylcholine. Pflügers Arch., 361, 207-213.

Toda, N. and West, T.C. (1967) Interactions of K, Na, and vagal stimulation in the S-A node of the rabbit. Am. J. Physiol., 212, 416-423.

Trautwein, W. and Dudel, J. (1958) Hemmende und "erregende" Weihungen des Acetylcholin am Warmblüterherzen. Zur Frage der spontanen Erregungsbildung. Pflügers Arch., 266, 653-664.

Tsien, R.W., Giles, W. and Greengard, P. (1972) Cyclic AMP mediates the effects of adrenaline on cardiac Purkinje fibres. Nature New Biol., 240, 181-183.

Vassort, G., Rougier, O., Garnier, D., Servait, M.P., Coraboeuf, E. and Gargouil, Y.M. (1969) Effect of adrenaline on membrane inward currents during the cardiac action potential. Pflügers Arch., 309, 70-81.

Yamagishi, S. and Sano, J. (1966) Effect of tetrodotoxin on the pacemaker action potential of the sinus node. Proc. Jap. Acad., 42, 1194-1196.

FADS AND FALLACIES IN CONTEMPORARY PHYSIOLOGY

Ernst Florey

Fakultät für Biologie
Universität Konstanz
D-7750 Konstanz, FRG

Ladies and Gentlemen, this is a splendid Congress, - so evidently splendidly organized. And this our science of physiology is a most splendid science. We are in the midst of an almost explosive phase of development. New findings are presented at a breathtaking rate, the boundaries of physiological experimentation and exploration are expanding in all directions and at a rate that makes it increasingly difficult to assess their full meaning. Physiology as a science can no longer be regarded as a fixed framework of reference. Traditional points of view, classical problems of physiology and its very terminology become inadequate. We realize this, of course, yet we are so tempted to exploit the new techniques which have made all these new discoveries possible that we rarely take the time to step back in order to assess the meaning of all the newly won knowledge.

This International Congress of Physiological Sciences offers an unusual opportunity to take stock of the current state of physiology. When the Programme Committee invited me to present a lecture on a topic of my own choosing, I happily accepted and supplied a list of five topics, among them the title of the present lecture. I was very pleased when the Committee encouraged me to speak on this rather general theme because this choice signifies and confirms a common concern for this our science of physiology as a developing body of knowledge.

In its abstract sense, the term "science" means systematized or ordered knowledge of natural phenomena; it implies an absolute truth for which we scientists strive. But what is truth? Is it the totality of being? Certainly not because such a definition would be an unnecessary tautology. Is truth then the "true" representation of the real world, is our science of physiology the true systematized knowledge of the functioning of real organisms? Truth thus would have an operant meaning and might be defined as a successful logical representation of organic function, success being defined as the achievement of predictability of - and hence control over - natural functions. Scientific truth thus becomes a utilitarian goal - and science becomes a moral issue because control is never purposeless but serves specific interests. Indeed, our science of physiology is usually said to have a very specific interest: that of "understanding" the functioning of the organs of animal organisms including, and foremost, those of man. To understand function means being able to explain function. The term "explanation" here signifies the reduction of a phenomenon to the underlying physical and chemical processes with the added possibilities provided by the application of cybernetics or systems theory.

Measured against the thousands of years of man's cultural development, the history of physiology covers a rather brief span of time. Physiology as we know it today began only in the last century, but if we take a look at recent textbooks, handbooks and monographs of Physiology, the impression is that it is really the last ten or twenty years of physiological research that are considered to be relevant. Our students, when asked to review a research paper older than five or ten years often feel reluctant to take its contents seriously because they regard it as coming straight from the stone age. This is the first fallacy I wish to attend to, the historical fallacy.

Indeed, progress in physiology has been staggering. The cause is twofold: an increase of well-trained physiologists, and a surge of new technological developments that have opened fields of inquiry previously out of reach, if not unimaginable. The results obtained with the new techniques are so fascinating, that preoccupation with technique overshadows the intellectual achievement made possible by them. This is the second fallacy I wish to consider: the technological fallacy.

The preoccupation with these new dimensions of observation restrains our vision of one of the most characteristic features of life on earth: the diversity of form and function, of life styles and life histories. Thus we tend to simplify, to unify, and to generalize. We invent prototypes of animals and speak of "mammals", of "vertebrates" and of "invertebrates". This is the third fallacy to which I want to direct your attention: the taxonomic fallacy.

The overemphasis of deductive reasoning also forces us to invent prototypes of organs and of cells, and tempts us to talk simply of "the heart", "the kidney", "smooth muscle", "the liver cell", "glial cells" and "neurones". The tendency to simplify concerns features of generalized functions, functions which are regarded as absolutes reminiscent of the ideas of Plato. Such idealized functions are "muscle contraction", "glomerular filtration", "synaptic transmission", "neurosecretion", "excitation-contraction coupling", "membrane transport" and numerous others. The real phenomena observable in specific animals, organs or cell types are thus regarded as models. This is a current fad much in vogue in contemporary physiology. The stomatogastric ganglion of Homarus has become a "model nervous system", kidney tubules of the flounder have been regarded as providing models of tubular function, and the electrical excitation of giant axons of squid have become the dominant model of the "nerve impulse". This is the fourth fallacy that is of concern, the simplification fallacy.

Immediately related to this fourth fallacy is a fifth one which I would like to call the anthropocentric fallacy. It is unavoidable, of course. After all we are humans and we cannot do otherwise than to view the world from a human point of view: man is the measure of all things. Its consequence is that we behave like clinical physiologists: "the heart" becomes the human heart, "muscle" is human muscle, the "central nervous system" is the human brain and spinal cord, and when we say "liver" we tend to have in mind our own digestive organ even if we study that of a rat.

There is another fallacy which, somehow, is likewise unavoidable. This sixth one may be referred to as a methodological fallacy. It results from the fact that physiologists like to work with so-called physiological preparations. What they hope to accomplish is an experiment akin to an experiment in physics where all parameters are perfectly controlled and preferably only one variable is altered. Work with isolated organs, and if at all possible with isolated cell groups, single cells, or even better with parts of cells, is preferred over experiments using an entire animal. Physiology is a highly analytical science - and "analysis" translates to mean "dissolution" or "separation". The fallacy arises when it is forgotten that parts of cells, whole cells, or even isolated organs are disconnected from their normal natural surroundings and that their behavior must necessarily be different from that which occurs within the context of the intact organism. Even an isolated receptor molecule like the now famous acetylcholine receptor, cannot be expected to behave like the receptor that is still embedded in its natural lipid matrix of the particular cell membrane from which it is taken.

Finally I wish to point out a seventh kind of fallacy which I will call the paradigmatic fallacy. This, perhaps, is the most serious one of all. All our concepts of animal organisms, and hence of the functioning of animal organisms, are based on historically derived and evolved ideas and doctrines that direct, and necessarily must limit our point of view. Permit me to enlarge further on the fallacies I have outlined.

THE HISTORICAL FALLACY

Why should we worry about the history of physiology, the historical derivation of its terms, its methods and its problems? Why be concerned about the philosophical implications of physiology, the consequences of physiological knowledge for our world view? Is it not enough to do good experiments, to search for new answers? Do we physiologists not have a satisfactory sense of direction to guide us?

The answer to these rhetorical questions must needs be a very complex one: of course it is true that every physiologist knows what he is doing. He may even be convinced that he knows why he does what he does. But does he really recognise the broad implications of his research? Do we think beyond the next publication, beyond the paper to be presented at the next meeting, beyond the critique of our presumed peers?

It is only when we view our scientific endeavours in the context of the history of science and in the context of the history of physiology that we can see them in their proper perspective - and whether we admit this or not, this perspective is a philosophical one. This is especially true for the field of neurophysiology which touches on such essential problems as the nature of memory, of perception and of consciousness itself.

Only in the historical perspective do we recognize that the very problems we study, and the methods we are employing to solve them, are determined by the usage introduced by past physiologists. The historical perspective encourages us to step out of traditions and gives us the courage to generate new ones.

THE TECHNOLOGICAL FALLACY

Physiological research, like the research in other fields of science, clearly follows current trends, current fads. Because so many other physiologists do the same kind of thing we feel confirmed in our own approach. There is positive reinforcement in the pursuit of currently established methodology. Indeed we feel inadequate if we do not apply the latest techniques. A neurophysiologist who does not wield a patch clamp is prone to suffer from an inferiority complex that compels him to search for a collaborator who can bring the new technique to his laboratory. And who can afford to admit that he is not yet using monoclonal antibodies?

It is interesting how certain approaches and techniques have come to dominate the work in physiological laboratories. About 40 years ago the first glass microelectrodes came into use; the so-called intracellular recording of membrane potentials and of potential changes then became the dominant feature of neurophysiological laboratories, so much so that a visitor to such a laboratory needed only a cursory glance at the set-ups to be satisfied that the usual equipment was at hand, and the host would say "you know what these labs look like". Indeed it became a prime challenge to succeed some day to stick a microelectrode into a human brain neurone, if only to prove that there was no "ghost in the machine", only potassium ions and a pump that held the sodium ions at bay until a nerve impulse permitted them to rush into the cell. The nerve impulse as an electrical signal came to be regarded as the essence of nervous activity, excitation and inhibition were defined in terms of increased or decreased probability of the firing of an action potential. The stage was set for the human brain to be regarded as a binary computer. Biophysics was considered to be progressive physiology, and many a department of physiology (and biophysics) prided itself on its ability to carry on successfully without the aid of chemistry!

How this has changed now! Neurochemistry, first only in the hands of homogenizing chemists, advanced to regional neurochemistry and finally arrived at the cellular level, only to join forces with immunocytochemistry and microscopy in a take-over bid for dominance on the neurophysiology scene. Suddenly the empty membranous bags called neurones became filled with exciting structures and chemicals, and the electrical events at the membrane came to be regarded as transitory signs of chains of events that reached deep inside the cells. What saved the biophysicists was the arrival of a most powerful tool, the patch clamp, which once again focused attention on the cell membrane because it enables the physiologist-biophysicist to observe directly the activity of single ionic channels. The way of neurophysiology went from the whole brain to brain centres to brain cells to single channels and transmitter molecules, and on to receptor and carrier molecules.

Suddenly the entire gamut of biochemistry has entered physiology, with the whole arsenal of chemical weaponry that is known as pharmacology - after all, what else is pharmacology but physiology with molecular tools? Even more technology has entered the field. Monoclonal antibodies, high pressure liquid chromatography, fluorescence microscopy, electronmicroscopy, backfilling with horseradish peroxidase, with nickel or cobalt, micro-gel-electrophoresis, radioimmune assays, and a host of computer programs now dominate the scene. The result is a bewildering mass of new findings, new data, new observations and the realisation that our physiological terminology has become outdated.

The new techniques are so sophisticated and precise that living systems with their inherent unpredictability are no longer adequate objects of investigation. Remember the biophysicists who resort to artificial membranes when they want to study membrane behaviour. The patch clampers are no different: they have to use cells in tissue culture when they wish to gain access to the membrane of the type of cell they want to investigate. Cells within living tissue are not suitable because they are covered by other cells, by connective tissue or by a basal lamina.

THE TAXONOMIC FALLACY

We recognize a staggering diversity of animal life and we have learned to appreciate the fact that each species is adapted to its environment in a special way and that all its body functions are thus adapted and specialized. A cat is above all a cat, and a mouse is a mouse, before it is a mammal. We know that taxonomically the domestic cat belongs to the family Felidae, and this family belongs like that of the Canidae, the Hyaenidae, the Ursidae and the Mustelidae to the order Carnivora. The mice, on the other hand, are related to the squirrels, the guinea pigs, the porcupines and the beavers - all part of the order Rodentia. Indeed mammals appear on this earth in a great variety of types: seals certainly strike one as different from elephants, the hippopotamus is quite distinct from a giraffe and a bat certainly differs from a chimpanzee. Generalizations, of course, are possible and the concept of "mammal" appears as logical as any concept of which our language is capable. Although it is difficult to imagine what a generalized mammal might look like, we find it not too difficult to write a list of common characteristics - or do we? Four legs? Think of bats! Viviparous? Think of the platypus! A fur-covered epidermis? Think of the armadillo! It is only when we study the internal anatomy that we can discover features that are not the common property of other animal groups like birds or reptiles: there is the secondary joint of the jaw, with the os articulare and os quadratum being relegated to the middle ear, there is the single right aortic arch and there is a telencephalon that has an immensely developed superficial neopallium, a true cerebral neocortex. But these general features do not make a living animal. It is only when we decide to concentrate our attention on the common laboratory mammals, that the task of defining a mammal becomes somewhat easier.

What has been said for the concept "mammal" is event more true for the concept "vertebrate". The appearance of a hagfish differs so strikingly from that of an ostrich and a rattlesnake looks so different from an orangutan that a composite picture of a "vertebrate" is out of the question. When a physiologist speaks of vertebrates, he has in his mind a picture that often strikingly resembles a cat! We simplify when we mean to generalize. What we observe in an adult cat is taken as valid for the whole lot of mammals just because we know the cat to be a mammal. But is anything that is true for another mammal always true for the cat? Certainly not!

What _is_ the characteristic of vertebrates? You might think it is the possession of vertebrae, but this is not so. The cyclostomes, so often studied by physiologists, have none! Of course everyone knows these agnathan fishes are primitive. But are we sure what "primitive" means? Systematists tell us that the cyclostomes are descendants of the earliest vertebrate types, the ostracoderm fishes. Of their internal organization we know little to nothing. The cyclostomes are parasites and their organization is highly specialized for this mode of life. Does the term "primitive" mean that we are dealing with an ancestral type of organism whose structural and physiological features represent the "original model" from which all later vertebrate functions can be derived? Parasites are certainly not ideal models of an ancestor. We ought to recognize that there is no such thing as a living fossil! The modern "primitives" cannot be assumed to have been bypassed by the process of evolution. Why they exist and how is however a most intriguing problem that challenges us as physiologists even more than it might please the evolutionist.

"Vertebrate", when placed in juxtaposition to "invertebrate" is an even more improper simplification. There is hardly anything so improbable as the construction of the category "invertebrate". Cyclostomes, as already pointed out, are without vertebrae, yet we do not regard them as invertebrates. What then is common to invertebrates? Can you find anything in common between a butterfly and a jelly fish, or between a starfish and a planarian? Indeed, for all we know, a sea cucumber is more closely related to a monkey than a lobster is to a squid. Insects are further removed from snails than are vertebrates from chaetognaths, echinoderms or tunicates. How can we generalize an invertebrate? We can not. There is a way, of course - this has nothing to do with science but is a matter of convenience: we can regard invertebrates as animals that are not vertebrates. Of the total number of recognized animal species, vertebrate animals account for less than 10% and current estimates of the actual number of animal species would give the vertebrates a figure below 5%.

How often have I read in the literature of "invertebrate muscle" or of the "invertebrate nervous system"! These terms are utterly meaningless. Try to draw a picture of such a nervous system and you will find out that you can not. Of course it is possible to sketch the nervous system of an octopus or of a spiny lobster. But what do these two animal forms have in common? Would you find a single muscle in an octopus that is homologous to, or even resembles in the slightest degree, a muscle of a lobster?

THE SIMPLIFICATION FALLACY

Generalization necessarily means simplification. Such simplifications are common in physiology. They are particularly apparent in our physiology texts and in our teaching. Simplification, like generalization is necessary. Indeed, we would be unable to go through life without it. Even our everyday communication by language is based on simplification. The fallacies arise when we become deceived by these simplifications and when we take these as the representation of the real world. Let me illustrate this by an example.

Mammals are usually regarded as homeotherms, that is they are thought of as being capable of regulating the body temperature so that is stays constant within very narrow temperature limits. This notion is false. In the first place it is the result of the taxonomic fallacy already mentioned: there is no such thing as a mammal. The term "mammal" refers not to any real animal but implies a concept, an idea - if you like, it is a common denominator of all animal forms we care to put into this category because of certain features they have in common and which set them apart from other animal forms. If in doubt what range of animal forms are included under this term "mammal", we can consult any text that is concerned with vertebrate taxonomy and will discover that the class Mammalia includes such diverse forms as the marsupials, the tenrecs, the whales, the bats, the pangolins and the camels. Are these forms homeotherms? It is known that the body temperature of the tenrec fluctuates widely over a range of some 15°C; even the camel permits its body temperature to vary over as much as 6°. When I say "the tenrec", or "the camel", I commit, of course, a fallacy: there is no such animal as "the tenrec", there is no such animal as "the camel". The tenrec I spoke of is a species called Centetes ecaudatus, and when I spoke of the camel, I meant a subspecies known as Camelus ferus bactrianus. Is it clear now what is meant by the word "tenrec" or the word "camel"? Perhaps you can form, in your mind, an image of a camel, even if you might have difficulties trying to compose a picture of a tenrec. But the camel you will be thinking of is an adult, while Camelus ferus bactrianus includes the young ones as well as the old senile ones and the embryos. We physiologists are a notoriously sloppy lot indeed!

So what about the homeothermy of mammals? Are we sure mammals are homeotherms? Of course, there are physiologists who know perfectly well that we cannot make a valid general statement like that - indeed, some suggest we use another term, that of endothermy: it is not the constancy of the body temperature that is regarded as the characteristic mammalian feature but the ability to generate enough metabolic heat to raise body temperature above ambient temperature without having to rely on radiant heat from the outside.

We see here an example of the negation of a perfectly useful physiological term that results from a conflict between simplification and the recognized diversity of the real world.

THE ANTHROPOCENTRIC FALLACY

Let me first return to the arguments of the taxonomic and the simplification fallacies: it is not only that we invented a generalized vertebrate animal, we even invented a more or less obligatory typology of organ systems, and functional definitions of these organ types that are based on an idealized understanding of the human organism. This typology is applied to non-human organisms, and the results of studies of any kind of animal, vertebrate or non-vertebrate, are considered subservient to this anthropocentric vision of idealized organ function. It is a current fad to refer to animals, or to their organs, as "model systems" or "animal models". The given animal or organ is not considered in its own right as a specialized system but as a representation of the idealized, anthropocentric type of organ function. The squid giant axon is regarded as a model of a generalized axon, and the ionic mechanism of its action potentials is considered to be generally valid. So dominant has this theory become that our textbooks do not even mention that in mammals there are neurones where the initial sodium current is not followed by an outward potassium current. But what is a "mammal" anyway?

The terms "heart" or "kidney" nowadays have a meaning that is clearly anthropocentric. Kidneys are thus regarded as organs of nitrogen excretion, even though in most vertebrates they are not! The proximal segment of kidney tubules is regarded as a structure capable of resorption of some 80% of the glomerular filtrate, even if in most vertebrates this is not so. Heart physiologists state cardiac output in terms of the minute volume pumped by the left ventricle and make it thereby impossible to compare heart performance of the mammalian heart with that, say, of a teleost fish.

If physiology is to be a true science, it must recognize the diversity of animal life before it claims generalizations based on a limited selection of convenient animal types. During the first decades of modern physiology whose beginning may be set at a time around the middle of the last century, physiology was indeed such a broad endeavour. Man was regarded as an offshoot of the animal kingdom. Physiology, insofar as it was not primarily medical physiology, was a comparative physiology, an inductive science that made it necessary to create a general physiology. Today, the term "comparative physiology" has a different meaning: it is thought of as a physiology of non-vertebrate animals, a most curious situation that demands our attention. It is the outcome of the anthropocentric fallacy which shapes the living world in the image of man and causes us to lose sight of the meaning and diversity of life. Physiology is a comparative science!

THE METHODOLOGICAL FALLACY

It is a curious fact that a physiologist interested in the function of liver cells - or hepatocytes, as they are called by the professionals - uses the methods of biochemistry to analyse the functions of these cells. His colleagues would presumably shake their heads if he would let them know that he is using microelectrodes to study the membrane potential of liver cells in order to study their mode of operation. Even if he reported that hepatocytes generate action potentials this would not impress them too much. The chemical machinery of these cells is what counts.

A neurophysiologist on the other hand can get away by pretending that his cells, the neurones, are membrane bags filled with an aqueous medium that predominantly contains inorganic ions! What satisfaction it is to be able to rip off a piece of membrane to achieve a giga-seal! The ion channels are of predominant interest. How wonderful it is to see them open and close - individually and in groups! Of course, this fun with channels is not the entire neurobiology. There are other approaches to the nervous system: there is the fascination of the cerebral cortex of the mammalian brain. Here the interest centres on the interaction of the neurones, on the reconstruction of neuronal circuits that supposedly represent the outside world as well as the mind. Neural assemblies are regarded as the equivalent of what the philosopher would call "ideas" - "to fire or not to fire, that is the question", as Shakespeare would phrase the physiologist's notion of neuronal function. The electrical nerve impulse is still regarded by many as the epitome of nervous function; the temporal and spatial patterns of nerve impulses generated by assemblies of neurones are regarded as the representation of the mind, indeed of consciousness itself. Again, this preoccupation with nerve impulses is not the entire concern of neurophysiologists. They are also interested in how neurones grow, how they differentiate, how they establish contact with other neurones, how and why they die. We are indeed in the midst of a major shift of emphasis: the development of the nervous system, the mechanisms by which neuronal circuits are established, the conditions a neurone requires for its survival -- these are major problems of present neurophysiology. Here the microelectrode and the patch clamp are not the primary tools. Instead it is tissue culture, the isolation of adhesion molecules, the use of mono- and polyclonal antibodies and the study of various chemical messengers, that have become the preoccupation of many a neurophysiological laboratory. However we must ask: what is the final goal of this research activity? Is it the understanding of how neuronal circuits arise within a given nervous system? If that is the problem, we must ask further: is the search for neuronal circuits still a legitimate concern of neurophysiology? Do we still have to believe in the utility of the concept of circuits? What indeed do we mean by the term "neuronal circuit"? Are the nerve impulses that travel these circuits the elementary units of neuronal "behaviour"?

There is increasing evidence that communication between neurones, and even between neurones and glia, involves chemical interactions far more subtle and complex than those permitted by the expedient of nerve impulses. The emerging picture of nervous function transcends the boundaries of prevailing concepts and models.

Model systems, generally, are selected because of a specialization; giant axons of squid because of their large size, the amphibian gall bladder because of the ease of preparation and because of its relatively simple structure. The gastrocnemius muscle of frogs has been selected because single electric shocks produce a large twitch contraction with a tension development that is similar in magnitude to the tension recorded during a tetanus. The sartorius muscle of rabbits and frogs has been so useful because its motor endplates are arranged in a neat row, and because the muscle is so thin and flat that its fibres and endplates can easily be visually observed and it is not difficult to separate innervated and non-innervated portions of the muscle. Torpedo electroplaques have served as model synapses because of the exceptionally large size of their motor endplates, and because of the ease with which pre- and postsynaptic membranes can be prepared. The cell bodies of neurones in the brain and ganglia of the marine snail Aplysia have been studied as model neurones because of their exceptionally large size and the ease with which they can be penetrated by one or even by several microelectrodes. Slices of the hippocampus of rat brains have been used as models of the cerebral cortex because of the simple and regular arrangement of the nerve cells. For similar reasons the heart ganglion and the stomatogastric ganglion of lobsters have been employed as even simpler model nervous systems because they consist of only a few readily identifiable neurones.

Of what are these interacting neurones models? How can something simple be a model of something far more complex? How can we base our understanding of the staggering diversity and complexity of nervous systems on such simple models? This is possible only if we reduce the problems at hand to the simple common denominator. Models do, of course, have tremendous heuristic value, but unless we realize also that what they teach us is indeed elementary, these models are dangerously deceptive.

Physiology is based on models of organs as well as of function. This is not only true for neurophysiology but for any other branch of physiology as well. Models underlie all physiological concepts and theories. When we discuss our hypotheses we have such models in mind. Conceptual models direct but also misdirect our research methods. Let me explain this by using the example of the study of what we commonly refer to as "muscle".

What do we mean by "muscle"? Most physiologists, when asked to draw a picture of a muscle come up with a structure that suspiciously resembles the gastrocnemius muscle of a frog: there is a tendon at each end, and the well known muscle belly in the middle. But anyone who has ever dissected a frog knows that most of the muscles of such an animal look quite different. Think of the sartorius, of the rectus abdominis, of the latissimus dorsi! We might perhaps argue that the shape does not matter, that the function, the contraction, is still the same. But the rectus abdominis does not contract like the gastrocnemius at all. We know that. Is it not curious that we do not include the rectus abdominis in our generalized concept of muscle?

Of course, physiologists do not think only in terms of shape and structure; they have also concepts of function - and when they study function they tend to follow established methodology. Indeed, how do we study muscle contraction? In our student laboratories we still make use of the centuries-old sciatic nerve-gastrocnemius preparation first used in the University of Bologna in the Eighteenth Century. We now use electrical stimulators to apply brief pulses of current to the motor nerve and record the resulting muscle twitch. What an extraordinary phenomenon: the entire set of muscle fibres contracts nearly synchronously. This gastrocnemius muscle is a jumping muscle, and indeed it is likely that in the live, intact animal all the motoneurones connected with this muscle fire in synchrony when the frog is ready to jump. But is this a "typical" nerve-muscle system? Is this mode of stimulation characteristic of the nervous activation of other skeletal muscles? Certainly not. Smooth movements and shifts of posture require asynchronous, perhaps even randomized nerve activity. No stimulator that I know of is capable of initiating such a randomized firing pattern in the stimulated nerve so that each nerve fibre fires more or less independently, perhaps in a unique pattern.

But who worries that the efferent nerve might contain efferent modulatory neurones, that amphibian or even mammalian skeletal muscle is sympathetically innervated? Our conceptual model does not include this feature. It is common to think of a nerve-muscle preparation as consisting of a bundle of motor axons synaptically connected with corresponding muscle fibres. The vascularization is usually ignored and what is more, the autonomic innervation of both vasculature and muscle is totally neglected. If this autonomic innervation is considered at all, it is regarded as being concerned with the control of the blood vessels. Yet there has been good evidence from work on cat skeletal muscle that the muscle fibres themselves receive a sympathetic innervation beside the motor innervation. Indeed, even if the sympathetic supply of such muscles involves no direct neuromuscular synapses it is not at all unreasonable to suppose that released transmitter can affect the muscle fibres and the motor terminals. There is well-documented evidence that noradrenaline reactivates fatigued neuromuscular transmission and enhances transmitter output. Is it not an oversimplification when we ignore such evidence?

Isolated cells or organs are placed into, or are perfused with an artificial medium of known composition. This composition is based on the known concentration of certain inorganic ions that are present in the animal's extracellular fluid compartment, with one or the other buffer added to maintain a constant hydrogen ion concentration or pH. This method was initiated by the systematic researches of the British pharmacologist Sidney Ringer who, mostly by trial and error, devised the famous "Ringer solution" for the maintenance of amphibian isolated organs, especially the frog's heart. Numerous other salines have been developed since and a literature search reveals that most experimental work in physiology still involves the use of such saline media. The control condition of the organs or cells under study is thus the behavior in an artificial, quite unnatural medium. In most cases there are no proteins present and the colloid osmotic pressure is far below that of the natural extracellular fluid. Although we are by now very well aware that cell membranes are beset with ionic pores, ion pumps, transport molecules and carriers of various kinds, and that there is a continuous exchange of ions and small molecules like CO_2, NH_3, HCO_3^-, H^+ and volatile acids going on at all times, we tend to ignore this - except in investigations specifically designed to study acid-base regulation. For all we know, extracellular fluid in the intact organism carries numerous organic humoural factors that effectively regulate cell activities. The use of artificial saline media deprives the cells and organs under study of these essential factors and may well make them behave unnaturally. Indeed some physiologists have noted that their preparations do not perform as well as they expect them to, and they have learned more or less secretly to add some blood or plasma to their saline in order to make their preparations "work". (Tissue culture experts have long known the advantages of adding a bit of "embryo extract" to their culture media!) It is naive to assume that saline-perfused physiological preparations are normal and in a control state. We may well have to recognize the need to reverse experimental procedures: not to add substances to study their effects on deprived preparations, but to remove components of normal extracellular media to study the change imposed by such deprivation. Our argument here leads on to the next fallacy.

THE PARADIGMATIC FALLACY

The use of the term "paradigm" has become a fad. It has become prominent through Thomas S. Kuhn's treatment of The Structure of Scientific Revolutions. Dictionaries generally define "paradigm" as accepted "model" or "pattern".

Kuhn uses the term in a wider sense when he refers to scientific practice. His usage includes theory, application, and instrumentation. His "paradigms" "provide models from which spring particular coherent traditions of scientific research". Paradigms thus become "conceptual boxes", and Kuhn regards "research as a strenuous and devoted attempt to force nature into the conceptual boxes supplied by professional education".

Paradigms may represent major concepts. In biology as well as in our specific science of physiology our thinking is dominated by the cell theory to such an extent that we neglect the extracellular matter of the organism under consideration. We do not, of course, ignore it but we do not regard it as part of the living system. From the physiologist's limited point of view what is alive or is living are the cells. Anything else might be regarded as a product of such cells but it is always viewed as something extraneous. The extracellular fluid is the milieu intérieur, the fluid environment of the body cells. The collagen and elastin matrix of connective tissue is regarded as nothing but a secretion product of fibroblasts. The basal lamina that is so conspicuous in the pictures provided by the electron microscopists is taken as a coating that has been somehow deposited by certain cells. Indeed anything outside the cell membranes is regarded as non-living matter, not part of the living system. I am fully aware, of course, that physiologists appreciate perfectly well the functional significance of extracellular matrix and fluids, but I feel equally sure that in their hearts and minds physiologists think of organismic functions in terms of cell function. When Max Verworn introduced the concept of a General Physiology, he expressly had in mind a cellular physiology.

Even if we do not worry about the definition of physiology, we do base our experimental approach on the cell theory to the extent that we focus our attention on cell types: the muscle physiologist on muscle cells, the neurophysiologist on nerve cells (or neurones as he prefers to call them), the endocrinologist on glandular cells, the kidney physiologist mostly on epithelial cells. So dominant is this outlook that muscle physiologists ignore even those cells that ensheathe the muscle fibres, and neurophysiologists by and large ignore the glial elements and the endothelium. The common talk about neuromuscular transmission refers simply to a presynaptic element, the nerve terminal, and to the postsynaptic element, the muscle fibre. Who bothers about the other cellular entities that form the structure of neuromuscular systems; who is concerned with the role of the extracellular matrix? Indeed, we arbitrarily set the cell boundaries at the cell membrane. Look at text book illustrations of nerve or muscle tissue and you will find this confirmed. Most neurophysiologists look at the brain as if it were composed exclusively of neurones. They try to elucidate brain function by experiments designed to discover neuronal circuits. Nervous systems are visualized in terms of the pictures emerging from Cajal's staining method. Patterns of nerve impulses are regarded as equivalent to elements of perception. To many neurobiologists the chemistry of the brain matters only to the extent that it influences neuronal firing. But in a human brain there are ten times more glial cells than neurones. The space between nerve cells is continuously filled by glia and blood vessels. Can we afford to ignore this?

Physiological research is indeed based on numerous paradigms which determine its various methods and techniques. A simpler, more common word for methods and techniques is "approach". Fallacies arise when the approach is too narrow or too selective. Take the concept of the "nerve impulse". We tend to regard it as an electrical event, a signal that is propagated along an axon and has the significance of being an element of information transfer. According to this paradigm the signal is an electrical process. The equipment applied to investigate it therefore, consists of electrical and electronic apparatus, and the results obtained by the application of theories of electricity and magnetism are of necessity refinements of an electro-physical theory of the operation of the nervous system. The limitation of methodology has already been referred to as the methodological fallacy. The paradigmatic fallacy is more serious because it provides a rationale for the methodological fallacy: the paradigm provides the theoretical justification for the choice of method to be applied in our research effort. If we had a chemical theory of the nerve impulse, or if the paradigm of nervous activity would include osmotic and mechanical processes, our approach to the nervous system would be very different. Indeed it looks now as if our views of the nervous system will become immeasurably wider and I am sure our Congress will provide numerous examples of this dramatic enlargement of neurophysiology. New paradigms will have to replace many an older one - no doubt, they will again and inevitably result in new fallacies.

It is not only the kind of global paradigm I wish to refer to but the large number of much more limited paradigms that dominate laboratory work and may set it off in the wrong direction even though they follow well established tradition. Indeed, it is the very fact that accepted paradigms include the acceptance of restricted methodology that can be seriously misleading.

Let me illustrate this by some examples. When studying ion fluxes across cell membranes we use standard techniques that include ion substitution. It is often forgotten, however, that such ion movements result in shifts of intracellular pH and in movement of water which, in turn, lead to pH and volume regulation. Measurements of membrane potential and membrane current can thus give rise to complex results and the interpretation in terms of a narrow paradigm may be quite inadequate.

The restricted application of only electrophysiological technique, so prevalent in many laboratories, can obscure important physiological processes. A similar critique can be applied to the use of pharmacological methods in research designed to study membrane transport or epithelial transport of small molecules or ions: just because a certain drug is known to inhibit a certain process, say chloride transport in one system, the results of its application in another context do not necessarily mean that in this system too, chloride transport has been affected. Too many cases of multiple action have already been observed to caution against such simple conclusions, even if the use of such a drug has become standard procedure and has been part of a paradigm.

Substitution of propionate ion for chloride ion is often made use of in experiments designed to study chloride permeability changes of cell membranes. It is sometimes forgotten, however, that propionic acid can penetrate cell membranes while propionate ions cannot. Since the ratio of the undissociated ion to the dissociated one is equal to the antilog of pK minus pH, but the extracellular pH demands a constant ratio, any propionic acid that penetrates into the cell will cause formation of more undissociated propionic acid from propionate ion. The consequence would be not only a loss of chloride ions from the cells but considerable swelling which, in turn, might lead to volume regulation, intracellular pH changes and a host of subsequent effects.

Let me turn once more to some paradigmatic fallacies of neurophysiology. The event of a nerve impulse that results from the integration of synaptic inputs is often taken as the decisive signal, and the central nervous system is hence regarded as a binary computer whose mode of operation is based on temporal and spatial impulse patterns. Such a view ignores the fact that even in the absence of nerve impulses, neurones affect each other synaptically and non-synaptically through discontinuous quantal, and through continuous non-quantal release of transmitter. The latter form of transmitter release may be quantitatively more important than impulse-coupled release. The effect of released transmitter may not result in a conductance change in the subsynaptic or postsynaptic membrane, yet it might be profound in as much as it alters intracellular chemical events of the affected cell. Communication between neurones is thus far more subtle than can be understood on the basis of impulse patterns. The recognition of the importance of interactions between neurones and glia will add a further dimension to our understanding of brain function.

It is a curious fact that neurophysiologists look at central nervous systems largely in terms of input/output relations and study them primarily as sensory-motor systems.

The autonomic nervous system is regarded as a special subdivision of the peripheral nervous system that is concerned with the control of vegetative functions, especially those of the various types of smooth muscle, particularly those of the gastrointestinal tract and of blood vessels. The nervous control of epithelial transport receives, by comparison, far less attention. The function of the massive peripheral ganglia of the autonomic nervous system remains largely unknown. There is intensive investigation at the cellular level but we have little idea of the significance of the complexity of this peripheral nervous system.

At the present time we are only dimly aware of the role of the nervous system, especially its autonomic division, in the control of the immune system and the control of haemopoesis.

It is a paradigmatic fallacy to stay within the bounds of traditional and well established sets of problems and associated methodology. We need to step over the confines of established paradigms if we wish to achieve major advances in our science. In numerous special areas of physiology, including neurophysiology, this is already being done - and to greatest advantage.

Ladies and Gentlemen, physiology is a splendid science, in spite of its many shortcomings. The spectre of the fads and fallacies I raised before your attentive minds is a necessary consequence of the fact that science by its very nature follows and creates paradigms, generates concepts, theories and hypotheses. These paradigms are, and must be, confined to a limited point of view and can represent only a highly selective aspect of the real world.

The methods we follow, the techniques we employ, must sooner or later present us with data and results that force us to recognize this limitation and the inadequacy of our initial framework of hypotheses and theories. If we recognize the nature of the processes inherent in scientific reasoning we can easily cope with the fallacies. Scientific progress is never continuous but proceeds through shifts of emphasis and perspective.

What do we expect of our science of physiology? We want it to be a body of knowledge which represents the mode of operation of animal organisms. The point of view is strictly Cartesian: organisms are to be explained as mechanisms or machines that operate without the necessary intervention of non-material principles. The human body, no doubt, is the prime object of physiology, but nature is more than man and our planet is still inhabited by millions of species of animals. The diversity of life cannot be ignored by physiologists and requires explanation. It is a major task of physiologists to increase our knowledge of the diversity of animal function, of the adaptation of these so varied functions to particular environments and life styles. Such knowledge will enhance our understanding of the meaning of life on this earth and will give us deeper insights into our own existence.

The fallacies I have exposed in this lecture are a hindrance to the achievement of a true science of physiology if they are not recognized as such. They are, as I explained, consequences of the methods of science, indeed of our language. Fallacies are the outcome of our use of paradigms that go unrecognized. To avoid them we must understand the method of science and the nature of its paradigms. If we recognize this we can happily go on with our experiments and, I think, we might even do better research in the service of this splendid science of physiology.

THE ARCHITECTURE OF NEURAL CENTRES AND UNDERSTANDING NEURAL ORGANIZATION

J. Szentágothai

Department of Anatomy
Semmelweis University Medical School
Budapest, Hungary

ARCHITECTURE

With the technical means available today it is possible to define and describe any neurone of the nervous system in exact quantitative terms (numerical: for numbers of various types of synapses received or given, distribution [spatial groups], etc.; geometrical: for size, shape, volume, orientation of both dendritic and axonal arborization; topological [edges, apexes]). This revolution in neuroanatomical techniques began with the development of new antero- and retrograde tracing methods (uptake and transport, both by nerve cells and by terminal arborizations of radiolabeled amino acids [occasionally other metabolites and/or mediators], of fluorescent dyes, enzymes - e.g. horseradish peroxidase [HRP] - cobalt compounds, etc). The next step was the combination of various classical and more recent histological procedures, like the Golgi precipitation, labeling with horseradish peroxidase, and by anterograde secondary axonal degeneration, performed simultaneously on the same neural structures and making possible the recovery under the electron microscope (in ultrathin section series) of any specific detail (especially of a given synapse) previously identified in the light microscope. Whole networks of mutually coupled neurones could be thus defined with hitherto unexpected clarity (Somogyi, Hodgson and Smith, 1979). These combined techniques were soon joined by further possibilities to detect the synaptic transmitters (both conventional mediators like acetylcholine, GABA, glycine [hopefully soon also glutamate and aspartate], and many of the newly discovered neuropeptides) in neuronal perikarya as well as in synaptic terminals by biochemical and especially by immunocytochemical techniques. Using several antibodies against various types of mediators simultaneously, modulators and/or their receptors can now be localized in the same or in synaptically interconnected neurones. Eventually the cross-correlation between anatomically and immunocytochemically identified neurones was joined by the direct intracellular (or intraaxonal) labeling by horseradish peroxidase of individual neurones that had previously been studied by unit recording for their physiological characteristics under various experimental conditions. Thus it is possible to correlate physiological, anatomical, and various biochemical properties of the same neurone both at the light and the electron microscopic level. By computer-aided reconstruction of the total dendritic and local axonal arborizations of physiologically identified neurones (or in other cases the terminal arborizations of axons of extraneous source) statistics of the synapses both given and received can now be made and their distribution in relatively large tissue spaces can be described in exact terms. Fig. 1 gives an example - stripped down to essentials - of the main elements of information upon which the understanding of such a cell is based - in this case a so-called "clutch cell" (The figure assembly was prepared by Peter Somogyi; the original data are quoted in the legend).

Understandably, this type of study can be most readily made in the cerebral neocortex, due to the many specific cell types present and due to the sophisticated experimental techniques by which the neurones, especially in the sensory cortices, can be studied. The classical observation of a vertical, so called "columnar", organization of the specific sensory cortices could gradually be matched, over the last 15-20 years, with new anatomical and mediator (modulator) architectonics, accumulated with the aid of these combined techniques. The concept of the "modular architectonics principle" of neural centres - although included implicitly in many earlier descriptions and illustrations - emerged and took shape gradually somewhat before the recent revolution in the neuroscientific techniques. This concept was based on the understanding that the larger organoid portions of neural centres were assembled from smaller repetitive modules of similar architecture and connectivity. This general principle had been forecast already by the columnar arrangement of neurones of similar receptive field properties, but it turned out that one way or another it can be observed in virtually all major organs of the CNS (Szentágothai, 1983; Szentágothai and Arbib, 1974). However, the cerebral neocortex is the most attractive paradigm of this architectonic principle,

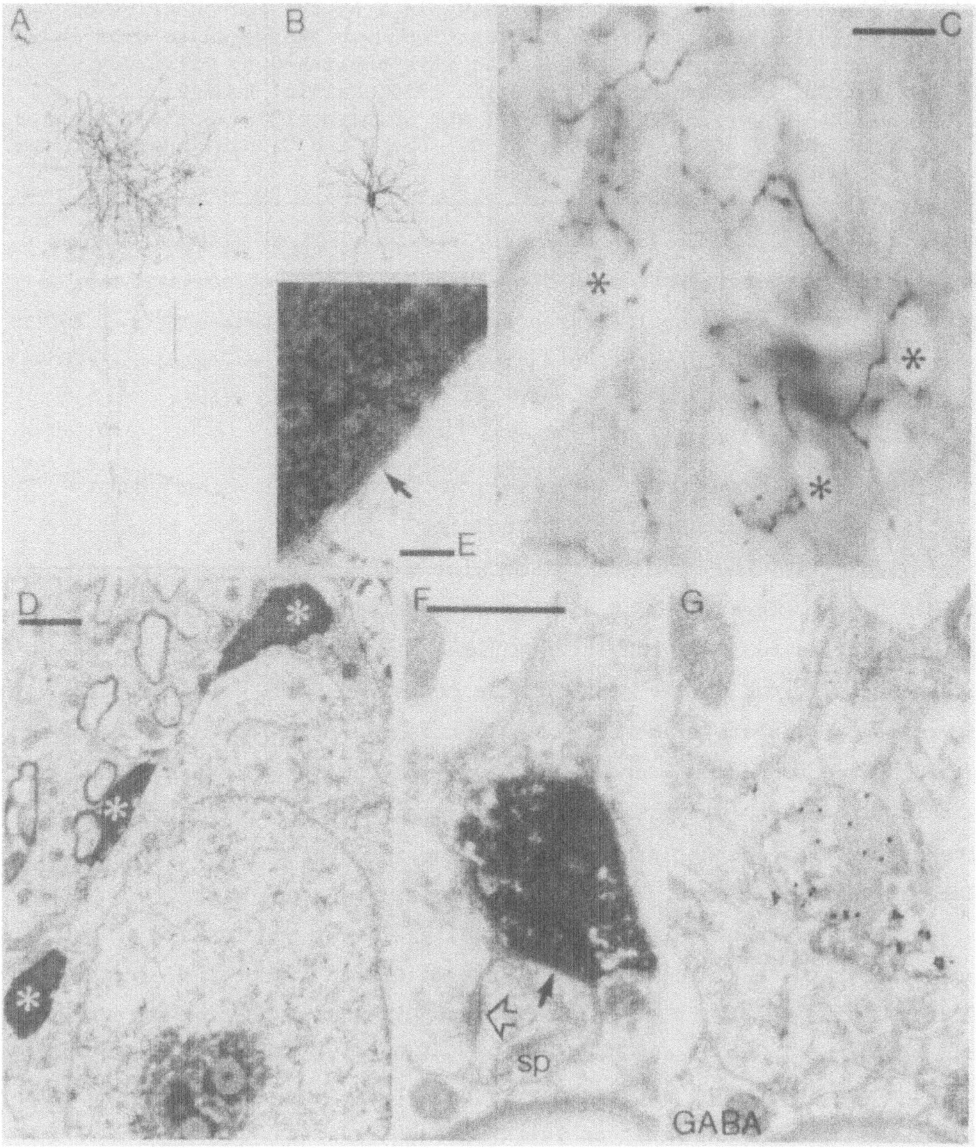

Fig. 1. Synaptic connections and putative transmitter of physiologically characterized HRP-filled clutch cells in layer IV of the cat's striate cortex. (A and B): Axonal and dendritic arborization (from a computer aided reconstruction turned into the coronal plane) of a clutch cell that could be monosynaptically activated by Y-type thalamic afferents and had a complex type of receptive field. (C): Light micrograph of a clutch cell axon in layer IV; some of the boutons contact neuronal somata (asterisks). (D): Electron micrograph of a soma that receives synapses from three boutons (asterisks) of the clutch cell shown above. (E): A synaptic contact (arrow) is shown at higher magnification. Clutch cells make symmetrical or type II contacts. (F and G): Serial sections of a clutch cell bouton making a type II synaptic contact (arrow) with a spine (sp) that also receives a type I synaptic contact (open arrow) from another bouton. Only the clutch cell bouton is (continued)

immunoreactive for GABA, as shown in G by the accumulation of colloidal gold, following incubation with antiserum to GABA in an immunogold procedure. Based on work published by Kisvárday, Martin, Whitteridge and Somogyi, 1985; Martin, Somogyi and Whitteridge, 1983; and Somogyi and Soltész, 1986. Scales: A and B, 100 μm; C, 20 μm; D, 1 μm; E, 0.1 μm; F and G, same magn., 0.5 μm. (Courtesy of P. Somogyi).

due to an amazing refinement of almost "seamless" (i.e. without overlap or empty spaces) interdigitation in the convergence of cortico-cortical afferents to the same cortical area from different distant cortical loci. The same principle of sophisticated interdigitation of terminal territories is observed in several other projections (Goldman-Rakic, 1984).

It would be impossible, but also unnecessary to enter here into details about the columnar cortical modules. Although other kinds of modules do exist, the most general modular units in the neocortex were defined by the convergence of a group of cortico-cortical afferents (Goldman and Nauta, 1977) having a minimal diameter of 150-300 μm width and cutting through the entire depth of the cortex (1-3mm). It is difficult to determine the maximal diameter of the cortico-cortical columns, because the long distance projections are generally - if viewed from the surface - zebra-stripe-shaped elongated territories that are regularly spaced and alternating with another projection fitting into the inter-stripe gaps (Goldman-Rakic and Schwartz, 1982). However, such long distance projections can be revealed only by labeling larger populations of connections; the individual terminal arborizations of cortico-cortical afferents have rarely if ever a larger span in the tangential direction than 3-500 μm, in most cases considerably less (Szentágothai, 1978). The stripe-shape projection pattern corresponds therefore, to the termination fields (or fields of origin) of larger groups of axons. It is remarkable that the number of cells in such a narrow cylindrical (or prismatic) column - the height of which is about ten times its width - is approximately constant over a wide phylogenetic scale (Rockel, Hiorns and Powell, 1980), apart from a larger number of cells in the primary visual area, especially of primates. The same is true for the types of cells found in any column, although regional differences may be considerable. The increase of cortical tissue volume and the number of neurones - on the phylogenetic scale - is not associated with or caused by any major change of cell shape, size, diversity, and distribution of various cell types (the major cell types are easily recognizable from mouse to man), but almost exclusively with a dramatic increase of the total number of modular units; a couple of

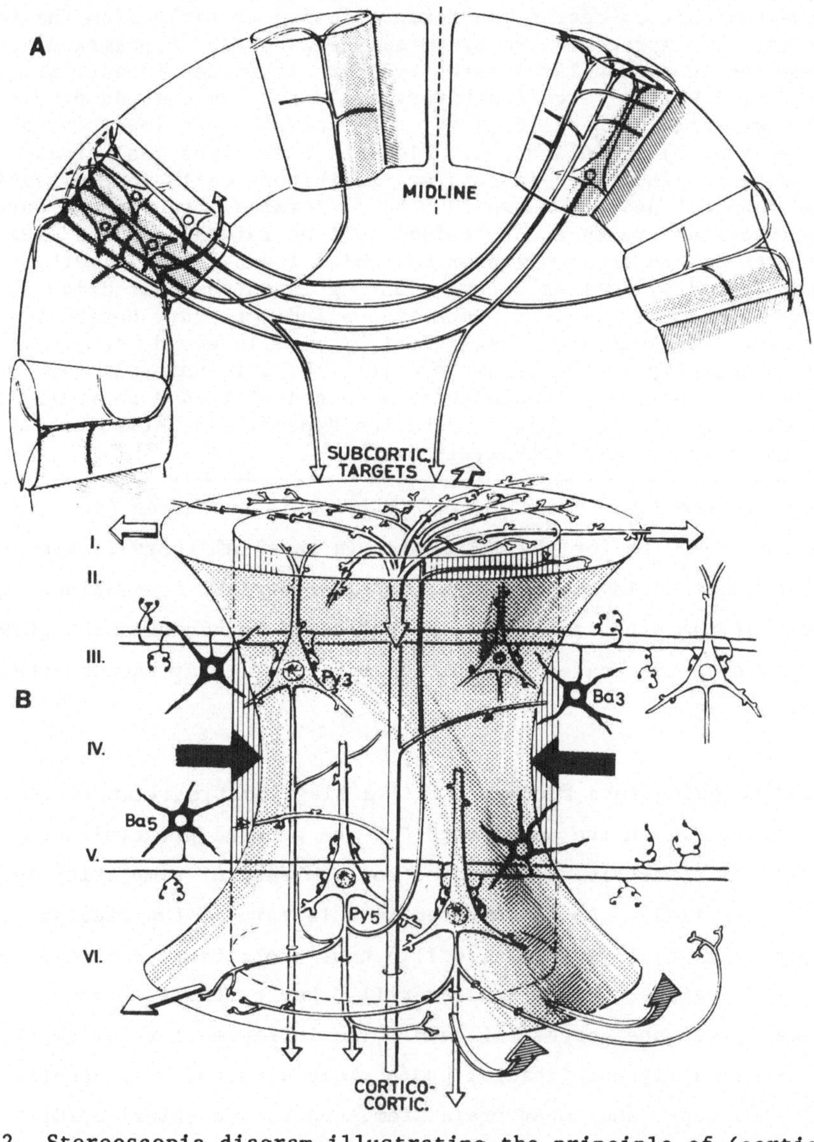

Fig. 2. Stereoscopic diagram illustrating the principle of (cortico-cortical) modular architecture of the neo-cortex. Part "A" shows cortico-cortical connectivity on the large scale: supragranular pyramidal cells (in laminae II and III; two cells of the outer row in left middle column) connect preferentially with columnar tissue spaces in the ipsilateral or in the contralateral cortex. Infragranular pyramidal cells (lower row of cells in middle column at left and one cell in one column at right) send their axons to subcortical targets, but some of them are addressed to - or give major collaterals to - contralateral cortical columns. Local (intracortical) connectivity extending over approximately ten neighbouring columns is neglected in this diagram. Part "B" shows the architecture of a single (cortico-cortical) column (vertically hatched inner cylinder) of 200-300 μm width and extending through the entire depth of the cortex, radically stripped down to a few essentials. A single cortico-cortical afferent arborization is

(continued)

placed into the centre of the drawing. (Representative for a few of the ten cortico-cortical afferents coming probably from the same source). Representative pyramidal cells of the supragranular (Py_3) and the infragranular layers (Py_5) are indicated in outline. Relatively long range inhibitory cells are represented by two large basket cells (Ba_3 and Ba_5) for the upper and the lower row each (see Fig. 3) in full black. Elements (and direction) of excitatory nature are indicated in outline, inhibitory cells (and directions of connections) are shown in black. Terminal branches of cortico-cortical afferents and pyramidal cell collaterals extend over the limits of the basic cylinder in lamina I and lamina VI; this is emphasized by outline arrows pointing in outward direction in the two layers. Conversely, dark arrows indicate that horizontally oriented basket (and other inhibitory) cells would tend to narrow down activity in the columnar unit. The original cylinder (vertical hatching) would hence become distorted dynamically into the shape of an hourglass (rotation hyperboloid; stippled). Adapted from Szentágothai, 1985.

---

thousand in smaller rodents and about 2 million in the human (assuming an average width of 300 µm for the columns). In parallel there is also a considerable increase in the number of cortico-cortical connections and in the richness of the arborizations, also reflected in a consequent growth - along the phylogenetic scale - in the neuropil: cell body (accumulated) volume ratio.

Instead of going into further detail a diagrammatic illustration can be given in Fig. 2. In the upper part "A" the general principle of cortico-cortical connectivity is shown (for the sake of simplicity in a lissencephalic animal). As illustrated in this part of the diagram, the upper (supragranular) layers (II and III) have projections exclusively to other cortical regions, mainly ipsilaterally, but partially also contralaterally via the corpus callosum. The infragranular layers (V and IV) have predominantly corticofugal projections directed to subcortical centres: to the upper and lower brain stem, and to the spinal cord; however a substantial fraction of the efferents from the infragranular layers - axons of pyramidal and other projecting cells of layer VI - are callosal fibres directed to the contralateral cortex. Probably a majority of these fibres are collateral side branches of the main fibres addressed to subcortical centres. The lower part "B" of Fig. 2 illustrates a single cortico-cortical column, neglecting all non-essential details. A single cortico-cortical afferent is placed here into the centre of the original cylindrical column (vertically hatched). The diagram contains only two inhibitory interneurone types, the larger basket cells (two rows of which have recently been identified, an upper in layer III-IV (Somogyi, Kisvárday, Martin and Whitteridge, 1983), and a lower mainly in lamina V

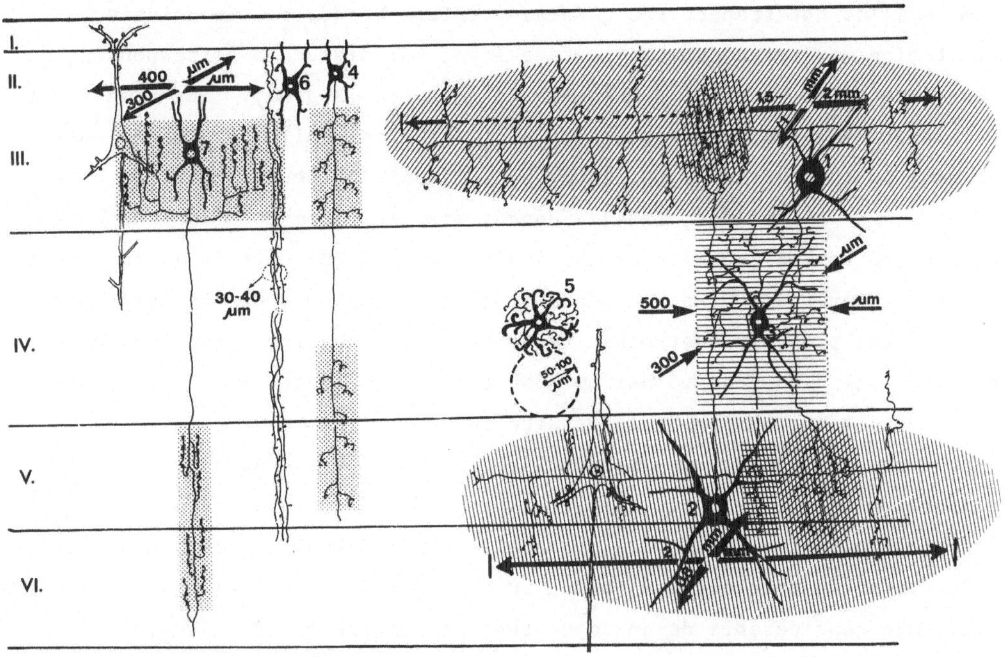

Fig. 3. Diagrammatic illustration of the seven types of well characterized (GABA-ergic) inhibitory interneurones. In order to facilitate recognition of dendrites and axons, the dendritic trees are simplified and drawn with exaggerated thickness. Fields of axonal arborization (where necessary) are indicated by hatching and stippling. Horizontal arrows indicate maximal (observed) extension of the axonal arborizations in the sagittal, and oblique arrows in the coronal directions (the extension of the axons in depth can be derived from the cortical layering indicated at left margin). (1) Large basket cell of the upper (supergranular) group (Somogyi et al., 1983) (2) large basket cell of the infragranular group (Kisvárday et al., in the press) (3) "clutch cell" terminating mainly in lamina IV (Kisvárday et al., 1985) (4) columnar basket cell (Szentágothai, 1983) (5) microglioform cell (both dendritic and axonal arborization most generally spherical (6) "cellule a double bouquet" of Ramón y Cajal (7) axo-axonic inhibitory interneurone (Szentágothai, 1978). There are more local interneurone types known that are probably also GABA-ergic, but they were not included, because their axonal arborizations have not yet been sufficiently well characterized. The spatial relations between cell types 1-3 have been adapted from a diagram of Somogyi and Soltész. (The earlier literature concerning the other cell types can be found in "Cerebral Cortex", Vol. 1; Peters, A. and Jones, E.G. eds. Plenum Press, London (1984)

and VI described by Kisvárday et al. (in the press). These GABA-ergic inhibitory interneurones have an unexpectedly wide range in the tangential direction (up to 1.5 mm and possibly more), but their synaptic terminal branches are oriented vertically. Since the terminal branches of cortico-cortical afferents and collaterals of pyramidal cells (especially from a type of lamina V pyramidal cells (Martin and Whitteridge, 1984) directed to lamina I) do not respect the general vertical border of the columns, excitation may spread in the surface layer for distances corresponding to the width of at least 10 successive columns (ca. 3-5 mm). The same is true for lamina VI, where the collaterals of pyramidal cells may reach - still staying in the cortex - and converge upon loci in the neighbouring columns (Martin and Whitteridge, 1984). Hence, excitation in a column can be assumed to spread (spill over the individual column) especially in layers I and VI. Conversely, the long inhibitory connections over the large basket cells - and probably over other types of inhibitory interneurones as well - can be assumed to tend to narrow down the column in the middle layers (III-V), thus corsetting the "waist of the column". This assumed dynamic "structure" of the column is symbolized in the lower diagram by the distortion of the cylinder into the shape of an hour-glass (rotation hyperboloid). The reality is certainly much more complicated than could be illustrated in this diagram (which in spite of such radical simplifications will make considerable demand upon the imagination of the viewer). As already forecast by most recent observations (Luhmann, Martinez-Millán and Singer, 1986), this "medium range" of tangential intercolumnar connectivity is vastly more complex than hitherto suspected.

For reasons of economy the several types of specific inhibitory interneurones will not be discussed in detail. Fig. 3 tries to give instead a comprehensive view of the seven main types so far identified (with the strict criteria mentioned), as GABA-ergic local interneurones. The different neuronal types are shown in simple semi-diagrammatic form, and in addition to their names the main references of the papers are indicated in the figure legend where the interested reader can find the most important recent analysis of the cell type made by the new synthetic approach. Both dendritic and axonal arborizations had to be compressed into a two-dimensional drawing, but whenever known the span of the arborization in the plane perpendicular to that of the figure is also indicated. Inhibitory interneurones are drawn in full black, the dendritic arborization is simplified and in order to indicate the targets of the axon terminals, a few pyramidal cells are indicated in outline. Statistics of the boutons given by various inhibitory interneurones have shown that contrary to earlier assumption the several neurones are much less selective in finding specific contact targets. For example the large basket cells (1 and 2) and the clutch cells (3) have only about one third of their synaptic contacts on the cell bodies of pyramidal cells and another third on large dendritic shafts of various - although mainly pyramidal - neurones.

Roughly one third contact dendritic spines, again probably mainly of pyramidal cells, but a few percent remain for identified inhibitory interneurones. Only a single neurone type of the seven indicated in Fig. 3, the earlier "chandelier" or more correctly "axo-axonic" cell, has practically 100% of its synapses on the initial segments of pyramid cell axons. Probably no pyramid cell exists in the neocortex and in the hippocampus that would lack this powerful inhibitory "choke" at the beginning of its axon, so that the axo-axonic (7) interneurone probably plays a crucial role in blocking the output of the pyramidal cells. One strange feature becomes obvious from Fig. 3: in addition to the wide tangential distribution of their axons, the cell types 1, 2, 3 and 7 have one (or two) vertically descending or ascending axon branch(es) that give synapses either vertically below (1,3,7) or above (2) its parent cell body. Two other cell types (4,6) have their own axonal arborization confined in narrow strictly vertical tissue spaces. It is logical to conclude from this that although inhibition has a certain tendency to spread in the tangential direction, there is a strong mutual inhibitory relation between the superficial and deeper layers of the cortex that is of small tangential range and is oriented vertically.

UNDERSTANDING THE MEANING OF ARCHITECTURE

One might think that such remarkable insights into cortical architectonics, in addition to a great body of correlated physiological evidence, and many attempts at bringing physiological facts into line with experimental psychology would already have given us a fair understanding of how the nervous system organizes behaviour. The harvest, alas, is meagre.

Sensory physiology and psychology can boast of numerous examples, in which the processing of sensory information correlates exceedingly well with established anatomical and physiological facts. In this respect the thoughts put forward by Barlow (1981; 1985a,b) are particularly stimulating. From this argument many known facts about the peripheral visual pathways become meaningful: their capacity to carry information, the exact topographic maps of the visual field in area 17, and less exact maps in the secondary and further visual regions of the cortex where progressively more "unique" and specific complex stimuli, like a hand or a human face are signalled by individual cells, etc. However, the progression over the hierarchy of visual centres from cells responding to relatively non-specific elementary stimuli (edges and orientations) towards progressively more specific cells: eventually so-called "cardinal" (or "pontifical", or "grandmother") cells, leaves us with greater frustration the more elegant the succession becomes of breaking down the entire visual field into details relevant to the survival of the animal. Involuntarily the question arises in one's mind: "so what?"; because somewhere at the end of the line there would always loom the "homunculus" (or "demon", or "ghost in the machine") for evaluating and making use of what is being perceived.

Leaving aside the problem of local neuronal architecture, the "wiring" of the brain on the gross anatomical scale is also miraculous with respect to - obviously genetically inbuilt - precision and preservation of topography and correct neighbourhood relations from the periphery to the centres. Various types of logical and coherent somatotopic maps are found virtually everywhere, from somatotopic arrangement of motoneurones in motor nuclei and of secondary cells for various sensory modalities in the sensory nuclei of the spinal cord and the brainstem, up to the spectacular organoid cell assemblies, the so-called "barrels" of the face (or snout) region of the somatosensory cortex of rodents (Woolsey and van der Loos, 1970), where each vibrissa has its exact corresponding "barrel" in correct order and topography in lamina IV of the somatosensory (I, and slightly less clearly in II) region. Cortico-cortical connectivity is hardly less sophisticated and precise (as already mentioned above) (Goldman-Rakic, 1984; Goldman-Rakic and Schwartz, 1982); but this is only what could be expected from the known connectivity between different visual cortical territories and their requirements as discussed by Barlow.

In view of these considerations, one question becomes significant. Is the whole blueprint of neuronal connections genetically preprogrammed, and if so, how far does this determination go?

The developmental mechanism is the same in all parts of the early embryonic medullary tube. Neurones and glial cells are produced in the germinal (ventricular) zone of the tube, where most of the cell divisions occur. Also the outward movement of prospective neurones from the germinal zone towards the outer "mantle zone" of the tube - guided by radiate glial cells that connect points of the base of the original epithelial cells in the pseudostratified cylindrical epithelium with the sites of their original contact with the outer surface of the tube (Rakic, 1984a) - proceeds in very much the same manner in virtually all parts of the brain. Even if rapidly proliferating cell masses are produced at special sites in the cephalic part of the tube, for example the rhombic lip, giving rise to the external granular layer of the cerebellum, the eventual mechanism of cell movement - although in this case from the surface towards the depth - will receive the same guidance by a special kind of radiate element, the Bergmann glia derived from the same original radiate glial cells (Rakic, 1984b). It is therefore not astonishing to find the basic cell types and interneurone connectivity virtually uniform all over the cerebellar cortex, and, albeit having very different cell types and a highly different connectivity, this applies also to all parts of the neocortex. (It would not be difficult to show how slight modifications of the same mechanisms in the more ancient parts of the hemispheres [the allocortex] lead to a uniformity of a somewhat different type). But how much of the later connectivity can be included in the original genetic program? (Needless to say that the genetic program is not considered here in the sense of a new molecular level "preformism", but rather in that of an epigenetic gradual unfolding in a tree-like arborization of various specific molecules attached to the cell surfaces [as suggested by the theory of Edelman (1984) by a system of "cell adhesion molecules"; CAM] that decide which part of which neurone makes contact with which part of other cell(s)). But how many of the 10^{14} synapses could be determined by genetic programming? We have seen in division I that, in spite of very characteristic arborization patterns of most cortical cells, the synaptic targets are in many cases almost as if selected at random. Conversely, at least in one cell type, the axo-axonic inhibitory interneurone, the localization of the synapses is 100% selective by contacting exclusively the initial axon segments of pyramidal cells. But even in this specific case it is beyond our present understanding how the convergence of about five different axo-axonic cell terminals per pyramidal cell initial axon segment is brought about during development. Similar questions could be repeated over and over again in an infinite variety, for practically every single type of neurone and synapse.

A certain degree of synaptic plasticity is very generally assumed to exist all over the CNS, especially in a "sensitive period" during relatively early postnatal life. This plasticity has been shown most convincingly in the visual cortex, where changes under specific experimental conditions - generally of functional deprivation - could be convincingly demonstrated in behavioural, unit level physiological tests, as well as in (hard wired) anatomy (Hirsch and Spinelli, 1979; Hubel and Wiesel, 1970; Pettigrew, 1974). Also the fact that, as a general rule, many more presynaptic elements are offered for final selection of the permanent synapses, is considered generally as evidence for a selection mechanism and the limits imposed by some "saturation factor" in synapse formation. (The classical example of this is the climbing fibre in the cerebellum, where several climbing fibres establish loose coils around each Purkinje cell body, but eventually only one of them climbs up along the tree of primary and secondary dendrites). It is highly probable that these plastic changes are intimately linked with function, but it is highly questionable whether such plasticity can serve as a paradigm for memory and learning - even if we were to assume that the "plasticity" involved in learning is shifted several orders of magnitude, from gross histological or synaptic changes towards something occurring on the molecular level.

A quite different aspect of neural - most specifically, of course, of cortical - organization might also merit serious consideration: virtually all projecting (pyramidal) cells of the supragranular layers of the cortex (laminae II and III, and probably partially IV) are addressed to other cortical areas. It is generally assumed, as a rule of thumb, that about 60% of all cortical cells are pyramidal cells, and since on 20% of the cortical cells are GABA-ergic inhibitory interneurones (Gabbott and Somogyi, 1986) and the number of excitatory local neurones could hardly contribute to much more than 10% of the total population, one is forced to the conclusion that over 70% of all nerve cells in the cortex are neurones the axons of which are addressed to targets at some distance from the cells of origin. Well over half of these are directed towards other regions of the cortex, say 50% of them. From the remaining cortical efferents again the majority belong to pathways that continue in chains of neurones that are eventually redirected to the cortex (over the upper brainstem nuclei, some over nuclei of the lower brainstem, and particularly over the cerebellar loop [the pontine nuclei, cerebellar nuclei and cortex, and the ventral-anterior thalamic nuclei]). The "re-entrance" repeated all over again has the sequel - as emphasized particularly by Mountcastle (1978) - that whatever occurs (or is stored) in the cortex must be distributed. The principle of re-entrance was used as the central piece of a new general theory of the brain by Edelman (1978) and Edelman and Finkel (1982).

It would be very difficult indeed to develop any generalized concept of the "neural" without incorporating such a fundamental architectural feature of the vertebrate brain. The essential similarity in local architecture of all cortical regions may be made palatable by Barlow's (1985a) statement that "it performs the same operation everywhere"; the difference is in the input - exactly as also in the cerebellar cortex. But what does this similarity mean in local processing and repeated re-entrance ad infinitum?

Elegant concepts have been proposed over the recent years that could claim to be considered as something like a "general brain theory". Although it is obviously impossible to discuss - in the framework of this paper - these theories in any detail, some of them will be mentioned briefly as options for further speculation, theoretical analysis, and eventually for being tested against biological, psychological (and other relevant) phenomena.

(1) The selective stabilization theory of Changeux and Danchin (1976) is characterized by being based on a host of well documented observations on the development, selection mechanisms of target finding by neurones, and a rational analysis of their possible biochemical and molecular biological substrates. More recently Changeux and co-workers (Heidmann, Heidmann and Changeux, 1984) have widened the scope of this original theory towards lines that show considerable convergence with my own trend of thought, to be outlined briefly below.

(2) The theoretical neurobiological approach - only some highlights can be mentioned: Wilson and Cowan (1973); von der Malsburg (1973); Grossberg (1982); Kohonen (1982); Hopfield (1982); - is a most fruitful one, but we cannot enter here into any details. Their mathematical approaches might give some hope for an emergent rational formalism for the exact modelling of such mystic properties of "the neural" as "memory and learning".

(3) The analysis of sensory-motor transformations by the aid of the "tensor network theory" by Pellionisz and Llinás (1979, 1985) is a most interesting approach - especially in light of such reflex mechanisms as the vestibulo-ocular and the vestibulo-neck muscle reflexes. It remains to be seen whether this theory can be extended over the limited scope of sensory-motor

coordination - even if the function of the cerebellum were included - towards wider applications to other phenomena of behaviour.

(4) Edelman's (Edelman, 1978; Edelman and Finkel, 1982) "Group (degenerate) selection and phasic re-entrant signaling theory" has the major advantage over all other variants in approach, that it may - potentially and in the future theoretically - incorporate the most elusive properties of "the neural", to which we attach the label "cognitive".

Although this author considers all of the mentioned theories most fruitful and the best options open to us for the time being, he would prefer two additional basic concepts for theorizing, that surface occasionally in the speculations, but still lack in attracting sufficient attention and serious consideration by theoreticians of the "neural". Both aspects are not new, but they ought to gain a central position in the basic issue at hand.

(1) The first aspect is "self-organization": I have recently drawn attention (Szentágothai 1984, 1985) to rather antiquated experiments with isolated neural centres which show very remarkable tendencies toward clearly recognizable functional "self-organization" in such centres (further communications - going into more detail - will be forthcoming in the near future).

Arguments supporting the self-organizing character of the nervous system do not come from embryological experiments only. The study of brain activity dynamics by EEG suggested that spatio-temporal patterns of cortical neural activity are internally generated representations of expected sensory input and not merely passive responses to such input. Following this line (Freeman, 1983; Freeman and Skarda, 1985) the statement can be made that the macroscopic co-operative activity of weakly but widely interconnected neurones emerges from the operations of individual neurones coupled by conventional synaptic transmission. Furthermore, the construction of a perceptual representation from the sensory input can be described in terms of self-organization.

Theoretical frameworks for analysing "dynamic activity patterns" in general terms, however, motivated by chemical physics, were given by the theories of dissipative structures (Nicolis and Prigogine, 1977) and of synergetics (Haken, 1978). In accordance with the spirit of these approaches the theory of "noise-induced transition" (Horsthemke and Lefever, 1984) suggests that small fluctuations may operate as "organizing forces" during structure formation and may modify drastically the qualitative dynamic behaviour of a macroscopic system. Specifically, the positive role of "neural noise" has been emphasized by Stryker (1982) and by the Changeux school (Heidmann et al. 1984). Simulation experiments of my co-workers (Érdi and Barna, 1984; Érdi and Szentágothai, 1985) support this view. It is my belief that here are the bases of the remarkable "spontaneity" of the neural activities that find their expression in behaviour. This spontaneity could hardly be accounted for by any "reflex-robot" or "gradual feature extracting and/or assembly" mechanisms envisaged by many physiologists engaged in the study of perception.

(2) The second aspect is the paramount principle of (repetitive) "re-entrance" - particularly in the cerebral cortex (but by no means only there) - that might secure the general distribution of all activity over the entire nervous system and weld its activity into a holistic process. It is therefore perhaps not too daring to visualize the essence of the neural in a system designed around the principle of "self-reference". This does not mean that nervous systems ought to be considered as purely "self-referential" (or autopoetic), because the reflex principle (and everything that it entails in the widest sense) cannot simply be thrown overboard.

It would probably be premature and grossly misleading to overemphasize - not to speak of giving any unique ontological status to - the two aspects mentioned; however, the facts discussed in some detail in the first part of this paper would forcefully suggest that these two aspects be incorporated into further theorizing about the "essence of the neural".

REFERENCES

Barlow, H.B. (1981) Critical limiting factors in the design of the eye and visual cortex, Proc Roy. Soc. B: 212, 1-34

Barlow, H.B. (1985a) Cerebral cortex as model builder, In: Models of the visual cortex Rose, D. and Dobson, V.G., eds., Wiley: New York. pp 37-46

Barlow, H.B. (1985b) The role of single neurones in the psychology of perception, Quart. exp. Psychol. 37A, 121-145

Changeux, J.P. and Danchin, A. (1976) Selective stabilization of developing synapses as a mechanism for the specification of neuronal networks, Nature 264, 705-712

Edelman, G.M. (1978) Group selection and phasic signaling: A theory of higher brain function. In: The Mindful Brain Edelman, G.M. and Mountcastle, V.B., eds., MIT Press: Cambridge, pp 51-100

Edelman, G.M. (1984) Modulation of cell adhesion during induction, histogenesis, and perinatal development of the nervous system, Ann. Rev. Neurosci. 7, 339-377

Edelman, G.M. and Finkel, L.H. (1982) Neuronal group selection in the cerebral cortex. 1st Symposium of the Neuroscience Institute, La Jolla, 3-8 October

Érdi, P. and Barna, G. (1984) Self-organizing mechanism for the formation of ordered neural mappings. Biol. Cybernetics 51, 93-101

Érdi, P. and Szentágothai, J. (1985) Neural connectivities between determinism and randomness. In: Dynamics of macrosystems. Aubin, J.P., Saari, D. and Sigmund, K., eds., Lect. Notes in Econ. and Math. Systems 257, 21-29, Springer: Berlin

Freeman, W.J. (1983) The physiological basis of mental images. Biol. Psychiatry 18, 1107-1125

Freeman, W.J. and Skarda, C. (1985) Spatial EEG patterns, non-linear dynamics and perception: the neo-Sherringtonian view. Brain Res. Rev. 10, 147-175

Gabbott, P.L.A. and Somogyi, P. (1986) Quantitative distribution of GABA-immunoreactive neurones in the visual cortex (area 17) of the cat, Exp. Brain Res. 61, 323-431

Goldman, P. and Nauta, W.J.H. (1977) Columnar distribution of cortico-cortical fibers in the frontal association, limbic and motor cortex of the developing Rhesus monkey, Brain Res. 122, 393-413

Goldman-Rakic, P.S. (1984) Modular organization of prefrontal cortex, Trends in Neurosci. 7, 419-424

Goldman-Rakic, P.S. and Schwartz, M.L. (1982) Interdigitation of contralateral and ipsilateral columnar projections to frontal association cortex in primates, Science 216, 755-757

Grossberg, S. (1982) Associative and competitive principle of learning and development: The temporal unfolding and stability of STM and LTM patterns. In: Competition and Cooperation in Neural Nets Amari, S.I. and Arbib, M.A., eds., Springer: Berlin, pp 295-341

Haken, H. (1978) Synergetics. An introduction. 2nd edition. Springer: Berlin

Heidmann, A., Heidmann, T. and Changeux, J.P. (1984) Stabilization selective de représentation neuronales par resonance entre "préreprésentations" spontanées du réseau cérébral et "percepts" evoqués par interaction avec le monde extérieur, C.R. Acad. Sci. Paris 299, 839-844

Hirsch, H.V.B. and Spinelli, D.N. (1979) Visual experience modifies distribution of horizontally and vertically oriented receptor fields in cats, Science 168, 869-871

Hubel, D.H. and Wiesel, T.N. (1970) The period of susceptibility to the physiological effects of unilateral eye closure in kittens, J. Physiol. (Lond.) 206, 419-436

Hopfield, J.J. (1982) Neural networks and physical systems with emergent collective computational abilities. Proc. Natl. Acad. Sci. USA 79, 2554-58

Horsthemke, W. and Lefever, R. (1984) Noise-induced Transitions: Theory and Applications in Physics, Chemistry and Biology. Springer: Berlin, 318p

Kisvárday, Z.F., Martin, K.A.C., Whitteridge, D. and Somogyi, P. (1985) Synaptic connections of intracellularly filled clutch cells: a type of small basket cell in the visual cortex of the cat, J. Comp. Neurol. 241, 11-137

Kohonen, T. (1982) Self-organized formation of generalized topological maps of observations in a physical system. Biol. Cyber. 43, 59-69

Luhmann, H.J., Martinez-Millán, L. and Singer, W. (1986) Development of horizontal intrinsic connections in cat striate cortex Exp. Brain Res. 63, 443-448

Martin, K.A., Somogyi, P. and Whitteridge, D. (1983) Physiological and morphological properties of identified basket cells in the cat's visual cortex Exp. Brain Res. 50, 193-200

Martin, K.A.C. and Whitteridge, D. (1984) Form, functions, and intracortical projections of spiny neurones in the cat's visual cortex, J. Physiol. (Lond.) 353, 463-504

Mountcastle, V.B. (1978) An organizing principle of cerebral functions: The unit module and the distributed system. In: The Mindful Brain. Edelman, G.M. and Mountcastle, V.B., eds., M.I.T. Press: Cambridge, pp 7-50

Nicolis, G. and Prigogine, I. (1977) Self-organization in non-equilibrium systems. Wiley: New York

Pellionisz, A. and Llinás, R. (1979) Brain modeling by tensor network theory and computer simulation. The cerebellum: Distributed processor for predictive co-ordination. Neuroscience 4, 323-348

Pellionisz, A. and Llinás, R. (1985) Cerebellar function and the adaptive feature of the central nervous system. In: Adaptive Mechanisms in Gaze Control, eds. A. Berthoz and G. Melvill-Jones, eds., Elsevier: Amsterdam, pp 223-232

Pettigrew, J.D. (1974) The effect of visual experience on the development of stimulus specificity by kitten cortical neurones, J. Physiol. (Lond.) 237, 49-74

Rakic, P. (1984a) Organizing principles for development of primate cerebral cortex, In: Organizing Principles of Neural Development, Sharma, S.C. ed. Plenum: New York, pp 21-48

Rakic, P. (1984b) Emergence of neuronal and glial cell lineages in primate brain, In: Cellular and Molecular Biology of Neuronal Development, Black, I.B., ed. Plenum: New York, pp 29-50

Rockel, A.J., Hiorns, R.W. and Powell, T.P.S. (1980) The basic uniformity in structure of the neocortex. Brain 103, 221-244

Somogyi, P., Hodgson, A.J. and Smith, A.D. (1979) An approach to tracing neurone networks in the cerebral cortex and basal ganglia. Combination of Golgi staining, retrograde transport of horseradish peroxidase and anterograde degeneration of synaptic boutons in the same material Neuroscience 4, 1805-1852

Somogyi, P., Kisvárday, Z.F., Martin, K.A.C. and Whitteridge, D. (1983) Synaptic connections of morphologically identified and physiologically characterized large basket cells in the striate cortex of cat. Neuroscience 10, 261-294

Somogyi, P. and Soltész, I. (1986) Immunogold demonstration of GABA in synaptic terminals of intracellularly recorded, horseradish peroxidase-filled basket cells and clutch cells in the cat's visual cortex, Neuroscience (in the press)

Stryker, M.P. (1982) Role of visual afferent activity in the development of ocular dominance columns. Neurosci. Res. Progr. Bull. 20, 540-549

Szentágothai, J. (1978) The neurone network of the cerebral cortex. A functional interpretation. The Ferrier Lecture 1977. Proc. Roy. Soc. B: 201, 219-248

Szentágothai, J. (1983) The "modular" architectonic principle of neural centres. Rev. Physiol. Biochem. Pharmacol. Vol. 98 Springer: Berlin

Szentágothai, J. (1984) Downward causation. Ann. Rev. Neurosci. 7, 1-11

Szentágothai, J. (1985) Theorien zur Organisation und Funktion des Gehirns, Naturwissenschaften 72, 303-309

Szentágothai, J. and Arbib, M. (1974) Conceptual models of neural organization, Neurosci. Res. Prog. Bull. 12, 307-510

von der Malsburg, Ch. (1973) Self-organization of orientation sensitive cells in the striate cortex. Kybernetik 14, 80-100

Wilson, H.R. and Cowan, J.D. (1973) A mathematical theory of the functional dynamics of cortical and thalamic nervous tissue. Kybernetik 13, 55-80

Woolsey, T.A. and van der Loos, H. (1970) The structural organization of layer IV in the somatosensory region (SI) of mouse cerebral cortex, Brain Res. 17, 205-242

THE ADRIAN-ZOTTERMAN LECTURE: CUTANEOUS SENSATION

Erik Torebjörk

Department of Clinical Neurophysiology
University Hospital
S-751 85 Uppsala, Sweden

> It is very simple: the (peripheral) nervous system is signalling with the simplest of all codes. It has only one sign -- the spike, which follows the all-or-nothing principle.... Thus there is no amplitude modulation in the nerve fibres.... It is the impulse frequency alone which signals the strength of the stimulus. These fundamental principles of nervous transmission were first conceived by Adrian and me in Cambridge on a raw November day in 1925.

Those are the words of Yngve Zotterman in his autobiography "Touch, Tickle and Pain" (1971). Everyone agrees with Adrian and Zotterman that the peripheral message to the brain occurs according to the principle of frequency modulation. But what is it in this laconic message that determines the qualities of sensation such as touch, tickle and pain? This very controversial issue is the topic of my lecture.

Two main theories have been advanced to explain the qualitative aspects of cutaneous sensation. One is the specificity theory, the other is the pattern theory. In 1882, Blix reported that stimulation of certain spots in the skin gave a sensation of warmth, whereas stimulation of other spots gave a sensation of cold. He assumed that these stimuli excited nerve endings which are specialized to respond to warm or cold stimuli, and that signals from these specialized receptors evoke specific qualities of sensation in the brain. That is the essence of the specificity theory.

Against this theory stands the pattern theory of Weddell (1955) and Sinclair (1955). They denied the existence of specialized nerve endings.

They attributed various sensations to the brain's capacity to decode various patterns of impulses in noncommitted nerve fibres. That is the essence of the pattern theory. However, the work of Zotterman, Hensel, Iggo, Perl and others clearly demonstrated the existence of highly specialized nerve endings which only respond to certain kinds of mechanical, thermal or noxious stimuli (see Iggo, 1985, for a review). Obviously this called for a revision of the original pattern theory, which did not recognize receptor specialization. In 1962 Melzack and Wall came up with a new version of the pattern theory in their paper "On the nature of cutaneous sensory mechanisms". They acknowledged receptor specialization but maintained that "every discriminably different somesthetic perception is produced by a unique pattern of nerve impulses". They admitted that their "knowledge of how it occurs is meagre", and yet they proposed that "more than one nerve impulse from a single afferent fibre, or more than one fibre carrying single nerve impulses, is essential for the central cells to detect the characteristics of a sensory stimulus". In a way they were safe, since at that time there was "no evidence that stimulation of a single fibre in isolation gives rise to a sensation of any kind, and until the experiment is done it is impossible to say what the result will be" (Sinclair, 1967). Sinclair went on, stating: "To stimulate a single fibre in an intact human subject, to prove satisfactorily that only that fibre and no others has been stimulated, and to record a meaningful sensory judgement is an almost incredibly difficult technical feat, and it will be a long time before unequivocal evidence can be obtained".

I have done this experiment in collaboration with Ochoa (Torebjörk and Ochoa, 1980). Independently, Vallbo has confirmed our results (Vallbo, 1981). We used the median nerve for our experiments. The median nerve supplies the glabrous skin area of the hand, which has a very rich cortical representation. This is important. The results that I will describe apply to the glabrous skin of the hand and we do not extrapolate the results to other regions of the body. We used the technique of microneurography, introduced by Vallbo and Hagbarth (1968). Microelectrodes were inserted through the skin into cutaneous fascicles of the median nerve in awake human subjects. We have previously shown that this technique allows recording from single sensory units of all types in intact human nerves (for a review, see Vallbo, Hagbarth, Torebjörk and Wallin, 1979). The new thing we did was to use the electrode for selective electrical stimulation of nerve fibres, whose receptive properties had been classified by recording, and we asked the subjects what they felt when these identified sensory units were stimulated.

We believe that it is possible, with this technique of intraneural microstimulation, to activate single myelinated fibres in isolation, and to stimulate groups of unmyelinated fibres without coactivation of myelinated fibres. There is abundant evidence to support this notion. First, the median current intensity used to evoke a threshold sensation in humans is of the order of 0.8 μamp (Vallbo, Olsson, Westberg and Clark, 1984) which is of the same order of magnitude as the current needed to stimulate a single myelinated fibre in the exposed saphenous nerve in the cat, using the same type of electrode as in the human experiments (Torebjörk, Vallbo and Ochoa, 1987). Thus we conclude that the current intensities used in humans to evoke threshold sensations are consistent with stimulation of single myelinated fibres.

A second evidence is the quantal nature of threshold sensations. As the current intensity is gradually increased from zero, there is typically a discontinuous recruitment of qualitatively different sensations, like tapping, pressure or stinging pain. The sensations have different temporal profiles: tapping is intermittent, pressure and pain are sustained sensations. The sensations are projected to separate small areas of skin. Each sensation is recruited according to the all-or-nothing principle. With recruitment of a new sensation, there is no change in quality, temporal profile, intensity or projection area of a previously recruited sensation. This quantal behaviour of threshold sensations appears to be a psychological counterpart to electrophysiological recruitment of single sensory units.

A third evidence for single unit stimulation is the remarkable coincidence between the projected field of a threshold sensation evoked by intraneural microstimulation, and the actual receptive field of a recorded unit. In a sample of 172 mechanoreceptive units, 10% of glabrous skin units had concentrically overlapping fields and 57% had some degree of overlap (Schady, Torebjörk and Ochoa, 1983). In fingertips, the overlap was almost perfect, whereas a slightly greater error was found for units innervating the palm.

The strongest evidence for single unit stimulation is the correspondence between quality of sensation evoked by intraneural microstimulation and the type of the largest amplitude unit recordable at the same electrode position. The degree of matching in a large population of mechanoreceptor units was highly significant (Schady and Torebjörk, 1983) and held true whichever was established first -- sensory unit type or sensation. There are four types of low threshold mechanoreceptor units in the glabrous skin of the human hand (Kniebestöl and Vallbo, 1970; Kniebestöl, 1973, 1975; Johansson, 1978). Two types are quickly adapting in the sense that they respond to changes of skin indentation. They are called RA (rapidly adapting) and PC (Pacinian corpuscle) units and correspond to Meissner and Paciniform nerve endings respectively (Johansson and Vallbo, 1979). Two other types are slowly adapting in the sense that they respond to static skin indentation. They are called "slowly adapting" (SA) type I and type II and correspond to Merkel and Ruffini nerve endings respectively (Chambers, Andres, von Duering and Iggo, 1972; Iggo and Muir, 1969; Johansson, 1978).

The typical sensation associated with intraneural microstimulation in sites where a single RA unit is dominating in the recording is a focal contact or tapping on a small area of skin (Torebjörk and Ochoa, 1980). A single impulse in a single RA unit can be felt in fingertips (Ochoa and Torebjörk, 1983; Vallbo et al., 1984). With increase in frequency of stimulation, the frequency of intermittent tapping is increased to evolve into flutter and buzzing at higher frequencies (Ochoa and Torebjörk, 1983). This result is consistent with the findings of Talbot, Darian-Smith, Kornhuber and Mountcastle, (1968) that the RA system signals flutter. The result is at variance with the proposition of the pattern theory that "more than one nerve impulse from a single afferent fibre, or more than one fibre carrying single nerve impulses, is essential for central cells to detect the characteristics of a sensory stimulus" (Melzack and Wall, 1962). For the RA system, a single impulse in a single fibre innervating the fingertip can be enough to allow the brain to detect, localize, delineate and qualify a sensory event.

A sensation of pressure is typically reported during stimulation in sites where a single SAI unit is recorded (Torebjörk and Ochoa, 1980). Pressure is projected to skin areas significantly larger than for tapping (Schady and Torebjörk, 1983). A single impulse is never felt; temporal summation of several impulses is always required for sensory detection (Torebjörk and Ochoa, 1983; Vallbo et al., 1984). Frequency of stimulation is not detected -- the sensation of pressure is sustained -- in contrast to tapping. Increase in stimulation frequency leads to increase in magnitude of sustained pressure. The result differs from the proposition of the pattern theory that "every discriminably different somesthetic perception is produced by a unique pattern of nerve impulses" (Melzack and Wall, 1962). If the same pattern is induced selectively in each of two different somesthetic systems, the RA and the SAI, two different sensations are reported which differ with respect to quality, temporal profile, projection area and need for temporal summation. If the stimulus pattern is varied, one basic sensation is not converted into the other -- they remain as separate and specific entities.

For PC units the available materials are small, but it seems that sensations of vibration or tickling are associated with stimulation of this unit type (Ochoa and Torebjörk, 1983; Schady and Torebjörk, 1983; Vallbo et al., 1984). This agrees with the findings of Talbot et al. (1968) that the PC system signals high frequency of vibration. No sensation is reported when SAII units are stimulated in isolation (Torebjörk and Ochoa, 1980; Ochoa and Torebjörk, 1983). If these units have a conscious sensory function, it must be through spatial summation.

The observed matching between quality of sensation and unit type is a strong piece of evidence against the proposition of the most recent update of the pattern theory that "the brain is abstracting elementary sensations from a diffuse barrage" (Wall and McMahon, 1985). Instead, our findings indicate that certain types of mechanoreceptor units innervating the human hand evoke specific qualities of sensation when stimulated in isolation.

An old question is whether different patterns of peripheral nerve fibre activity are required for the central nervous system to differentiate between itch and pain. According to the pattern theory "the sensation of itching is dependent upon the establishment of a specific, orderly pattern of neural activity within the central nervous system. This pattern is associated with a low frequency in peripheral nerves resulting from the stimulation of pain endings in the skin at an intensity below pain threshold" (Chapman, Goodell and Wolff, 1960). In their discussion of the neuronal basis of pain and itch, Wall and Cronly-Dillon (1960) also support the idea that "a temporal pattern of discharge would signal the (qualitative) nature of the stimulus".

It is now possible to test this hypothesis by intraneural microstimulation in humans. In certain intraneural electrode positions weak electrical shocks may produce a pure sensation of pain and in other positions a pure sensation of itch is evoked. Changes in stimulus frequency induce changes in perceived magnitude of pain or itch but modulation of stimulus pattern does not transform one quality of sensation into the other (Torebjörk and Ochoa, 1981). Thus, a less intense pain is perceived at lower stimulus frequencies but does not become itch, and itch is felt stronger at higher stimulus frequencies but is not converted to pain. Our observations indicate that temporal pattern _per se_ does not determine the qualitative differences between pain and itch, and we must conclude that the pattern theory is also wrong on this point.

Even though some of the propositions of the pattern theory obviously do not hold water, the pattern theory must not sink altogether. Clearly, the brain can combine and reorganize temporal and spatial inputs from many afferent units of different types to form much more complex percepts than the simple sensations I have discussed so far. Rather than conflicting specificity against pattern I find it more fruitful to ask questions about how patterns influence specific attributes of sensations. Let me give just two examples from our experimental work.

The first example is post-ischaemic paraesthesiae. We know from direct microneurographic recordings in human subjects that myelinated fibres engage in ectopic impulse generation in the post-ischaemic period (Torebjörk, Ochoa and McCann, 1979). Repeated bursts with very high impulse frequencies appear asynchronously in multiple sensory units of different types (Ochoa and Torebjörk, 1980). Because the brain can recognize quality, time course, magnitude and location of sensations from certain types of single afferent units in the hand, this exceptional pattern of impulses causes a chaotic percept consisting of an assortment of intense sensations, like "tingling", "buzzing", "pins and needles" (Merrington and Nathan, 1949) which are projected in irregular succession to multiple spots of the skin. The unphysiological spatio-temporal pattern of the total afferent discharge makes the sensory experience highly unusual and in that sense abnormal, and yet the sensations evoked from the discharging units are specifiable in terms of quality, time course, magnitude and location. Thus accepting the influence of pattern does not challenge specificity.

The next example is the influence of pattern in the RA system. I have already mentioned that a single impulse in a single RA unit can be felt as a single tap. What happens if two impulses are so closely packed in time that a subject can only feel one tactile event? A stronger tap is felt (unpublished observations). This is the solution to a puzzling problem, discussed by Mountcastle many years ago (1967): is it conceivable that one highly specialized system can encode two dimensions of sensory attributes, i.e. frequency and intensity of tactile events? Indeed, our results suggest that the time constant in the RA system is sufficiently short to allow for frequency detection when intervals between spikes are long and regular, but is sufficiently long to allow for temporal summation and intensity detection when the spike intervals are short. The patterned response to frequency and amplitude of mechanical oscillations was described by Talbot et al. (1968) for primary RA units, and by Mountcastle, Talbot, Sakata and Hyvarinen, (1969) for RA-specific cortical cells in the primary somatosensory area in awake monkeys. I have now explained the significance of this pattern (Fig. 1). The result was not predicted by the pattern theory which states: "The time course of the afferent barrage is poorly related to the time course of perception" (Wall and McMahon, 1985). As just indicated, this generalization does not apply for the RA tactile system.

What went wrong with the pattern theory? One explanation is given by Sinclair (1967): "The pattern theory of cutaneous sensation ... was originally based not upon human experimentation or clinical observations but upon electrophysiological findings in cats and frogs". It is salutary, in this context, to remember the words of Adrian (1931): "There may be many pitfalls in an argument which equates sensation in Man with nervous discharges from the frog's skin". Another explanation is given by Sherlock Holmes: "It is a capital mistake to theorize before one has data. Insensibly one begins to twist facts to suit theories instead of theories to suit facts". Perhaps the experimental basis for the pattern theory was not solid enough to carry the hypothetical constructions.

It is a hope for the future that theories on somatosensory mechanisms will be based on solid experimental data, and that scientists will realize that the dichotomy between pattern theory and specificity theory is highly artificial. In my view, the brain uses a variety of patterns from specific sensory receptors to form sensations which can be integrated, enhanced or suppressed to form complex percepts in everyday experience. It would be erroneous to deny the existence of specificity for the most elementary of these sensations, or to deny the modulation by pattern of integrated sensory experience. Also, it would be foolish to argue against the obvious matter of fact that sensation is influenced by "a number of factors such as attention, distraction, expectation, immediate and past experience, significance, etc." (Wall, 1971). Furthermore, it is quite conceivable

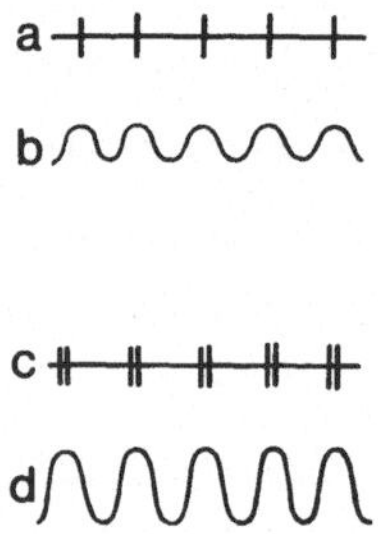

Fig. 1. Schematic representation of neural discharges in a single RA unit (a and c) in response to sinusoidal mechanical stimulation applied at low (b) and high (d) amplitude to the receptive field. It is seen that the frequency of the mechanical oscillations is signalled by the periodicity of the neural discharge, and increase in amplitude is signalled by grouping of two impulses per cycle. This patterned discharge from an individual RA unit has psychological counterparts in terms of frequency <u>and</u> intensity of tactile events.

that normal specificity and normal pattern integration are upset under pathological conditions. However the possibility that pathology deviates from normality is no proof against normality itself. Therefore I find no logical support for the assumption that "the hyperpathic syndrome [would be] a challenge to specificity theory" (Wall, 1984).

CONCLUSIONS

1. Recent experimental evidence from the combined use of microneurography and intraneural microstimulation in normal human subjects suggests that there is a high degree of specificity in the somatosensory system.

2. Accepting specificity does not preclude accepting an influence of pattern on sensory experience.

3. Accepting normal specificity and normal pattern integration in a normal nervous system does not preclude acceptance of abnormal signal processing in an abnormal nervous system.

4. Afferent inputs are processed differently in different somatosensory subsystems. This observation warrants caution in generalizing results from one system into a common theory of the origin of sensory modality, as has been done so indiscriminately in the past.

ACKNOWLEDGEMENT

Supported by the Swedish Medical Research Council Grants B87-14X-05206 and B87-14P-6153.

REFERENCES

Adrian, E.D. (1931) The messages in sensory nerve fibres and their interpretation. Proc. Roy. Soc. B:109, 1-18

Blix, M. (1882) Experimentela bidrag till lösning af frågan om hudnervenas specifika energi. Upsala Läkareförenings Förhandlingar 18, 87-102

Chambers, M.R., Andres, K.H., von Duering, M. and Iggo, A. (1972) The structure and function of the slowly adapting type I mechanoreceptor in the hairy skin. Quart. J. exp. Physiol. 57, 417-445

Chapman, L.F., Goodell, H., and Wolff, H.G. (1960) Structures and processes involved in the sensation of itch. In: Advances in Biology of Skin, Vol. I, Cutaneous Innervation. Montagna, W., ed. Pergamon, Oxford, pp 161-188

Iggo, A. (1985) Sensory receptors in the skin of mammals and their sensory functions. Rev. Neurol. 141, 599-613

Iggo, A. and Muir, A.R. (1969) The structure and function of a slowly adapting touch corpuscle in hairy skin. J. Physiol. (Lond.) 200, 763-796

Johansson, R.S. (1978) Tactile sensibility in the human hand: receptive field characteristics of mechanoreceptive units in the glabrous skin area. J. Physiol. (Lond.) 281, 101-123

Johanson, R.S. and Vallbo, Â.B. (1979) Tactile sensibility in the human hand: relative and absolute density of 4 types of mechanoreceptive units in glabrous skin. J. Physiol. (Lond.) 286, 283-300

Kniebestöl, M. (1973) Stimulus-response functions of rapidly adapting mechanoreceptors in the human glabrous skin area. J. Physiol. (Lond.) 232, 427-452

Kniebestöl, M. (1975) Stimulus-response functions of slowly adapting mechanoreceptors in the human glabrous skin area. J. Physiol. (Lond.) 245, 63-80

Kniebestöl, M. and Vallbo, Å.B. (1970) Single unit analysis of mechanoreceptor activity from the human glabrous skin. Acta Physiol. Scand. 80, 178-195

Melzack, R. and Wall, P.D. (1962) On the nature of cutaneous sensory mechanisms. Brain 85, 331-356

Merrington, W.R. and Nathan, P.W. (1949) A study of post-ischaemic paraesthesiae. J. Neurol. 12, 1-18

Mountcastle, V.B. (1967) The problem of sensing and the neural coding of sensory events. In: The Neurosciences Quarton, G.C., Melnechuk, T. and Schmitt, F.O., eds. Rockefeller: New York, pp 393-408

Mountcastle, V.B., Talbot, W.H., Sakata, H. and Hyvarinen, J. (1969) Cortical neuronal mechanisms in flutter-vibration studied in unanaesthetized monkeys. Neuronal periodicity and frequency discrimination. J. Neurophysiol. 32, 452-484

Ochoa, J.L. and Torebjörk, H.E. (1980) Paresthesiae from ectopic impulse generation in human sensory nerves. Brain 103, 835-853

Ochoa, J.L. and Torebjörk, H.E. (1983) Sensations evoked by intraneural microstimulation of single mechanoreceptor units innervating the human hand. J. Physiol. (Lond.) 342, 633-654

Schady, W.J.L. and Torebjörk, H.E. (1983) Projected and receptive fields: a comparison of projected areas of sensations evoked by stimulation of mechanoreceptive units, and their innervation territories. Acta Physiol. Scand. 199, 267-275

Schady, W.J.L., Torebjörk, H.E. and Ochoa, J.L. (1983) Cerebral localization function from the input of single mechanoreceptive units in man. Acta Physiol. Scand. 119, 277-285

Sinclair, D. (1955) Cutaneous sensation and the doctrine of specific energy energy. Brain, 78, 584-614

Sinclair, D. (1967) Cutaneous Sensation Oxford: London, pp 1-306

Talbot, W.H., Darian-Smith, I., Kornhuber, K.H. and Mountcastle, V.B. (1968) The sense of flutter-vibration: comparison of the human capacity with response patterns of mechanoreceptive afferents from the monkey hand. J. Neurophysiol. 31, 301-334

Torebjörk, H.E. (1981) Human microneurography and the problems of pain. Jap. J. EEG EMG (Suppl.), pp 169-175

Torebjörk, H.E., and Ochoa, J.L. (1980) Specific sensation evoked by activity in single identified sensory units in man. Acta Physiol. Scand. 110, 443-447

Torebjörk, H.E. and Ochoa, J.L. (1981) Pain and itch from C fiber stimulation. Soc. Neurosci. Abstracts 7, p. 228

Torebjörk, H.E., Ochoa, J.L. and McCann, F.V. (1979) Paresthesiae: abnormal impulse generation in sensory nerve fibres in man. Acta Physiol. Scand. 105, 518-520

Torebjörk, H.E., Vallbo, Å.B. and Ochoa, J.L. (1987) Intraneural microstimulation in humans. Its relation to specificity of tactile sensation. Brain. In press

Vallbo, Å.B. (1981) Sensations evoked from the glabrous skin of the human hand by electrical stimulation of unitary mechanosensitive afferents. Brain Res. 215, 359-363

Vallbo, Å.B. and Hagbarth, K.-E. (1968) Activity from skin mechanoreceptors recorded percutaneously in awake human subjects. Exp. Neurol. 21, 270-289

Vallbo, Å.B., Hagbarth, K.-E., Torebjörk, H.E. and Wallin, B.G. (1979) Somatosensory, proprioceptive and sympathetic activity in human peripheral nerves. Physiol. Rev. 59, 919-957

Vallbo, Å.B., Olsson, K.A., Westberg, K.-G. and Clark, E.J. (1984) Microstimulation of single tactile afferents from the human hand. Brain 107, 727-740

Wall, P.D. (1971) Somatosensory mechanisms. In: Handbook of Electroencephalography and Clinical Neurophysiology. Cobb, W.A., ed. Elsevier: Amsterdam. pp 9-1 to 9-6

Wall, P.D. (1984) The hyperpathic syndrome: A challenge to specificity theory. In: Somatosensory Mechanisms. von Euler, C., Franzén, O., Lindblom, U. and Ottoson, D., eds. Macmillan: London. pp 327-337

Wall, P.D. and Cronly-Dillon, J.R. (1960) Pain, itch and vibration. Arch. Neurol. 2, 365-375

Weddell, G. (1955) Somesthesis and chemical senses. Ann. Rev. Psychol. 6, 119-136

Zotterman, Y. (1971) Touch, Tickle and Pain. An Autobiography. Part Two. Pergamon: Oxford. pp 1-293

NOVEL NEUROTRANSMITTERS AND THE CHEMICAL CODING OF NEURONES

J.B. Furness, M. Costa, J.L. Morris and I.L. Gibbins

Centre for Neuroscience and Departments of
Anatomy and Histology and of Physiology
School of Medicine, Flinders University
Bedford Park, S.A. 5042, Australia

SUMMARY

It took many years, from 1904 to the middle 1950's, firmly to establish chemical transmission as a fact and to demonstrate the involvement of acetylcholine and noradrenaline as transmitter substances. There followed a period of re-evaluation of the pharmacology of transmission using antagonists of cholinergic and noradrenergic transmission. This led to a stage, in the 1960's, where a number of transmission processes that did not seem to depend on either of these substances were recognized. Then, in the late 1970's, immunochemical and other methods led to the discovery of numerous potential transmitter substances in central and peripheral neurones. A major further discovery has been that neurones frequently contain two or more substances (neuronal markers) that are, or may be, involved in the transmission process. Furthermore, the patterns of association of neuronal markers indicate that there is a chemical coding of neurones that functionally subdivides similar neurones according to classes and species of animals, and according to the targets they supply. The interactions between substances that are involved together in neurotransmission suggests that neurotransmission is a plurichemical process in which the substances that are released may cause acute changes in excitability, may enhance each other's effectiveness, or may prolong or curtail events in the target cells.

Du Bois Reymond (1877) suggested that two theories of neurotransmission should be entertained: transmission might be chemical, due to a stimulatory secretion, or it might be electrical in nature. The idea that chemical transmission was the more likely began with the thorough considerations of the actions of adrenaline and of the effects of stimulation of sympathetic nerves made by Elliott. In 1904, he published a short summary in which he concluded that sympathetic nerves acted upon smooth muscle by the liberation of adrenaline. The next year there appeared a complete analysis of the experimental data that lead to this conclusion (Elliott, 1905). A year later, Dixon (1906) proposed that parasympathetic nerves acted by the liberation of a muscarine-like substance, which Dale (1914) suggested to be acetylcholine. Dale (1954) recalls that Loewi had visited Cambridge and had discussions with Elliott in 1903, at the time that Elliott was formulating his ideas on the action of adrenaline. It was then Loewi and his colleagues who finally demonstrated the chemical nature of transmission, in experiments performed on the vagal nerve pathways to the frog heart (Loewi, 1921; Loewi and Navratil, 1926). Loewi demonstrated his classical experiment to the Twelfth International Congress of Physiology in Stockholm in 1926, almost exactly 60 years prior to this Thirtieth Congress (Holmstedt and Liljestrand, 1963). This was followed by experiments demonstrating that an adrenaline-like substance was, as Elliott predicted, a transmitter released by sympathetic nerves supplying the intestine (Finkelman, 1930) and the heart (Cannon and Bacq, 1931). It was later shown that, in mammals, this substance is noradrenaline (von Euler, 1956). A series of elegant experimental investigations showed that acetylcholine was also a transmitter at synapses in sympathetic ganglia (Feldberg and Gaddum, 1934) and at the motor end plate (Dale, Feldberg and Vogt, 1936; Brown, Dale and Feldberg, 1936). The idea that transmission from neurones was chemical did not go unchallenged, and it was not until the 1950's that it was accepted as true for the majority of neuroeffector junctions (Bacq, 1975), there being a few specialized junctions where electrical transmission occurs.

Following the acceptance of chemical transmission as a fact, there was a period of consolidation during which a number of drugs were developed that could block transmission from noradrenergic neurones. Drugs such as atropine and nicotine that blocked cholinergic transmission had been available since the last century and so the stage was now set to re-examine neuroeffector transmission at various sites. Thus in the 1960's came a number of important papers that pointed to the existence of transmission that was not explainable by the release of acetylcholine or of noradrenaline. Transmission from a class of inhibitory neurones to gastrointestinal muscle was found to be maintained in the presence of antagonists of noradrenergic and cholinergic transmission (Burnstock, Campbell, Bennett and Holman, 1964; Burnstock, Campbell and Rand, 1966; Martinson, 1965; Campbell, 1966). Other instances of transmission with an unexplained pharmacology also were discovered, for example excitatory transmission to intestinal muscle (Ambache and Freeman, 1968) and slow potentials in sympathetic ganglia (Nishi and Koketsu, 1968; see Dun, 1983). Once some of these instances had been documented it was possible to look back over earlier literature and identify more cases where both acetylcholine and noradrenaline seemed inadequate candidates as neurotransmitters (Campbell, 1970).

A further stage in this history was reached when immunohistochemical methods were used to demonstrate a number of small peptides in neurones (e.g. Hökfelt, Efendic, Johansson, Luft and Arimura, 1974; Hökfelt, Elde, Johansson, Luft and Arimura, 1975a; Hökfelt, Johansson, Efendic, Luft and Arimura, 1975b; Hökfelt, Kellerth, Nilsson and Pernow, 1975c; Hökfelt, Johansson, Ljungdahl, Lundberg and Schultzberg, 1980a). These peptides were known stimulants of various tissues, for example smooth muscle and glands, as well as having actions within the central nervous system. Within a few years, several hundred papers relating to the presence of peptides in neurones and to the possibility of their being transmitters had been published and the number of neuropeptides identified had increased to over 30 (see reviews by Cuello, 1978; Hökfelt et al., 1980a; Snyder, 1980; Furness and Costa, 1982).

Even while scientists were wrestling with the problem of the possible transmitter roles for the many peptides that had been discovered in neurones a further complication came to light: individual neurones contained not one but two, or even three, substances that could be proposed to be neurotransmitters (Lundberg, Hökfelt, Anggard, Terenius, Elde, Markey, Goldstein and Kimmel, 1982a; Lundberg and Hökfelt, 1983).

We are thus faced with dual problems: do the small peptides contained in neurones participate in neurotransmission; and what is the significance of there being several substances with the potential to be neurotransmitters in the one neurone? The two parts to the title of this lecture indicate our contention that many of these substances (and also non-peptides) do act as neurotransmitters and that their patterns of co-localization follow some system suggestive of a chemical coding of neurones.

There has been a huge volume of literature published on neuropeptides that it is not possible to review in these pages. Instead we have drawn on a few examples to highlight some of the ideas that have been developing in the last few years.

OTHER NEUROTRANSMITTER MOLECULES

Although much of the emphasis here is on peptides, there are a number of other molecules that also participate in neurotransmission, for example, amino acids (such as glycine, gamma-aminobutyric acid and aspartate), aromatic amines (such as histamine, dopamine, 5-hydroxytryptamine and adrenaline) and nucleotides (adenosine and adenosine triphosphate).

ARE PEPTIDES NEUROTRANSMITTERS?

We cannot attempt to answer this question for all peptides in all the neurones where they are found. However, in Table 1 we provide a tabulation of some places where an involvement of peptides in the transmission process is well documented. We will discuss an example where we have personal experience: substance P as an excitatory transmitter to intestinal muscle.

Substance P was first detected by von Euler and Gaddum (1931) who found this active compound in extracts of both brain and intestine. Much later, immunohistochemical studies showed substance P to be in nerve fibres in the intestine (Nilsson, Larsson, Håkanson, Brodin, Pernow and Sundler, 1975; Pearse and Polak, 1975). In the guinea-pig small intestine, where we now focus our attention, substance P immunoreactivity is in nerve cell bodies and nerve fibres but not in other cell types such as endocrine cells (Costa, Cuello, Furness and Franco, 1980). Chromatographic separation and radioimmunoassay showed that the immunoreactive material was indeed authentic substance P (Murphy, Furness, Beardsley and Costa, 1982). Substance P contracts intestinal muscle and also exhibits the interesting property of desensitizing its own receptors, that is, exposure to substance P in sufficient concentration renders the muscle refractory to further exposure (Gaddum, 1953). Franco, Costa and Furness (1979) exploited this desensitization in an examination of transmission from non-cholinergic

TABLE 1

Examples of Known Involvement of Peptides in Neurotransmission

Peptide	Site	Reference
LHRH	Excitatory transmitter with acetylcholine in amphibian sympathetic ganglia	Jan and Jan (1982,1983)
Somatostatin	Cardioinhibitory transmitter with acetylcholine in an amphibian	Campbell et al (1982)
VIP	Transmitter of gastrointestinal inhibitory neurones along with unknown compound(s)	Furness and Costa (1986)
VIP	Transmitter with acetylcholine in cat salivary gland	Lundberg (1981)
GRP	Excitatory transmitter to gastrin cells of the mammalian stomach	Schubert et al. (1985)
SP	Transmitter for neurogenic inflammation (antidromic vasodilatation and extravasation)	Lembeck and Gamse (1982)
SP	Excitatory transmitter to intestinal muscle	see text
Proctolin	Excitatory transmitter with glutamate in motoneurones to cockroach skeletal muscle	Adams and O'Shea (1983)
NPY	Vasoconstrictor transmitter with noradrenaline to cat spleen	Lundberg et al. (1984)

excitatory neurones to the longitudinal muscle of the guinea-pig small intestine. They showed that substance P caused desensitization of its own contractile effect, but did not restrict the effectiveness of other agonists (carbachol, DMPP, 5-HT or bradykinin). Substance P desensitization abolished transmission from the non-cholinergic neurones

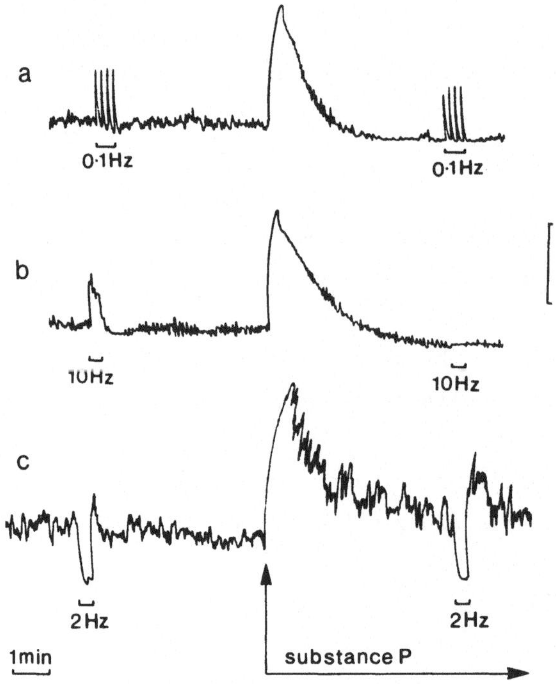

Fig. 1. Effects of substance P desensitization on (a) cholinergic excitatory transmission, (b) non-cholinergic excitatory transmission and (c) transmission from enteric inhibitory neurones to the longitudinal muscle of the guinea-pig intestine. The non-cholinergic transmission, evoked by transmural stimulation at 10Hz with hyoscine present to block cholinergic transmission was selectively blocked by desensization using a substance P concentration of 7.5×10^{-8}M. The vertical scale is 1g for (a) and (b), 2g for (c). Modified from Franco et al. 1979.

and, at the same concentration, had no effect on transmission from cholinergic or enteric inhibitory neurones (Fig. 1). It is interesting that Paton and Zar reported similar experiments in a presentation to the International Pharmacology Congress in 1966 but appear not to have published the results in full. Analogues of substance P that block its receptors without having significant agonist action have now become available and these also antagonize transmission from the non-cholinergic excitatory neurones to the longitudinal muscle of the guinea-pig small intestine as well as to the circular muscle and to the taenia coli (Leander, Håkanson, Rosell, Folkers, Sundler and Tomqvist, 1981; Leander, Hakanson, Horig and Folkers, 1983; Björkroth, Rosell, Xu and Folers, 1982; Yokoyama and North, 1983; Costa, Furness, Pullin and Borstein, 1985).

Intracellular microelectrodes can be used to examine the electrical correlates of neuromuscular transmission. Using such techniques, Niel, Bywater and Taylor (1983) and Taylor and Bywater (1986) have described transient depolarizations elicited in the muscle of the guinea-pig small intestine when non-cholinergic neurones are stimulated; these depolarizations are blocked by desensitization of substance P receptors or by a substance P antagonist. Similar pharmacological tests of transmission point to an involvement of substance P in non-cholinergic excitatory transmission to the feline pyloric sphincter (Lidberg, Lundberg, Dahlstrom, Rosell, Folkers and Ahlman, 1982; Lidberg, Dahlstrom, Lundberg and Ahlman, 1983), to the muscularis mucosae in both the opossum oesophagus (Domoto, Jury, Berezin, Fox and Daniel, 1983) and in the canine colon (Angel, Go and Szurszewski, 1984) and to the feline lower oesophageal sphincter (Reynolds, Ouyang and Cohen, 1984).

In most of the examples referred to above, electrical stimulation was used to activate excitatory neurones. There is however, evidence for the involvement of substance P in excitatory transmission when enteric neurones are activated reflexly. Fluid distension of the guinea-pig ileum causes a non-cholinergic reflex excitation of the circular muscle that is blocked or substantially reduced by substance P desensitization or substance P antagonists (Bartho, Holzer, Donnerer and Lembeck, 1982a,b; Yokoyama and North, 1983). Reynolds et al. (1984) showed that the reflex contraction of the lower oesophageal sphincter when the lumen of the distal oesophagus is acidified involves the activation of non-cholinergic neurones to the muscle. The reflexly-induced contractions were mimicked by substance P and were abolished or substantially reduced by desensitization of substance P receptors or by substance P antagonists. Costa et al. (1985) also examined reflexes, in this case reflex contractions of the circular muscle of the small intestine that occur oral to a point of distension, and found that the non-cholinergic components of these reflexes were also blocked by antagonists of receptors for substance P.

When substance P neurones in the intestine are stimulated electrically or reflexly some of the peptide escapes into the fluid bathing the organ or into the vasculature and can be detected by radioimmunoassay (Baron, Jaffe and Gintzler, 1983; Angel et al., 1984; Donnerer, Holzer and Lembeck, 1984; Holzer, 1984).

In summary, there is conclusive evidence for the participation of substance P in non-cholinergic excitatory transmission to gastrointestinal muscle. The peptide is present in nerve fibres supplying the muscle and it is released when enteric neurones are stimulated; the muscle is contracted by substance P; and when non-cholinergic neurones supplying the muscle are stimulated electrically or reflexly the subsequent depolarizations or contractions are antagonized if substance P receptors on the muscle are blocked by desensitization or by substance P analogues.

There are numerous other instances where the release of peptides from neurones has been detected: in fact, as far as we are aware in all places where a peptide is present and evidence for release has been sought, such evidence has been found. It is probably reasonable to assume that all peptides found in nerve endings can be released when the nerve fibres are activated. The peptides all seem to affect the excitability of effector cells when they are applied artificially which implies that they all have the potential to be involved in the process of neurotransmission.

CHEMICAL CODING AND CHEMICAL MARKERS

We have begun to look for patterns of co-localization of potential transmitters and have encountered some unusual combinations of substances that have obliged us to hesitate in the use of the term neurotransmitter for a substance when first encountered in a neurone. In the submucous ganglia there are nerve cell bodies that contain immunoreactivity for each of the following substances: calcitonin gene-related peptide (CGRP), cholecystokinin (CCK), cholineacetyltransferase (ChAT), galanin (GAL), neuropeptide Y (NPY) and somatostatin (SOM) (Furness, Costa and Keast, 1984; Furness, Costa, Gibbins, Llewellyn-Smith and Oliver, 1985 and unpublished; see Fig. 2). As a shorthand for these neurones we have referred to them as CCK/CGRP/ChAT/GAL/NPY/SOM neurones; this is their chemical code based on their content of potential transmitters. It is our presumption that immunoreactivity for ChAT means that these neurones are cholinergic. It seems likely that each of the other substances somehow participates in the process of transmission, but it seems unlikely that each is a transmitter in its own right in these neurones, for reasons elaborated below. It is probable that each substance has a different role and that each contributes to a different extent to the transmission process. We therefore take the view that we should refer to each substance as a chemical marker, that is, a substance whose presence, either alone or

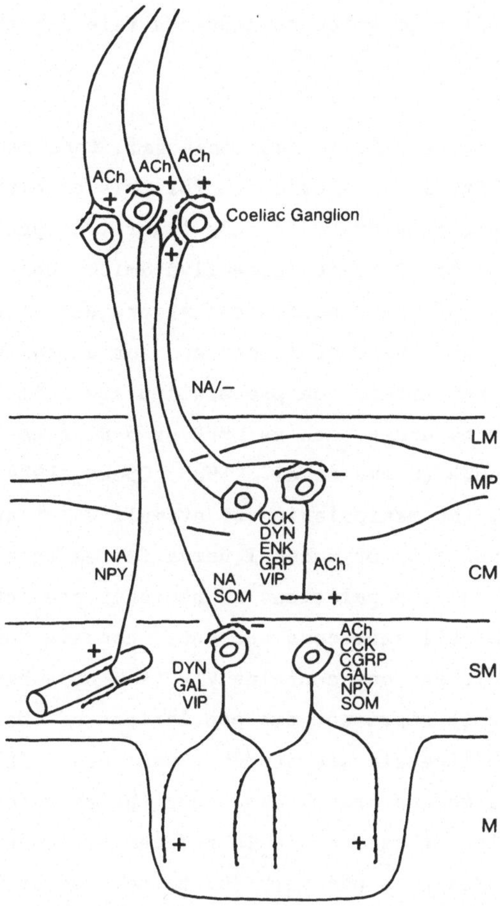

Fig. 2. Examples of chemical coding in nerve pathways associated with the guinea-pig small intestine. There are three chemical codes for noradrenaline neurones in the coeliac ganglion (NA/SOM, NA/NPY and NA alone). The axons from these neurones join common nerve trunks but run selectively to different targets within the small intestine: NA/NPY to arterioles, NA/SOM to submucous ganglia and NA/- to cholinergic neurones within the myenteric plexus. Within submucous ganglia, NA/SOM neurones selectively innervate DYN/GAL/VIP secretomotor neurones but do not supply another set of secretomotor neurones coded by ACh/CCK/CGRP/GAL/NPY/SOM. The separate coding of coeliac ganglion noradrenaline neurones has allowed us to discover that intestinofugal neurones containing CCK/DYN/ENK/GRP/VIP selectively supply noradrenaline neurones that reduce motility and secretion, but do not innervate noradrenaline vasoconstrictor neurones.

in combination with other substances, can distinguish one class of neurones from another. Use of the term chemical marker is a admission that we may have insufficient data from which to deduce a role for the substance so designated.

The nerve cell bodies of the submucosa fall into four groups, two of which are shown in Fig. 2. The evidence for this conclusion, and for the projections and connections drawn in this figure, is published elsewhere (Furness and Costa, 1986; Furness, Llewellyn-Smith, Borstein and Costa, 1986). The functions of those neurones that project to the mucosa can be explored by setting up a sheet of mucosa and submucosa in an Ussing chamber. When neurones within the preparation are stimulated with brief electrical pulses or by drugs such as DMPP or 5-HT a net secretion of water and electrolytes occurs (Cooke et al., 1983; Cooke, 1984; Keast, Furness and Costa, 1985a,b). No manipulation of stimuli or of antagonist drugs is able to reveal a population of mucosal nerve fibres that enhance absorption. Of the three populations of neurones projecting to the mucosa two, totalling 40% of all submucous neurones, contain ChAT and are presumably cholinergic and one contains VIP (45% of submucous neurones). These neurones probably represent two populations of cholinergic and one population of non-cholinergic secretomotor neurones. If receptors for acetylcholine in the mucosa are blocked then the secretory response to electrical stimulation of the nerves is reduced by about 50%. The remainder of the response is possibly due to release of VIP and of other secretory stimulants in the neurones. This brings us to an interesting puzzle: if the neurones just stimulate secretion, what is the reason for the combination of substances in the CCK/CGRP/ChAT/GAL/NPY/SOM neurones? Secretion is increased by acetylcholine (Tapper, 1983) and by CCK (Gardner et al., 1967; Matachansky et al., 1972) but is decreased by neuropeptide Y (Miller, 1985; Saria and Beubler, 1985) and by somatostatin (Dharmsathaphorn et al., 1980; Keast et al., 1986). Thus in the same neurone there are both stimulants and inhibitors of secretion. This is not the only example of this type: there are nerve fibres supplying the guinea-pig uterine artery in which dynorphin (DYN), NPY and VIP are co-localized (Morris et al., 1985). NPY constricts and VIP dilates this artery. Moreover, Lundberg et al. (1985) have shown that CGRP and substance P released from the peripheral ends of the same sensory neurones exert different effects. These examples underscore the need to be circumspect about the roles of the peptides and justifies referring to them simply as chemical markers when no clear role can be attributed.

TABLE 2

Examples* of Inconstancy of Association of Chemical Markers

Combination of Markers	Neurones
Example 1: Somatostatin neurones	
ACh/SOM	Cardiodepressor neurones in an amphibian
NA/SOM	Inhibitory fibres to inteetinal secretomotor neurones
ACh/CCK/CGRP/GAL/NPY/SOM	Intestinal secretomotor neurones
SOM + ?	Primary sensory neurones
Example 2: VIP neurones	
ACh/VIP	Salivary gland motoneurones
DYN/GAL/VIP	Non-cholinergic intestinal secretomotor neurones
DYN/NPY/VIP	Nerve fibres to uterine artery
Example 3: Substance P neurones	
ACh/SP	Intestinal secretomotor neurones
5-HT/SP/TRH	Raphe neurones
CCK/CGRP/DYN/SP	Primary sensory neurones

* Further details relevant to the examples and the species in which they occur are provided in the text.

Inconstancy of Association of Chemical Markers

It might be expected that there would be some constancy of association of chemical markers, but this is not the case. From Table 2 it can be seen that substances come together in a wide variety of combinations and few rules have emerged that would allow the combinations present to be predicted in any novel situation. It is as if a range of substances can equally well fill roles in a variety of functionally distinct neurones. This should not come as a surprise. After all, acetylcholine participates in transmission to skeletal muscle, to sweat glands, to the constrictor of iris and to the heart and has different actions and roles in these different organs. Thus we must be cautious about predicting associations of chemical markers in novel situations.

Chemical Coding of Sub-groups of Neurones

We have just warned that, in general, there is no predictability in the associations of substances that occur within classes of neurones. Nevertheless, within the one system, the associations that exist seem to provide an exquisite coding that divides neurones into functional and anatomical subgroups. The coeliac ganglion is composed almost entirely of noradrenergic neurones that innervate the abdominal viscera, with the small intestine being a major target. We have examined the chemical coding of nerve fibres that reach the intestine from these ganglia in the guinea-pig (Costa and Furness, 1984). It was found that noradrenaline neurones containing NPY (NA/NPY neurones) run exclusively to intestinal arterioles, NA/SOM neurones supply the submucous ganglia but never the arterioles, and noradrenaline neurones containing neither of these peptides supply myenteric ganglia. Studies of the distributions of the nerve cell bodies in the ganglia showed that small clumps of each type are dispersed throughout (Macrae et al., 1986). Thus three separately coded sets of noradrenergic cell bodies in the one ganglion send their processes out via mixed nerve trunks to innervate targets in the one organ precisely according to their chemical coding.

A similar target-related chemical coding exists for small diameter primary sensory neurones (Gibbins et al., 1986a; Fig. 3). In the dorsal root ganglia of the guinea-pig there are neurones containing various combinations of the peptides CCK, CGRP, DYN and SP. In individual ganglia there are mixtures of cell bodies with the peptide combinations

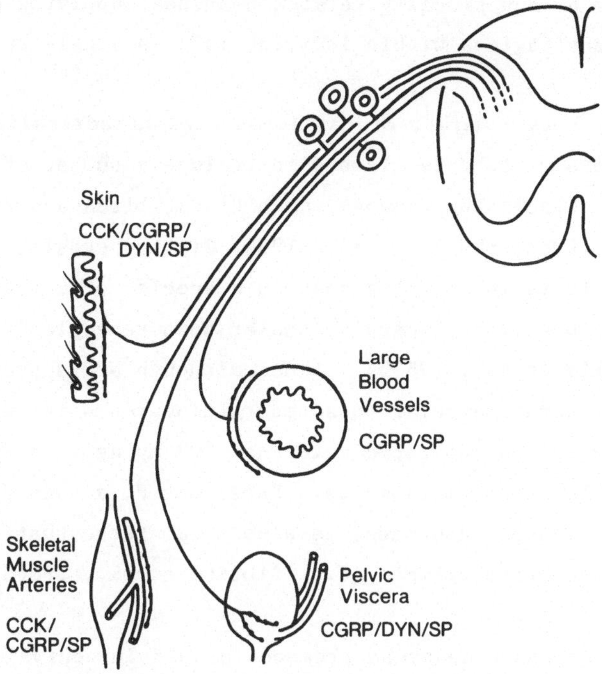

Fig. 3. Chemical coding in small diameter sensory neurones of the guinea-pig. Neurones that contain different combinations of CCK, CGRP, DYN and SP selectively supply different peripheral targets, although the cell bodies are in the same ganglia and the fibres run in the same nerve trunks.

CCK/CGRP/DYN/SP, CCK/CGRP/SP (without DYN), CGRP/DYN/SP (without CCK) and CGRP/SP (without CCK or DYN). Processes of these differently coded nerve cells leave the ganglia in the same nerve trunks. Their endings, however, selectively supply different fields: CCK/CGRP/DYN/SP endings are in the skin; CCK/CGRP/SP endings are around the vessels of skeletal muscle; CGRP/DYN/SP endings supply the pelvic viscera and airways; and CGRP/SP endings run to the heart and great vessels. Thus there are preferential peripheral distributions of these primary afferent fibres based on their chemical coding.

Hierarchies of Chemical Coding

Present data suggest that there exist hierarchies of coding, such that certain chemicals can be markers of broad groups of functionally related neurones while other markers or patterns of markers define subpopulations. Some coding molecules extend across several classes of vertebrates, others appear to be confined within classes or subclasses, while others appear to define sub-groups of functionally related neurones supplying an organ, or supplying different regions within individuals of a single species.

Neuropeptide Y is a marker of cardiovascular noradrenaline neurones that extends across vertebrate classes to include a number of species of placental mammals (including humans), amphibians, birds and reptiles (Lundberg et al., 1983; Ekblad et al., 1984; Gibbins et al., 1986b; Morris et al., 1986a). It is interesting that in a species of marsupial mammal, the brush-tailed possum, most cardiovascular noradrenergic fibres do not contain NPY (Morris et al., 1986b). Thus, although NPY appears to be a general marker of cardiovascular noradrenergic neurones in vertebrates, whose lines of evolution separated more than 300 My ago, there may have been a secondary loss in some species. Substance P, as one of the markers of small diameter sensory neurones, is also a substance that is present across several vertebrate classes (e.g. Gibbins et al., 1986b).

In contrast to the widespread presence of NPY in equivalent neurones of species in different vertebrate classes, there are examples of restriction of chemical codes. For example, somatostatin participates with acetylcholine in transmission from vagal postganglionic neurones to the heart in some amphibian species (Campbell et al., 1982), but not in mammals. Also in an amphibian, LHRH has been shown to be a neurotransmitter in the cholinergic preganglionic fibres supplying some neurones of sympathetic ganglia, although it is not found in equivalent fibres in mammals (Jan and Jan, 1983).

Within the one vertebrate class there are sometimes restricted codes. For example, VIP appears to be in cholinergic neurones supplying the salivary glands in the cat and to participate in vasodilator transmission (Lundberg, 1981), whereas this transmission appears to be entirely cholinergic in the rabbit (Morley et al., 1966). In one mammalian species, the guinea-pig, nerve cell bodies of a specific subgroup of noradrenaline neurones in the coeliac ganglion contain somatostatin and send their processes to submucous ganglia (Costa and Furness, 1984). These are noradrenergic neurones that inhibit secretomotor neurones (Furness and Costa, 1986). Although this same functional group is known to exist in the cat (Sjövall, 1984), they do not share this chemical coding because there are no somatostatin containing noradrenaline neurones in the feline coeliac ganglion (Lundberg et al., 1982a).

Within individuals of the one species there can be regional differences in coding. As explained earlier, in the case of small diameter sensory fibres to blood vessels in guinea-pig, CCK/CGRP/DYN/SP nerve fibres supply cutaneous blood vessels, CCK/CGRP/SP fibres supply vessels of skeletal muscle beds and CGRP/SP endings supply large systemic arteries.

Consequences and Significance of Chemical Coding

The existence of a chemical coding of neurones has provided the means to unravel neuronal circuits and to provide maps of the distributions and connections of neurones such as that in Fig. 2. The coding also allows conclusions to be made about patterns of neuronal branching. For example, sensory nerve fibres to skeletal muscle and cutaneous vascular beds are unlikely to be collaterals of each other because they are differently coded. So, just as the advent of immunohistochemical staining applied for single substances has allowed circuits to be more fully analyzed, the presence of a chemical code involving several substances in the one neurone is giving us the opportunity to analyze finer details of neurocircuitry.

The discovery that there may be several substances in the one neurone has led to considerable speculation about their relative roles (e.g. Burnstock, 1976; Dismukes, 1979; Kupferman, 1979; Hökfelt et al., 1980b; Cuello, 1982). Now we have examples with four or more substances present, submucous CCK/CGRP/ChAT/GAL/NPY/SOM neurones (see above); CCK/DYN/ENK/GRP/VIP neurones projecting from the intestine to the coeliac ganglion (Macrae et al., 1986); and CCK/CGRP/DYN/SP sensory neurones (Gibbins et al., 1986a). If all these substances are capable of being released, as seems likely, then it is perhaps sensible that we regard transmission as a plurichemical process. At each site of transmission different substances may have different roles in the transmission process. For example, in transmission to amphibian sympathetic ganglia, acetylcholine causes a rapid excitation of neurones (30-50msec) while LHRH, released from the same neurones, causes a sustained increase in neuronal excitability (the late slow e.p.s.p.), lasting 5-10 min (Jan and Jan, 1982, 1983). The late slow e.p.s.p. is associated with a sustained increase in membrane resistance and therefore considerably augments the excitability of the neurone (Schulman and Weight, 1976). In the salivary gland of the cat, where VIP and ACh are involved in transmission from the same neurones (Lundberg, 1981), VIP in doses that are themselves ineffective greatly enhance the effectiveness of ACh (Lundberg et al., 1982b). In contrast to these examples, two cases where substances in the neurones cause opposite effects have been noted earlier in this review. Transmitters may also have metabolic effects on target cells. For example, Ip and Zigmond (1984) have shown that the enzyme tyrosine hydroxylase is activated in nerve cells of the superior cervical ganglion when nerve fibres supplying these cells are stimulated. Thus chemical transmission is a process that may often involve the combined actions of several substances, each of which, acting alone, would not mimic transmission and some of which might have contrasting effects on the excitability of the target cell.

ACKNOWLEDGEMENTS

The original work and the thought behind it owes much to collaboration and discussion with a number of colleagues, in particular Joel Bornstein, Graeme Campbell, Rony Franco, Stephen Johnson, Janet Keast, Ida Llewellyn-Smith, Mhairi Macrae, Roger Murphy, Ray Papka, Terry Smith and Alan Wilson. Excellent technical support has been provided by Venetta Esson, Janine Falconer, Julie Giles, Sue Graham, Charles Humphreys, Fiona Renton, Di Trussell, Rae Tyler and Pat Vilimas. These endeavours have been supported financially primarily by the National Health and Medical Research Council, the National Heart Foundation and Flinders University.

REFERENCES

Adams, M.E. and O'Shea, M. (1983) Peptide co-transmitter at a neuromuscular junction. Science 221, 286-289

Ambache, N. and Freeman, M.A. (1968) Atropine resistant longitudinal muscle spasms due to excitation of non-cholinergic neurons in Auerbach's plexus. J. Physiol. (Lond.) 199, 705-727

Angel, F., Go, V.L.W. and Szurszewski, J.H. (1984) Innervation of the muscularis mucosae of canine proximal colon. J. Physiol. (Lond.) 357, 93-108

Bacq, Z.M. (1975) Chemical Transmission of Nerve Impulses. A Historical Sketch. Pergamon: Oxford, 106p

Baron, S.A., Jaffe, B.M. and Gintzler, A.R. (1983) Release of substance P from the enteric nervous system: direct quantitation and characterization. J. Pharmacol. exp. Ther. 227, 365-368

Bartho, L., Holzer, P., Donnerer, J. and Lembeck, F. (1982A) Evidence for the involvement of substance P in the atropine-resistant peristalsis of the guinea-pig ileum. Neurosci. Lett. 32, 69-74

Bartho, L., Holzer, P., Donnerer, J. and Lembeck, F. (1982b) Effects of Substance P, cholecystokinin octapeptide, bombesin and neurotensin on the peristaltic reflex of the guinea-pig ileum in the absence and in the presence of atropine. Arch. Pharmacol. 321, 321-328

Björkroth, U., Rosell, S., Xu, J.-C. and Folkers, K. (1982) Pharmacological characterization of four related Substance P antagonists. Acta Physiol. Scand. 116, 167-173

Brown, G.L., Dale, H.H. and Feldberg, W. (1936) Reactions of the normal mammalian muscle to acetylcholine and to eserine. J. Physiol. (Lond.) 87, 394-424

Burnstock, G. (1976) Do some nerve cells release more than one transmitter? Neuroscience 1, 239-248

Burnstock, G., Campbell, G., Bennett, M. and Holman, M.E. (1964) Innervation of the guinea-pig taenia coli: are there intrinsic inhibitory nerves which are distinct from sympathetic nerves? Int. J. Neuropharmacol. 13, 163-166

Burnstock, G., Campbell, G. and Rand, M.J. (1966) The inhibitory innervation of the taenia of the guinea-pig caecum. J. Physiol. (Lond.) 182, 504-526

Campbell, G. (1966) The inhibitory nerve fibres in the vagal supply to the guinea-pig stomach. J. Physiol. (Lond.) 185, 600-612

Campbell, G. (1970) Autonomic nervous supply to effector tissues. In: Smooth Muscle. Bulbring, E., Brading, A.F., Jones, A.W. and Tomita, T., eds. Arnold: London, pp 451-495

Campbell, G., Gibbins, I.L., Morris, J.L., Furness, J.B., Costa, M., Oliver, J.R., Beardsley, A.M. and Murphy, R. (1982) Somatostatin is contained in and released from cholinergic nerves in the heart of the toad Bufo marinus. Neuroscience 7, 2013-2023

Cannon, W.B. and Bacq, Z.M. (1931) Studies on the conditions of activity in endocrine organs, XXVI. A hormone produced by sympathetic action on smooth muscle. Am. J. Physiol. 96, 392-412

Cooke, H.J. (1984) Influence of enteric cholinergic neurons on mucosal transport in guinea-pig ileum. Am. J. Physiol. 246, G76-G82

Cooke, H.J., Shonnard, K. and Wood, J.D. (1983) Effects of neuronal stimulation on mucosal transport in guinea pig ileum. Am. J. Physiol. 245, G290-G296

Costa, M., Cuello, A.C., Furness, J.B. and Franco, R. (1980) Distribution of enteric neurons showing immunoreactivity for substance P in the guinea-pig ileum. Neuroscience 5, 321-331

Costa, M. and Furness, J.B. (1984) Somatostatin is present in a subpopulation of noradrenergic nerve fibres supplying the intestine. Neuroscience 13, 911-920

Costa, M., Furness, J.B., Pullin, C.O. and Bornstein, J. (1985) Substance P enteric neurons mediate non cholinergic transmission to the circular muscle of the guinea-pig intestine. Arch. Pharmacol. 328, 446-453

Cuello, A.C. (1978) Immunocytochemical studies of the distribution of neurotransmitters and related substances in CNS. In: Handbook of Psychopharmacology, Vol. 6. Iversen, L.L., Iversen, S.D. and Snyder, S.H., eds. Plenum: New York pp 69-137

Cuello, A.C. (1982) Co-transmission, Macmillan: London.

Dale, H.H. (1914) The action of certain esters and ethers of choline, and their relation to muscarine. J. Pharmacol. exp. Ther. 6, 147-190

Dale, H.H. (1954) The beginnings and the prospects of neurohumoral transmission. Pharmacol. Rev. 6, 7-14

Dale, H.H., Feldberg, W. and Vogt, M. (1936) Release of acetylcholine at voluntary motor nerve endings. J. Physiol. (Lond.) 86, 353-380

Dharmsathaphorn, K., Binder, H.J. and Dobbins, J.W. (1980) Somatostatin stimulates sodium and chloride absorption in the rabbit ileum. Gastroenterology 78, 1559-1564

Dismukes, R.K. (1979) New concepts of molecular communication among neurons. Behav. Brain. Sci. 2, 409-448

Dixon, W.E. (1906) Vagus inhibition. Brit. Med. J. 2, 180-181

Domoto, T., Jury, J., Berezin, I., Fox, J.E.T. and Daniel, E.E. (1983) Does substance P comediate with acetylcholine in nerves of opossum esophageal muscularis mucosa? Am. J. Physiol. 245, G19-G28

Donnerer, J., Holzer, P. and Lembeck, F. (1984) Release of dynorphin, somatostatin and substance P from the vascularly perfused small intestine of the guinea-pig during peristalsis. Br. J. Pharmac. 83, 919-925

Du Bois Reymond, E.R., (1877) Gesammelte Abhandlung der allgemeinen Muskel- und Nervenphysik. Vol. 2, 700p.

Dun, N.J. (1983) Peptide hormones and transmission in sympathetic ganglia. In: Autonomic Ganglia Elfvin, L.-G., ed. Wiley: London, pp 345-366

Ekblad, E., Edvinsson, L., Wahlestedt, C., Uddman, R., Håkanson, R. and Sundler, F. (1984) Neuropeptide Y co-exists and co-operates with noradrenaline in perivascular nerve fibres. Reg. Peptides 8, 225-235

Elliott, T.R. (1904) On the action of adrenaline. J. Physiol. (Lond.) 31, 20P

Elliott, T.R. (1905) The action of adrenalin. J. Physiol. (Lond.) 32, 401-467

Feldberg, W. and Gaddum, J.H. (1934) The chemical transmitter at synapses in a sympathetic ganglion. J. Physiol. (Lond.) 81, 305-319

Finkleman, B. (1930) On the nature of inhibition in the intestine. J. Physiol. (Lond.) 70, 145-157

Franco, R., Costa, M. and Furness, J.B. (1979) Evidence for the release of endogenous substance P from intestinal nerves. Arch. Pharmacol. 306, 195-201

Furness, J.B. and Costa, M. (1982) Identification of gastrointestinal neurotransmitters. Handbook of Exp. Pharmacol. 59, 383-460

Furness, J.B. and Costa, M. (1986) The Enteric Nervous System, Livingstone: Edinburgh.

Furness, J.B., Costa, M., Gibbins, I.L., Llewellyn-Smith, I.J. and Oliver, J.R. (1985) Neurochemically similar myenteric and submucous meurons directly traced to the mucosa of the small intestine. Cell Tiss. Res. 241, 155-163

Furness, J.B., Costa, M. and Keast, J.R. (1984) Choline acetyltransferase and peptide immunoreactivity of submucous neurons in the small intestine of the guinea-pig. Cell Tiss. Res. 237, 328-336

Furness, J.B., Llewellyn-Smith, I.J., Bornstein, J.C. and Costa, M. (1986) Neuronal circuitry in the enteric nervous system. In: Handbook of Chemical Neuroanatomy. Owman, C., Björklund, A. and Hökfelt, T., eds. Elsevier: Amsterdam (in press)

Gaddum, J.H. (1953) Tryptamine receptors. J. Physiol. (Lond.) 119, 363-368

Gardner, J.D., Peskin, G.W., Cerba, J.J. and Brooks, F.P. (1967) Alterations of in vitro fluid and electroyte absorption by gastrointestinal hormones. Am. J. Surg. 113, 57-64

Gibbins, I.L., Furness, J.B. and Costa, M. (1986a) pathway-specific patterns of co-existence of substance P, calcitonin gene-related peptide, cholecystokinin and dynorphin in dorsal root ganglion neurons of the guinea-pig. Cell Tiss. Res. (submitted)

Gibbins, I.L., Morris, J.L., Furness, J.B. and Costa, M. (1986b) Innervation of systemic blood vessels. In: Non-adrenergic Innervation of Blood Vessels, Burnstock, G. and Griffith, S., eds. CRC Press: Florida.

Hökfelt, T., Efendic, S., Johansson, O., Luft,, R. and Arimura, A. (1974) Immunohistochemical localization of somatostatin (growth hormone release-inhibiting factor) in the guinea pig brain. Brain Res. 80, 165-169

Hökfelt, T., Elde, R., Johansson, O., Luft, R. and Arimura, A. (1975a) Immunohistochemical evidence for the presence of somatostatin, a powerful inhibitory peptide, in some primary sensory neurons. Neurosci. Lett. 1, 231-235

Hökfelt, T., Johansson, O., Efendic, S., Luft, R. and Arimura, A. (1975b) Are there somatostatin-containing nerves in the rat gut?: Immunohistochemical evidence for a new type of peripheral nerves. Experientia 31, 852-854

Hökfelt, T., Kellerth, J.O., Nilsson, G. and Pernow, B. (1975c) Substance P: Localization in the central nervous system. Science 190, 889-890

Hökfelt, T., Johansson, O., Ljungdahl, A., Lundberg, J.M. and Schultzberg, M. (1980a) Peptidergic neurones. Nature 284, 515-521

Hökfelt, T., Lundberg, J.M., Schultzberg, M., Johansson, O., Ljungdahl, A. and Rehfeld, J. (1980b) Coexistence of peptides and putative transmitters in neurons. Adv. Biochem. Psychopharmacol. 22, 1-23

Holmstedt, B. and Liljestrand, G. (1963) Readings in Pharmacology. Pergamon: Oxford, 395p

Holzer, P. (1984) Characterization of the stimulus-induced release of immunoreactive substance P from the myenteric plexus of the guinea-pig small intestine. Brain Res. 297, 127-136

Ip, N.Y. and Zigmond, R.E. (1984) pattern of presynaptic nerve activity can determine the type of neurotransmitter regulating a postsynaptic event. Nature 311, 472-474

Jan, L.Y. and Jan, Y.N. (1982) Peptidergic transmission in sympathetic ganglia of the frog. J. Physiol. (Lond.) 327, 219-246

Jan, Y.N. and Jan, L.Y. (1983) Coexistence and corelease of cholinergic and peptidergic transmitters in frog sympathetic ganglia. Fed. Proc. 42, 2929-2933

Keast, J.R., Furness, J.B. and Costa, M. (1985a) Different substance P receptors are found on mucosal epithelial cells and submucous neurons of the guinea-pig small intestine. Arch. Pharmacol. 329, 382-387

Keast, J.R., Furness, J.B. and Costa, M. (1985b) Investigations of nerve populations influencing ion transport that can be stmulated electrically, by serotonin and by a nicotinic agonist Arch. Pharmacol. 331, 260-266

Keast, J.R., Furness, J.B. and Costa, M. (1986) Effects of noradrenaline and somatostatin on basal and stimulated mucosal ion transport in the guinea-pig small intestine. Arch. Pharmacol. (in press)

Kupferman, I. (1979) Modulatory actions of neurotransmitters. Ann. Rev. Neurosci. 2, 447-465.

Leander, S., Håkanson, R., Horig, J. and Folkers, K. (1983) Screening of substance P antagonists. Proceedings of the International Symposium on

Substance P, Skranbanek, P. and Powell, D., eds. Boole Press: Dublin, pp 12-13

Leander, S., Håkanson, R., Rosell, S., Folkers, K., Sundler, F. and Tomqvist, K. (1981) A specific substance P antagonist blocks smooth muscle contractions induced by non-cholinergic, non-adrenergic nerve stimulation. Nature 294, 467-469

Lembeck, F. and Gamse, R. (1982) Substance P in peripheral sensory processes. In: Substance P in the Nervous System, Ciba Foundation Symposium 91, Pitman: London, pp 35-54

Lidberg, P., Dahlström, A., Lundberg, J.M. and Ahlman, H. (1983) Different modes of action of substance P in the motor control of the feline stomach and pylorus. Reg. Peptides 7, 41-52

Lidberg, P., Edin, R., Lundberg, J.M., Dahlstrom, A., Rosell, R., Folkers, K. and Ahlman, H. (1982) The involvement of substance P in the vagal control of the feline pylorus. Acta Physiol. Scand. 114, 307-309

Loewi, W. (1921) Über humorale Übertragbarkeit der Herznervenwirkung. Pflügers Arch. 189, 239-242

Loewi, W. and Navratil, E. (1926) Über humorale Übertragbarkeit der Herznervenwirkung. X. Über das Schicksal des Vagusstoffes. Pflügers Arch. 214, 678-688

Lundberg, J.M. (1981) Evidence for coexistence of vasoactive intestinal polypeptide (VIP) and acetylcholine in neurons of cat exocrine glands. Acta Physiol. Scand. Suppl. 496, 1-57

Lundberg, J.M., Anggard, A. and Fahrenkrug, J. (1982b) Complementary role of vasoactive intestinal polypeptide (VIP) and acetylcholine for cat submandibular gland blood flow and secretion. III. Effects of local infusions. Acta Physiol. Scand. 114, 329-337

Lundberg, J.M., Anggard, A., Theodorsson-Norheim, E. and Pernow, J. (1984) Guanethidine-sensitive release of neuropeptide Y-like immunoreactivity in the cat spleen by sympathetic nerve stimulation. Neurosci. Lett. 52, 175-180

Lundberg, J.M., Franco-Cereceda, A., Hua, X., Hökfelt, T. and Fischer, J.A. (1985) Co-existence of substance P and calcitonin gene-related peptide-like immunoreactivities in sensory nerves in relation to cardiovascular and bronchoconstrictor effects of capsaicin. Eur. J. Pharmacol. 108, 315-319

Lundberg, J.M. and Hökfelt, T. (1983) Coexistence of peptides and classical nurotransmitters. Trends in Neuroscience 6, 325-333

Lundberg, J.M., Hökfelt, T., Anggard, A., Terenius, L., Elde, R., Markey, K., Goldstein, M. and Kimmel, J. (1982a) Organizational principles in the peripheral sympathetic nervous system: Subdivision by coexisting peptides (somatostatin-, avian pancreatic polypeptide-, and vasoactive intestinal polypeptide-like immunoreactive materials). Proc. Natl. Acad. Sci. USA 79, 1303-1307

Lundberg, J.M., Terenius, L., Hökfelt, T. and Goldstein, M. (1983) High levels of neuropeptide Y in peripheral noradrenergic neurons in various mammals including man. Neurosci. Lett. 42, 167-172

Macrea, I.M., Furness, J.B. and Costa, M. (1986) Distribution of subgroups of noradrenaline neurons in the coeliac ganglion of the guinea-pig. Cell Tiss. Res. 244, 173-180

Martinson, J. (1965) Vagal relaxation of the stomach. Experimental reinvestigation of the concept of the transmission mechanism. Acta Physiol. Scand. 64, 453-462

Matachansky, C.P., Juet, P.M., Mary, J.Y., Rambaud, J.C. and Bernier, J.J. (1972) Effects of cholecystokinin and metoclopramide on jejunal movements of water and electrolytes and on transit time of luminal fluid in man. Eur. J. Clin. Invest. 2, 169-175

Miller, R.J. (1985) Control of eithelial ion transport by neuropeptides. Reg. Peptides (Suppl. 4) 203-208

Morley, J., Schachter, M. and Smaje, L.H. (1966) Vasodilation in the submaxillary gland of the rabbit. J. Physiol. (Lond.) 187, 595-602

Morris, J.L., Gibbins, I.L., Campbell, G., Murphy, R., Furness, J.B. and Costa, M. (1986a) Innervation of the large arteries and heart of the toad (Bufo marinus) by adrenergic and peptide-containing neurons. Cell Tiss. Res. 243, 171-184

Morris, J.L., Gibbins, I.L., Furness, J.B., Costa, M. and Murphy, R. (1985) Co-localization of NPY, VIP and dynorphin, in non-noradrenergic axons of the guinea-pig uterine artery. Neurosci. Lett. 62, 31-37

Morris, J.L., Gibbins, I.L. and Murphy, R. (1986b) Neuropeptide Y-like immunoreactivity is absent from most perivascular noradrenergic axons in a marsupial, the brush-tailed possum. Neurosci. Lett. (submitted)

Murphy, R., Furness, J.B., Beardsley, A.M. and Costa, M. (1982) Characterization of substance P-like immunoreactivity in peripheral sensory nerves and enteric nerves by high pressure liquid chromatography and radioimmunoassay. Reg. peptides 4, 203-212

Niel, J.P., Bywater, R.A.R. and Taylor, G.S. (1983) Effect of substance P on non-cholinergic fast and slow post-stimulus depolarization in the guinea-pig ileum. J. Autonom. Nerv. Syst. 9, 573-584

Nilsson, G., Larsson, L.I., Håkanson, R., Brodin, E., Pernow, B. and Sundler, F. (1975) Localization of substance P-like immunoreactivity in mouse gut. Histochemistry 43, 97-99

Nishi, S. and Koketsu, K. (1968) Early and late after discharge of amphibian sympathetic ganglion cells. J. Neurophysiol. 31, 33-42

Paton, W.D.M. and Zar, A.M. (1966) Evidence for transmission of nerve effects by substance P in guinea-pig longitudinal muscle strip. Abstr. III Int. Pharmac. Congr., Sao Paulo, Brazil, p 9.

Pearse, A.G.E. and Polak, J.M. (1975) Immunocytochemical localization of substance P in mammalian intestine. Histochemistry 41, 373 375

Reynolds, J.C., Ouyang, A. and Cohen, S. (1984) A lower esophageal reflex involving substance P. Am. J. Physiol. 246, G346-G354

Saria, A. and Beubler, E. (1985) Neuropeptide Y (NPY) and peptide YY (PYY) inhibit prostaglandin E_2-induced intestinal fluid and electrolyte secretion in the rate jejunum in vivo. Eur. J. Pharmacol. 119, 47-52

Schubert, M.L., Saffouri, B., Walsh, J.H. and Makhlouf, G.M. (1985) Inhibition of neurally-mediated gastrin secretion by bombesin antiserum. Am. J. Physiol. 248, G456-G462

Schulman, J.A. and Weight, F.F. (1976) Synaptic transmission: long-lasting potentiation by a postsynaptic mechanism. Science 194, 1437-1439

Sjövall, H. (1984) Sympathetic control of jejunal fluid and electrolyte transport. An experimental study in cats and rats. Acta Physiol. Scand. Suppl. 535, 63p

Snyder, S.H. (1980) Brain peptides as neurotransmitters. Science 209, 976-983.

Tapper, E.J. (1983) Local modulation of intestinal ion transport by enteric neurons. Am. J. Physiol. 244, G456-G468

Taylor, G.S. and Bywater, R.A.R. (1986) Antagonism of non-cholinergic excitatory junction potentials in the guinea-pig ileum by a substance P analogue antagonist. Neurosci. Lett. 63, 23-26

von Euler, U.S. (1956) Noradrenaline. Thomas: Springfield, 382p

von Euler, U.S. and Gaddum, J.H. (1931) An unidentified depressor substance in certain tissue extracts. J. Physiol. (Lond.) 72, 74-87

Yokoyama, S. and North, R.A. (1983) Electrical activity of longitudinal and circular muscle during peristalsis. Am. J. Physiol. 244, G83-G88

CENTRAL PATTERN GENERATION: A CONCEPT UNDER SCRUTINY

K.G. Pearson

Department of Physiology
University of Alberta
Edmonton, Canada

SUMMARY

What are the mechanisms that generate the motor patterns for rhythmic movements in animals? Until recently it has widely been accepted that the relative timing of motor activity is determined by central neuronal circuits termed central pattern generators (CPGs). The validity of this view is examined for four motor systems in which there is a high degree of sensory regulation: the walking systems of the cat, cockroach and stick insect, and the flight system of the locust. In none of these systems is the evidence sufficient to conclude that a CPG is primarily responsible for establishing the timing of motor activity. In the stick insect there is no evidence for the existence of a CPG for walking at all, while in the cockroach the deafferented motor pattern differs from the normal walking pattern to such an extent that phasic afferent input must play an important part in establishing the timing of motor activity. Reflex pathways have been identified which appear to function to produce some aspects of the normal motor pattern for walking in insects. No firm conclusion can be reached regarding the role of central pattern generators in establishing the timing of motor activity in the walking system of the cat because the motor patterns in completely deafferented preparations have not been analysed in sufficient detail to allow an adequate comparison with the intact motor pattern. Finally, in the flight system of the locust it has recently been found that one entire phase of the normal flight cycle depends on phasic afferent input from wing receptors, and that sensory feedback is involved in the generation of rhythmicity. It is now clear that CPGs as defined in deafferented preparations do not form the basis for establishing the timing of activity in all rhythmic motor systems. Thus central pattern generation cannot be regarded as a universal functional principle for rhythmic motor systems.

INTRODUCTION

Most natural movements are produced by a precise, often quite complex, pattern of motor activity. Not only is the timing of the onset and termination of activity in different muscles highly regulated but the profile of the amplitude of activity in each muscle usually varies in a stereotyped manner for repetitions of the same movement. A fundamental problem is to determine the mechanisms which establish these patterns of activity. To date the mechanisms for patterning motor activity have been analysed in most detail in rhythmic motor systems. These systems have two obvious advantages. The first is that for most rhythmic movements the pattern of motor output remains relatively constant from cycle to cycle. The second is that these patterns can usually be readily initiated in restrained and extensively dissected preparations. The latter often allows a cellular analysis of patterning with the use of intracellular recording techniques.

A common feature of virtually all rhythmic motor systems is that a rhythmic motor pattern can be generated in the absence of phasic sensory feedback from peripheral proprioceptors. This general finding has led to the idea that the "timing of ... repetitive movements that constitute any rhythmic behaviour is regulated by intrinsic properties of the central nervous system rather than sensory feedback from moving parts of the body" (Delcomyn, 1980). In neurophysiological terms this means that the motor patterns for rhythmic movements are established by central mechanisms within the nervous system. In this article I refer to this idea as the concept of central pattern generation. Delcomyn (1980) has suggested that this concept constitutes a "general principle of neuronal organization applicable to all animals with central nervous systems".

Today it is common to refer to the neuronal systems generating the motor patterns in deafferented systems as central pattern generators (CPGs). Grillner and Wallen (1985) have defined a central pattern generator as the system of "...neurons responsible for creating a particular motor pattern (in a nervous system deprived of all sensory feedback)... regardless of whether all aspects of the motor pattern of the intact animal are produced or some part is missing". The latter part of this definition acknowledges that the motor patterns observed in deafferented preparations are usually not identical to those observed in intact behaving animals. Thus in most systems sensory feedback must play a role in establishing some details of the intact motor patterns but it is not regarded as being responsible for the generation of the basic rhythmicity or as a primary factor in establishing the relative timing of activity in different muscles. Exactly how sensory feedback interacts with a CPG has not yet been established in any rhythmic motor system.

An important aspect of the concept of central pattern generation is that each CPG is regarded as a unit within the central nervous system for generating one part of the overall motor pattern. The notion of CPGs as functional units has led to the proposal that complex behaviours may result from the linking of a number of CPGs (Grillner, 1985). For example, in the locomotory system of the cat the CPGs associated with the four legs may be combined in various ways to produce different patterns of coordination. A related proposal is that the CPG for a single leg can itself be fractionated into different functional units for independent control of movements at different joints. Independent action at each joint can therefore be achieved by activating just part of the CPG network for stepping. For example, "if one wants to wiggle the big toe, one may merely call on the unit CPG for the big toe, and so forth" (Grillner, 1985). One general view, therefore, is the existence in motor systems of a large number of functional units (CPGs) which can be combined in a variety of ways to produce a wide variety of movements. Two obvious questions are raised by the proposal that the basic motor patterns for complex movements can be explained by combining a variety of CPGs: 1) does the proposal of combinatorial sets of CPGs really further our understanding of how movements are produced since knowledge of how various combinations are selected and their interactions specified is presumably required for a satisfactory explanation of motor patterning, and 2) what criteria are we to use to identify a unitary CPG (e.g. a CPG for wiggling the big toe) and how would we establish the characteristics of the coupling between different CPGs?

Recently the concept of central pattern generation has been strongly criticized (Lundberg, 1980; Pearson, 1985; Bässler, 1986a). The most fundamental criticism of the concept is that it can lead to non-falsifiable hypotheses (Lundberg, 1980; see CONCLUSIONS). This raises the issue of whether the concept as currently formulated can be regarded as being scientifically valid. A less severe criticism is that the CPGs as defined in deafferented preparations may not always act as functional units in the intact animal, i.e. the CPGs may in some cases be so strongly influenced by proprioceptive feedback that these circuits lose their functional unity (Pearson, 1985). Related to this is Bässler's criticism that it is not necessarily the case that intact motor patterns can be explained in terms of the interaction of two well defined subsystems, the CPG and the sensory structures (Bässler, 1986a). Thus he concludes that the functional principles for pattern generation may be quite different in intact and deafferented systems. In light of these criticisms it seems worthwhile to review some of the evidence for central pattern generation in motor systems in which sensory feedback is known to be important in controlling the behaviour. The systems I have chosen are the walking systems of cats and insects and in the flight system of locusts. The main issue I consider is whether there is sufficient evidence to conclude that in each of these systems the CPG (as defined in deafferented preparations) is the basis for determining the overall timing of motor activity.

WALKING IN THE CAT

One of the first findings clearly indicating that the basic reciprocity of activity in flexors and extensors in the hindleg of a stepping cat is determined by central mechanisms was Graham Brown's observation that rhythmic alternating contractions of the deafferented ankle flexor and extensor muscles could be evoked by severing the spinal cord (Graham Brown, 1911). Graham Brown proposed that this rhythmic reciprocal activity could be generated by two mutually inhibiting centres -- he termed these half-centres -- with the generation of the rhythmicity depending on a fatigue process in each of the inhibitory pathways following activation of the corresponding half-centre. This finding had little influence on the views of how stepping movements are generated until the discovery in the 1960's that a locomotor-like rhythm could be generated in immobilized spinal cats following the intravenous injection of DOPA (Jankowska, Jukes, Lund and Lundberg, 1967a,b). This finding confirmed the existence of circuits within the spinal cord for programming alternating bursts of activity in flexors and extensors. However, EMG studies in walking animals had shown that the activity patterns were considerably more complex than a simple alternation of flexor and extensor activity, the most notable being a different pattern of activation of flexors acting at the hip and knee. Thus Lundberg (1969) raised the question "whether this differential activation is "built into" the central program or whether the central program consists of a simple alternating activation of extensors and flexors and the differential movement may be provided by proprioceptive reflexes". After a careful consideration of known reflex pathways Lundberg (1969) concluded that the latter is more likely. This conclusion has not been widely accepted (Lundberg, 1981). Subsequent experiments by Grillner and his colleagues found that the motor patterns generated in immobilized and deafferented preparations were more complex than a simple alternation of flexor and extensor activity (Grillner and Zangger, 1974, 1979; Grillner, 1981). Instead of attempting to modify the half-centre hypothesis to account for the features of the deafferented motor patterns not consistent with it, Grillner and his colleagues proposed that a central pattern generator (CPG) was responsible for establishing the relative timing of motor activity (Grillner, 1981; Grillner and Wallen, 1985).

The existence of central neuronal circuits which can generate many features of the motor pattern for walking is suggested most strongly from experiments on mesencephalic treadmill-walking cats in which motor patterns of the hindleg muscles were recorded before and after deafferentation of the hindlegs (Grillner and Zangger, 1984). The important finding was that following deafferentation the motor pattern was remarkably similar to that in intact animals. From this observation it appears that afferent feedback from proprioceptors in the hindlegs is not necessary for patterning activity in hindleg motoneurones. This finding is sufficient to exclude some of the proposals made by Lundberg (1969) concerning segmental reflex modulation of centrally programmed activity in flexors and extensors. But is it sufficient to allow the conclusion that proprioceptive feedback has little direct role in patterning motor activity? The confounding aspect of this study is that the forelegs were stepping normally and presumably providing considerable phasic afferent input to the spinal cord. It is conceivable, therefore, that these signals could be involved in patterning hindleg activity either by 1) direct intersegmental reflex pathways from foreleg afferents to hindleg motoneuronal and interneuronal pools, or 2) foreleg afferent input influencing the foreleg pattern generating network and the latter interacting strongly with the hindleg pattern generating network. Since no results on the effects of perturbation of the forelegs on the motor pattern in the deafferented hindlegs were reported it is impossible to assess the extent to which phasic proprioceptive signals from foreleg afferents are involved in patterning hindleg activity. Thus the results reported by Grillner and Zangger (1984) on the effects of deafferentation on treadmill-walking mesencephalic cats, although suggestive, do not conclusively demonstrate that CPGs establish the basic motor pattern for walking.

What we really need to know is 1) what are the motor patterns in animals deprived of all phasic input from leg receptors and 2) how closely do these patterns resemble those observed in intact animals? Unfortunately both these questions are difficult to answer due to the fact that the motor patterns have not been extensively analysed in a quantitative manner, and that results vary depending on the preparation used. First let us consider data from the work of Perret and his colleagues on the motor patterns in immobilized (curarized) or deafferented decorticate cats (Perret and Cabelguen, 1976, 1980; Perret, 1983). In these animals there is a strict alternation of activity in pure flexor and extensor muscles of the hip, knee and ankle. The patterns in bifunctional muscles, however, are variable, depend on the nature of tonic afferent input, and usually do not resemble those seen in intact animals. For example, in immobilized animals the efferent pattern in semitendinosus motoneurones (knee flexor/hip extensor) can vary between two extremes: one in which activity is present during the extensor phase, and one in which activity occurs during flexor activation. Tonic stimulation of the ipsilateral limb can cause a transition from the former to the latter pattern. Following deafferentation, semitendinosus activity is more likely to occur throughout the period of extensor activity. This is quite unlike the pattern in intact walking animals where semitendinosus becomes active for a short period early in the flexor phase and sometimes briefly at the beginning of the extensor phase.

The second set of data is that from the work of Grillner and Zangger (1979) on the motor patterns in acute spinal cats following the injection of DOPA. In acute deafferented preparations strong rhythmic patterns of motor activity can be generated. But are these the patterns of walking? Because only a small number of motoneurone pools were analysed in this study it is difficult to make direct comparisons with the data from intact animals. Even so, the published records of activity in deafferented or immobilized DOPA-injected spinal cats give the impression that the motor pattern is not much different from a simple alternation of flexor and extensor activity with almost all the subtle details of timing differences and intensity variation typical of intact walking being absent. Forssberg, Grillner and Halbertsma (1980) state "that the output pattern of a cat with isolated spinal cord (i.e. with dorsal roots and spinal cord transected) has so far not been shown to produce as complex a pattern (as in the intact cat) although it is more complicated than a flexor-extensor alternation". The precise details of the deviations from a strict alternation have not been described in any study of DOPA-injected spinal cats. Furthermore, it appears that following deafferentation the patterns generated in spinal DOPA-treated animals differ from those in decorticate animals. For example, in the former semitendinosus activity occurs throughout the entire flexor phase (Figs. 1 and 2 in Grillner and Zangger, 1979), while in the latter it usually occurs during the extensor phase (Perret and Cabelguen, 1980). One aspect of the deafferented pattern which was found by Grillner and Zangger to be similar to the intact pattern was that the dependence of the duration of flexor bursts on cycle time was less than that for the extensor bursts. More recently, however, other studies have failed to duplicate this finding (Jordan, Brownstone, Kriellaars and Noga, 1986).

Given the lack of quantitative analyses of the motor patterns in immobilized and deafferented preparations, the variability in these patterns, and the different patterns in decorticate and spinal animals, it is not possible to state the properties of the CPG with precision, nor is it possible to judge the assertion that a CPG is the basis for patterning the motor activity for walking. On the other hand there is no question about the involvement of afferent input in patterning motor activity. The most interesting sensory effect found so far is that input from hip afferents near the end of the stance phase is the primary causal event in initiating the stance-to-swing transition (Grillner and Rossignol, 1978). The influence of leg proprioceptors on the central rhythm has also been demonstrated by the fact that imposed rhythmic movements of the leg will entrain the central rhythm over a wide range (Andersson, Forssberg, Grillner and Wallen, 1981). Cutaneous input can also strongly influence the centrally generated pattern (Perret and Cabelguen, 1980) and can readily alter the motor output in a phase dependent manner in intact, mesencephalic and spinal cats (Grillner, 1981). Given these strong sensory influences the issue is whether the CPG is entirely responsible for establishing the relative timing of motor activity, or whether some aspects of timing are established by phasic sensory feedback. It is conceivable that the effects of sensory input can be explained by the influence this input has on the CPG (Grillner, 1985; Grillner and Wallen, 1985) but before this can be accepted we need a more precise description of the CPG, clear proposals on how these properties can be influenced by afferent input, and a description of exactly how afferent input interacts with the CPG to produce the normal motor pattern.

WALKING IN INSECTS

Walking in insects has often been regarded as a behaviour in which the basic motor pattern is generated centrally (Delcomyn, 1980). This conclusion arose from early studies on the cockroach in which it was found that motor patterns recorded in deafferented preparations resembled those in intact animals (Pearson, 1972; Pearson and Iles, 1970, 1973). However, these initial studies on the cockroach also demonstrated that afferent feedback was involved in establishing the timing of motor activity and was directly responsible for the generation of some features of the intact motor pattern (Pearson, 1972; Pearson and Duysens, 1976). Two questions have arisen concerning the early data on the cockroach. The first is whether they constitute sufficient evidence to conclude that a CPG is the basis for patterning walking motor activity (Pearson, 1985), and the second is whether the motor patterns in the deafferented preparation are really associated with walking (Bässler and Wegner, 1983; Zill, 1985). In regard to the first question it is important to note that there are major differences in the intact and deafferented patterns as follow. 1) There is no relationship between the intensity of a flexor burst and the subsequent extensor burst in deafferented preparations. In some preparations a flexor burst can occur in the complete absence of extensor activity while in others the extensor can discharge anything from a few spikes following a flexor burst to a high intensity burst (Pearson and Iles, 1970). In intact animals the discharge rate of flexors and extensors is highly correlated (Pearson, 1972). 2) In deafferented preparations extensor activity is highest at the beginning of each burst while in walking animals it is highest near the end of the burst (Pearson, 1972). 3) There is no coordination of rhythmic motor output in the two hindlegs in deafferented preparations. Usually rhythmic activity is generated on one side only compared to the pattern of strict alternation occurring in intact walking animals. 4) In intact animals flexor bursts in ipsilateral middle and hindleg segments follow each other with a significant delay. This delay is abolished following deafferentation (Pearson and Iles, 1973). Given these differences in a single pair of muscles and their homologues we would expect even greater differences if more muscles were analysed. Even so, the known differences indicate that it is unlikely that the motor pattern for walking is centrally generated. Nevertheless the published data do suggest that even if the overall pattern is not centrally generated then some features are established centrally, namely the reciprocity of activity in flexors and extensors and the generation of flexor bursts (Pearson and Iles, 1973; Pearson and Duysens, 1976). On the other hand extensor activity can be strongly influenced by proprioceptors excited during the stance phase and in slowly walking animals it is likely that phasic feedback from these proprioceptors is the main factor involved in the production of the extensor activity (Pearson, 1972; Zill, 1985).

The second criticism of the early work on the cockroach is that the motor patterns generated in deafferented preparations are unrelated to walking. Instead they could be associated either with searching movements (made by legs when they lose contact with the substrate) or with grooming movements (made by the hindlegs to remove irritating stimuli from the abdomen and cerci). Many of the early experiments were done in either headless animals or in preparations with the meso-metathoracic connectives cut. Headless animals rarely walk in a coordinated manner but they do display searching and grooming movements, and grooming movements are very easily evoked following severance of the meso-metathoracic connectives. Also consistent with the notion that the patterns are associated with searching or grooming is that the rhythmicity in deafferented preparations is usually generated in only one deafferented hindleg at a time. However, in the absence of more extensive data on motor patterns in deafferented animals it is not possible to assess which behaviour is most likely to be associated with the deafferented pattern. It may well be that certain features of the deafferented pattern are shared by all three behaviours. One of these features could be the bursts of activity in flexor motoneurones. EMG recordings in walking, grooming and righting animals show that flexor bursts are similar in all three behaviours and similar to bursts generated in deafferented preparations (Reingold and Camhi, 1977; Zill, 1985). A combination of afferent signals and central inputs may establish how these flexor bursts are expressed in each behaviour.

An important investigation on the mechanisms of walking in insects was made by Bässler and Wegner (1983) in the stick insect. In this study recordings were taken from motor axons to leg promotor and remotor muscle in animals subject to varying degrees of deafferentation. Coordinated walking-like patterns of activity occurred in these neurones provided at least one leg was intact and stepping in a normal manner (all other hemisegments including the one in which recordings were made were deafferented). However, following complete deafferentation of all legs no motor pattern resembling walking was produced. Promotor and remotor motoneurones could still be induced to generate rhythmic reciprocal activity but the characteristics of this activity were quite different from those of walking. A quantitative analysis of these burst patterns revealed that less than 15% of the reciprocal activity was vigorous regular alternation and during these periods there was no coordination of activity with motoneurones in the ipsilateral adjacent leg. These patterns resembled those occurring during leg searching movements. Weak regular alternation occurred even less frequently and during these periods there was coordination of activity in adjacent half-ganglia. The latter pattern of activity resembled that associated with rocking and not with walking. Thus Bässler and Wegner failed to find any evidence for central patterning of motor activity for walking. It might be argued that a CPG does exist but that total deafferentation leads to an inability to activate it. This argument is impossible to disprove, and consequently is of no value (see CONCLUSIONS). A more positive explanation for the differences between the intact walking and the deafferented patterns is that the motor patterns for walking are timed by phasic sensory input (Bässler and Wenger, 1983). Consistent with this is the existence of powerful reflexes which are appropriate for generating parts of the walking pattern (Bässler, 1986b).

In summary we can conclude that there is no compelling evidence that the overall motor pattern for walking in the stick insect and cockroach is established by central mechanisms. In both animals deafferentation leads to a significantly different motor pattern than that seen in walking animals, and, in the stick insect at least, these patterns are more like those for behaviours other than walking. In addition, numerous afferent pathways have now been identified in both animals which are appropriate for generating certain aspects of the walking motor pattern. These observations indicate that sensory feedback plays an essential role in establishing the motor pattern for walking in insects, and suggest that the oscillating systems observed in deafferented preparations are not the basis for establishing the timing of motor activity in intact walking animals.

FLIGHT IN THE LOCUST

The flight system of the locust has generally been regarded as a very good example of a centrally generated behaviour, despite criticism from those analysing sensory regulation of this behaviour (Wendler, 1983; Altman, 1983; Pearson, Reye and Robertson, 1983). There is no doubt that a rhythmic reciprocal pattern of activity in flight motoneurones can be generated in the absence of sensory input from wing receptors (Wilson, 1961; Robertson and Pearson, 1983). Superficially this pattern looks similar to that in the intact animal except that the repetition rate is about half normal. Furthermore electrical stimulation of sensory afferents can speed up the rhythm resulting in a pattern looking even more like that of the intact animal (Wilson and Gettrup, 1963). These early results led to the hypothesis that the flight motor pattern was produced by a tonic sensory influence on a central oscillator (Wilson, 1961; Wilson and Gettrup, 1963). More recent data have shown this to be incorrect (Wendler, 1983; Pearson et al., 1983; Wolf and Pearson, 1986). First, a reexamination of the effects of deafferentation on the motor pattern has shown major qualitative changes in the pattern (Pearson and Wolf, 1986). In other words the intact pattern is not simply a speeded-up version of the deafferented pattern. Second, afferent feedback has strong phase-dependent influences on the centrally generated rhythm (Pearson et al., 1983) and on the rhythm in intact animals (Wendler, 1983).

Until recently there has been no detailed analysis of the motor patterns in elevator motoneurones, and in particular how the relative timing of activity in elevators and depressors depends on flight frequency. Pearson and Wolf (1986) made a direct comparison of the motor patterns in flight motoneurones before and after deafferentation. A quantitative analysis of these patterns revealed that in addition to slowing the rhythm, deafferentation produced a qualitative change in the motor pattern. First, the activity in the elevators became more variable and second, the onset of elevator activity following depressor activity was delayed and this interval became strongly dependent on cycle time. In intact animals the depressor to elevator interval is short (ca. 20 msec) and relatively independent of cycle time. Thus sensory input appears intimately involved in the generation of the normal motor pattern. The question is how is the phasic sensory feedback utilized in the production of the intact motor pattern?

One approach to an answer to this question has been to compare the intracellularly recorded pattern of synaptic input to identified neurones in intact flying animals and in deafferented preparations (Wolf and

Pearson, 1986). These recordings have revealed quite different profiles of synaptic input to elevator motoneurones in the two preparations. In intact animals the elevators are briefly and rapidly depolarized following depressor activity, whereas in deafferented preparations the depolarization develops slowly, is longer lasting and reaches a peak immediately preceding the onset of depressor activity. These two quite different profiles of depolarization in elevator motoneurones in intact and deafferented preparations appear to be produced by two different mechanisms (Wolf and Pearson, 1986). For instance, both components can be seen within each cycle in intact animals when the rhythm slows down. At the normal flight frequency (ca. 20 cycles/sec) only the early rapid component is observed but as the rhythm slows the second later component progressively increases and there is little change in the first component. Since the first component occurs only in intact flying animals it is important to know how this component is generated. Because the first component disappears entirely following deafferentation the simplest hypothesis is that it is generated by phasic afferent input from wing proprioceptors. Recently we have identified a set of proprioceptors with properties appropriate for generating the main part of elevator depolarizations in intact animals. These receptors are the tegulae, one of which is located at the base of each wing. The afferents from the tegulae are strongly excited during wing depression (Neumann, 1985), they make strong excitatory connections to elevator motoneurones (Kien and Altman, 1979), and their removal abolishes the early rapid depolarization in elevator motoneurones (Pearson and Wolf, unpublished).

In contrast to the observations on the elevator motoneurones, no data indicate that afferent input is responsible for the depolarization of depressor motoneurones. In fact, deafferentation produces only slight changes in the patterns of synaptic input to depressor motoneurones indicating that depressor activity is largely centrally generated. Therefore, a recent proposal is that the intact flight pattern is produced by a combination of central and afferent mechanisms: a central mechanism generating the depressor bursts and phasic sensory feedback generating the elevator bursts (Wolf and Pearson, 1986). This proposal is radically different from the idea that the basic flight pattern is generated by a CPG and the details of the pattern are established by afferent feedback.

Apart from their role in activating flight motoneurones, wing proprioceptors also contribute to the generation of the rhythmicity in intact animals. One important system contributing to rhythm generation involves the tegulae and a central pathway coupling elevator activity to the following depressor burst (Hedwig and Pearson, 1984; Wolf and Pearson, unpublished). The sequence of events is as follows: wing depression, excitation of the tegulae, reflex excitation of elevators, delayed central excitation of depressors, wing depression, excitation of the tegulae, etc. The contribution of feedback from the tegulae in maintaining rhythmicity can be readily demonstrated by comparing the flight performance of animals with and without tegulae. In the former a brief wind stimulus usually leads to sustained flight activity which can last many minutes. By comparison brief wind stimuli in animals lacking tegulae result in only short flight sequences usually lasting no longer than a few seconds.

Another group of proprioceptors which appears to be involved in the generation and maintenance of the rhythmicity are the wing stretch receptors. Three observations indicate their importance in rhythm generation: 1) their removal decreases the frequency from about 20 to about 10 cycles/sec. (Wilson and Gettrup, 1963), 2) they discharge in high frequency bursts during flight in intact animals (Möhl, 1985), 3) stimulation in deafferented preparations can reset the central rhythm (Pearson et al., 1983), and 4) repeated phase-locked stimulation in deafferented animals increases the frequency and prolongs the duration of rhythmic activity (Pearson et al., 1983). These four properties of the stretch receptors are those usually used for classifying an element of a rhythm generating system. Simply because they are peripherally located is no reason for excluding them from this system. The details of how stretch receptor feedback increases wingbeat frequency have not been determined.

In summary, recent work on the flight system of the locust has shown that phasic afferent feedback is causally involved in generating activity in elevator motoneurones in intact animals, and that proprioceptors are integral elements in the rhythm generating network. Although some features of the motor pattern depend largely on central mechanisms (such as the generation of depressor bursts) the overall patterning and rhythm generation depend on interactions of central and peripheral mechanisms. Since these interactions do not involve the CPG (as it is defined in the deafferented preparations) it is our opinion that the concept of central pattern generation should be abandoned for the flight system of the locust.

CONCLUSIONS

In this article I have considered some of the evidence for central pattern generation in the walking systems of cats and insects, and in the flight system of the locust. In none of these systems is the evidence sufficient to conclude that a CPG provides all the timing cues for the generation of the intact pattern. In the walking system of the cat the CPG in deafferented preparations has not been precisely defined and it remains unclear whether all the temporal features of the intact pattern are centrally programmed. In the walking systems of insects there is now good evidence that phasic sensory feedback establishes some aspect of the intact pattern, and in the stick insect at least there is no indication of a CPG for walking. Phasic sensory feedback has also been found to be a major component in the production of elevator activity in intact flying locusts and is intimately involved in the generation of the basic rhythmicity for flight. Thus, one of the main conclusions of this article is that afferent feedback is involved in establishing the timing of motor activity in some rhythmic motor systems. The degree of sensory involvement in patterning motor activity probably varies from system to system depending on the exact functional requirements. Rhythms associated with automatic internal functions such as leech heartbeat and lobster pyloric movements may be entirely centrally programmed. Less so may be those behaviours involving movements of body segments, and for these it may prove that sensory interaction with a CPG is the correct concept for explaining the normal motor pattern. Finally, behaviours involving the movements of appendages which can be easily perturbed by environmental events will probably have a high degree of sensory involvement and for these the concept of central pattern generation may not always be appropriate.

Apart from the lack of empirical support for central patterning of activity in all rhythmic motor systems, the concept of central pattern generation has been criticised for leading to non-falsifiable hypotheses (Lundberg, 1980). Consider the definition of a central pattern generator provided by Grillner and Wallen (1985): "the neurons responsible for creating the particular motor pattern (by a nervous system deprived of all sensory feedback) constitute the central pattern generator (CPG), <u>regardless of whether all aspects of the motor pattern of the intact animal are produced or some parts missing</u>" (my emphasis). With this definition how could it be shown that a central pattern generator does not exist in any particular motor system? Presumably the only circumstance is when no motor pattern is generated in the deafferented system. But even then it might be argued that a central pattern generator exists and the failure to observe a motor pattern is due to inappropriate means of activating the CPG. Since there is no way to disprove this argument this is a clear example of how the concept of central pattern generation leads to a non-falsifiable hypothesis. Now let us consider the more typical situation in which deafferentation leads not to an abolition of motor patterning but to a significant change in the motor pattern. Under these circumstances it has often been proposed that the intact motor pattern can be explained by sensory interaction with the CPG. The problem with many of these proposals is that the CPG and the mechanisms by which it is influenced by sensory feedback are so poorly defined that it is impossible to conceive of experiments which might disprove the proposal. One notable exception is in the flight system of the locust. Wilson initially proposed that afferent input acted in a tonic manner on the central oscillator (CPG) to speed it up and so generate the intact flight pattern. This proposal has now been disproved by the demonstration that the intact pattern is not a speeded-up version of the deafferented pattern (Pearson and Wolf, 1986) and the finding that afferent input has strong phasic effects on the motor pattern (Pearson et al., 1983; Wendler, 1983). On the other hand for the walking system of the cat it is difficult to conceive of experiments that might falsify the hypothesis that the CPG continues to provide the timing cues when all the normal afferent signals are present. This is because the properties of the CPG for cat walking have not been clearly established and there are no detailed proposals on how these properties are modified by afferent input.

Another problem with the concept of central pattern generation is the assumption that the CPGs identified in deafferented preparations retain their functional identities in intact systems (Pearson, 1985). An hypothetical example illustrates this problem. Suppose a rhythmic reciprocal pattern is generated by a mutually inhibiting pair of interneurones (the CPG) in a deafferented preparation. Now suppose in the intact animal that the movement produced by activity in one interneurone leads to reflex activation of the other interneurone and reflex inhibition of itself and vice versa to yield a chain-reflex system. In this situation the timing cues provided by the CPG are completely overridden by afferent feedback signals, and reflexes are responsible for the generation of the motor pattern. Even though a centrally generated pattern of activity can be generated in this case the primary basis for the motor pattern is not the CPG but the reflexes. For this example, it makes no sense to consider the CPG as a functional unit and attempt to explain pattern generation in terms of sensory interaction with the CPG. On the other hand it is easy to imagine situations in which sensory modulation of a well defined CPG does provide a satisfactory explanation of the motor pattern, such as one similar to that originally proposed by Wilson for locust flight. The problem is this: how are we to conceive of systems in which there is a clear sensory involvement but not in the form of chain reflexes or a simple modulation of a CPG? Perhaps each system will have to be judged individually for its unique properties. Although it is an admirable aim to seek general principles in the study of motor systems, as in any other area, we have to face the fact that there may be a variety of methods for patterning rhythmic motor activity, and afferent feedback may be utilized in many different ways in the generation of the normal motor patterns.

ACKNOWLEDGEMENTS

I thank Ian Gynther and Harald Wolf for their critical comments on early drafts of this article and their suggestions for improvement. Supported by the Medical Research Council of Canada.

REFERENCES

Altman, J. (1983) Sensory inputs and the generation of the locust flight motor pattern: from the past to the future. Biona Report 2, Nachtigall, W., ed. Fischer: Stuttgart. pp 126-137.

Andersson, O., Forssberg, H., Grillner, S. and Wallen, P. (1981) Peripheral feedback mechanisms acting on the central pattern generators for locomotion in fish and cat. Can. J. Physiol. Pharmacol. 59, 713-726.

Bässler, U. (1986a) On the definition of central pattern generator and its sensory control. Biol. Cybern. 54, 65-69.

Bässler, U. (1986b) Afferent control of walking movements in the stick insect Cuniculina impigra. II. Reflex reversal and the release of the swing phase in the restrained foreleg. J. Comp. Physiol. 158, 351-362.

Bässler, U. and Wegner, U. (1983) Motor output of the denervated thoracic ventral nerve cord in the stick insect Carausius morosus. J. Exp. Biol. 105, 127-145.

Delcomyn, F. (1980) Neural basis of rhythmic behaviour in animals. Science 210, 492-498.

Forssberg, H., Grillner, S. and Halbertsma, J. (1980) The locomotion of the low spinal cat. I. Coordination within a hindlimb. Acta Physiol. Scand. 108, 269-281.

Graham Brown, T. (1911) The intrinsic factors in the act of progression in the mammal. Proc. Roy. Soc. B:84, 308-319.

Grillner, S. (1981) Control of locomotion in bipeds, tetrapods, and fish. In: Handbook of Physiology. Vol. III. Motor Control. Brooks, V.B., ed. American Physiological Society: Bethesda, pp 1179-1276.

Grillner, S. (1985) Neurobiological bases of rhythmic motor acts in vertebrates. Science 228, 143-149.

Grillner, S. and Rossignol, S. (1978) On the initiation of the swing phase of locomotion in chronic spinal cats. Brain Res. 146, 269-277.

Grillner, S. and Wallen, P. (1985) Central pattern generators for locomotion, with special reference to vertebrates. Ann. Rev. Neurosci. 8, 233-262.

Grillner, S. and Zangger, P. (1975) How detailed is the central pattern generator for locomotion? Brain Res. 88, 367-371.

Grillner, S. and Zangger, P. (1979) On the central generation of locomotion in the low spinal cat. Exp. Brain Res. 34, 241-261.

Grillner, S. and Zangger, P. (1984) The effect of dorsal root transection on the efferent motor pattern in the cat's hindlimb during locomotion. Acta Physiol. Scand. 120, 393-405.

Hedwig, B. and Pearson, K.G. (1984) Patterns of synaptic input to identified flight motoneurons in the locust. J. Comp. Physiol. 154, 745-760.

Jankowska, E., Jukes, M.G.M., Lund, S. and Lundberg, A. (1967a) The effect of DOPA on the spinal cord. V. Reciprocal organization of pathways transmitting excitatory action to alpha motoneurones of flexors. Acta Physiol. Scand. 70, 369-388.

Jankowska, E., Jukes, M.G.M., Lund, S. and Lundberg, A. (1967b) The effect of DOPA on the spinal cord. VI. Half-centre organization of interneurones transmitting effects from flexor reflex afferents. Acta Physiol. Scand. 70, 389-402.

Jordan, L.M., Brownstone, R.M., Kriellaars, D.J. and Noga, B.R. (1986) Spinal modules for walking movements revealed in fictive locomotion experiments. Soc. Neurosci. Abstr. 12, 877.

Kien, J. and Altman, J.S. (1979) Connections of the locust wing tegula with metathoracic flight motoneurons. J. Comp. Physiol. 133, 199-210

Lundberg, A. (1969) Reflex control of stepping. The Nansen Memorial Lecture to the Norwegian Academy of Sciences and Letters. Universitetsforlaget: Oslo.

Lundberg, A. (1980) Half-centres revisited. In: Regulatory Functions of the CNS Motion and Organization Principles. Szentáothai, J., Palkovits, M. and Háori, J., eds. pp 155-167.

Möhl, B. (1985) The role of proprioception in locust flight control. II. Information signalled by forewing stretch receptors during flight. J. Comp. Physiol. 156, 103-116.

Neumann, L. (1985) Experiments on tegula function for flight coordination in the locust. In: Insect Locomotion. Gewecke, M. and Wendler, G., eds. Parey: Hamburg. pp 149-156.

Pearson, K.G. (1972) Central programming and reflex control of walking in the cockroach. J. Exp. Biol. 56, 321-330.

Pearson, K.G. (1985) Are there central pattern generators for walking and flight in insects? In: Feedback and Motor Control in Invertebrates and Vertebrates. Barnes, W.J.P. and Gladden, M., eds. Croom Helm: London. pp 307-316.

Pearson, K.G. and Duysens, J.D. (1976) Function of segmental reflexes in the control of stepping in cockroaches and cats. In: Neural Control of Locomotion. Herman, R.M., Grillner, S., Stein, P.S.G. and Stuart, D.G., eds. Plenum: New York. pp 519-538.

Pearson, K.G. and Iles, J.F. (1970) Discharge patterns of coxal levator and depressor motoneurones in the cockroach, Periplaneta americana. J. Exp. Biol. 52 139-165.

Pearson, K.G. and Iles, J.F. (1973) Nervous mechanisms underlying intersegmental coordination of leg movements during walking in the cockroach. J. Exp. Biol. 58, 725-744.

Pearson, K.G., Reye, D.N. and Robertson, R.M. (1983) Phase-dependent influences of wing stretch receptors on flight rhythm in the locust. J. Neurophysiol. 49, 1168-1181.

Pearson, K.G. and Wolf, H. (1986) Comparison of motor patterns in the intact and deafferented flight system of the locust. 1. Electromyographic analysis. J. Comp. Physiol. In press.

Perret, C. (1983) Centrally generated pattern of motoneuron activity during locomotion in the cat. In: Neural Origin of Rhythmic Movements. Roberts, A. and Roberts, B., eds. Symp. Soc. Exp. Biol. 37, 405-422.

Perret, C. and Cabelguen, J.M. (1976) Central and reflex participation in the timing of locomotor activations of bifunctional muscle, the semi-tendinosus, in the cat. Brain Res. 106, 390-395.

Perret, C. and Cabelguen, J.M. (1980) Main characteristics of the hindlimb locomotor cycle in the decorticate cat with special reference to bifunctional muscles. Brain Res. 187, 333-352.

Reingold, S.C. and Camhi, J.M. (1977) A quantitative analysis of rhythmic leg movements during three different behaviours in the cockroach Periplaneta americana. J. Insect Physiol. 23, 1407-1420.

Robertson, R.M. and Pearson, K.G. (1983) Interneurons in the fight system of the locust: distribution, connections and resetting properties. J. Comp. Neurol. 215, 33-50.

Wendler, G. (1983) The locust flight system: functional aspects of sensory input and methods of investigation. In: Biona Report 2. Nachtigall, W., ed. Fischer: Stuttgart. pp 113-125.

Wilson, D.M. (1961) The central nervous control of flight in a locust. J. Exp. Biol. 38, 471-490.

Wilson, D.M. and Gettrup, E. (1963) A stretch reflex controlling wingbeat frequency in grasshoppers. J. Exp. Biol. 40, 171-185.

Wolf, H. and Pearson, K.G. (1986) Comparison of motor patterns in the intact and deafferented flight system of the locust. 2. Intracellular recordings from flight motoneurons. J. Comp. Physiol. In press.

Zill, S.N. (1985) Proprioceptive feedback and the control of cockroach walking. In: Feedback and Motor Control in Invertebrates and Vertebrates. Barnes, W.J.P. and Gladden, M.H., eds. Croom Helm: London. pp 187-208.

SYMPATHETIC NERVE RECORDINGS IN MAN

B.G. Wallin

Department of Clinical Neurophysiology
Sahlgren's Hospital
University of Göteborg
S-413 45 Göteborg, Sweden

INTRODUCTION

For a long time a student of sympathetic physiology in man had to use sympathetic effector responses as markers of the nervous activity. This meant that he had to record changes of blood flow, sweat production, heart rate, blood pressure etc. and from such data try to draw conclusions about the sympathetic drive. There are obvious drawbacks with this approach. Sympathetic effector organs are sluggish and respond not only to nerve impulses but also to mechanical, chemical and hormonal stimuli and therefore the sympathetic component may be difficult to separate out. Another problem is that the basal level of sympathetic activity is difficult to estimate.

The situation suddenly changed when Hagbarth and Vallbo (1968) found that their newly developed microneurographic technique could also be used to record sympathetic impulses. With this important methodological development it became possible to monitor sympathetic action potentials to skin and muscle unobscured by the sluggishness and unspecificity of the effector organs. Visceral sympathetic and parasympathetic activity is still inaccessible in man but in spite of the limitation a body of new data has been obtained not only about specific sympathetic reflexes to skin and muscle but also more generally about the organization of the sympathetic system (Wallin and Fagius, 1986). Recently the method has also been reversed i.e. instead of recording spontaneous sympathetic activity the electrode has been used for electrical stimulation of different nerve fibres including the sympathetic ones with simultaneous monitoring of the resulting effector responses (Blumberg and Wallin, 1986).

METHODS

The majority of sympathetic fibres going to the extremities run in the peripheral nerves which are made up of fascicles surrounded by connective tissue. Distally in the extremities, where most recordings take place, the fascicles are "pure" in the sense that each fascicle is connected either with a muscle or a skin area. When the tungsten micro-electrode tip penetrates a fascicle it can record action potentials from many different types of fibre within the fascicle but fortunately enough there is no crosstalk between fascicles. To identify a fascicle physiological stimuli are delivered to sensory receptors in the extremity and the experimenter observes the afferent action potentials in mechanoreceptive fibres on their way up to the spinal cord. The identification rests upon the fact that receptors in skin and muscle are activated by different stimuli. Once a fascicle is identified the next step is to find the sympathetic fibres. These are not distributed evenly in the fascicles, but lie clustered in groups in Schwann cells. Only when the electrode tip comes close to such a bundle can spontaneous and induced sympathetic activity be recorded. Technical details and evidence of the sympathetic nature of the impulses have been published previously (Vallbo, Hagbarth, Torebjörk and Wallin, 1979).

As in the motor system there are sympathetic spinal and brainstem reflexes which are controlled from higher centres. In human experiments one can only record from the postganglionic nerves which means that the activity is influenced from many different functional levels proximal to the recording site. Even if this limits the possibilities to draw conclusions, some of the difficulties can be overcome by suitable experimental design and occasionally by an accident of nature. For example, spinal sympathetic reflexes, which normally are concealed by supraspinal influences can be studied in patients with traumatic spinal cord lesions (see below).

In this review I will first discuss what the human recordings have taught us about the functional organization of the sympathetic system and

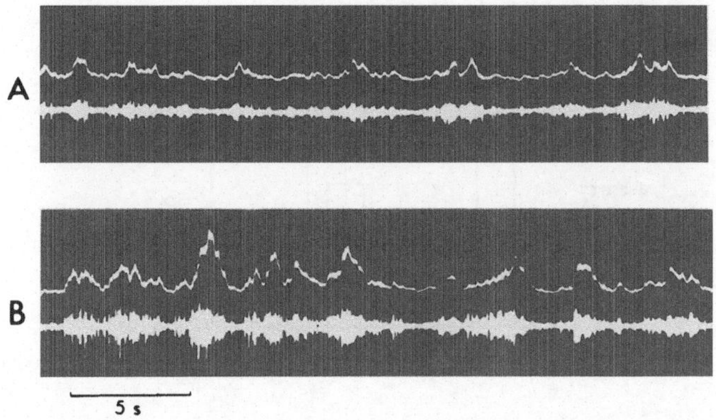

Fig. 1. Skin sympathetic activity recorded from the median nerve at elbow after Lidocaine block of afferent impulses distal to the recording electrode. Note irregularity of discharge both in single- and multi-unit activity. Nerve activity displayed both in mean voltage (upper traces) and original neurogram (lower traces). From Hallin and Torebjörk (1974) with permission.

then turn to muscle sympathetic activity and discuss both its reflex control and its importance for plasma concentrations of noradrenaline.THE

TEMPORAL PATTERN OF SYMPATHETIC ACTIVITY

In the first recordings of sympathetic activity in animals Adrian, Bronk and Philips, (1932) noted that the impulses were grouped in synchronized discharges, or bursts. The pattern is also similar in human recordings. Fig. 1 shows a typical example of the irregular pattern of discharge in multiunit recordings of skin sympathetic activity in man. The neurogram also contains a single unit which stands out above the background and which shows the same irregular firing. At rest, the average firing frequency of such a unit is rarely above 1-2 Hz, but because of the irregularity of firing the instantaneous rate, i.e. the rate calculated from the interval between 2 successive spikes, may reach 30-40 Hz (Iggo and Vogt, 1960; Hallin and Torebjörk, 1974). The question I want to ask is the following: Does this irregularity of discharge have physiological importance? Previously one did not think so but recently new evidence has accumulated which has started to change our view.

In an attempt to answer the question we selected a neurogram similar to that in Fig. 1 in which a few units stood out above the noise and

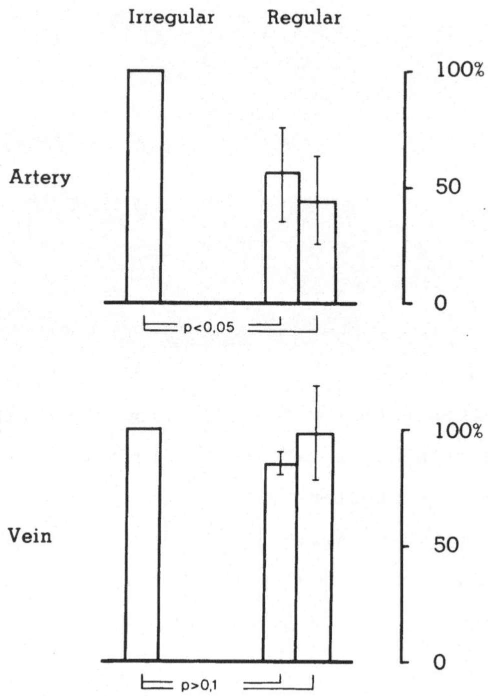

Fig. 2. Comparison of integrated contractile responses (mean ± SE, n=6) to electrical field stimulation of rat mesenteric vessels when the same number of stimuli were delivered at irregular and regular intervals. Two irregular stimulation frequencies were used with average frequencies of 1.6 and 1.8 Hz, respectively. Contractile responses at irregular stimulation set to 100%. Data from Nilsson et al. (1985).

transferred the units to a tape (Nilsson, Ljung, Sjöblom and Wallin, 1985). The tape was then used to stimulate the nerves to isolated mesenteric blood vessels of approximately 150 μm diameter which were mounted in an organ bath. Two stimulation sequences with average frequencies of 1.6 and 1.8 Hz were used and the aim of the experiment was to see if the average contractile responses of arteries and veins differed when the same number of impulses were given evenly and when given with the normal irregular discharge pattern.

As illustrated in Fig. 2 there were clear differences between the responses of arteries and veins. The venous contractions were quantitatively similar with the two modes of stimulation but the arterial contractions were about twice as large with the irregular as with the regular stimulations. The findings can be explained by the different character of the stimulus response curves to even stimuli for the two types of vessels. For the vein the initial part of the stimulus-response curve was linear, i.e. increases and decreases of frequency caused linear changes of the response. For the artery, however, the stimulus-response curve was sigmoid. Below 2 Hz the contractile response was very small but already at slightly higher frequencies significant contractions occurred. With the irregular impulse pattern higher frequencies occurred every now and then and therefore the mean contractile response became greater than with even stimulation. Similar results have been obtained for several parasympathetically innervated effectors (Andersson, 1983) and I think it is fair to say that today when one wants to evaluate autonomic effector responses one must consider not only average firing frequencies in the nerve but also the pattern of firing.

DIFFERENTIATION OF SYMPATHETIC OUTFLOW

In recordings from skin and muscle nerves there are clear differences in the patterns of sympathetic activity. Muscle sympathetic activity (MSA) is made up of sequences of fairly regular bursts of impulses occurring in the cardiac rhythm whereas skin sympathetic activity (SSA) consists of a much more irregular fluctuating activity with no obvious cardiac rhythmicity. When we started these recordings in the late 1960s this simple finding of different patterns in MSA and SSA was already of some significance. The reason was that at that time opinions differed as to how the sympathetic system was organized. Cannon's model of a diffusely organized system was the commonly accepted one and seemed to get support from sympathetic recordings in animals. According to this view there was a central rhythm generator from which activity descended to the sympathetic motoneurones in the spinal cord and then out in parallel in different sympathetic nerves. The contrasting view emanated from haemodynamic data which suggested that different vascular beds could receive completely different neural commands during various manoeuvres. On the basis of such evidence Folkow became an early advocate of a differentiated sympathetic system (Folkow, 1960). Against this background we felt that the different patterns in MSA and SSA gave some support to the latter hypothesis and this view has been strengthened over the years: not only the resting patterns differ but most stimuli and manoeuvres also give rise to different responses in the two types of nerves. For example, in SSA an arousal stimulus evokes a generalized distinct reflex discharge (Hagbarth, Hallin, Hongell, Torebjörk and Wallin, 1972) whereas in MSA there is no response

(Fig. 3B)(DeLius, Hagbarth, Hongell and Wallin, 1972a, Stjernberg, Blumberg and Wallin, 1986). In contrast, the Valsalva manoeuvre leads to an increase of MSA (DeLius, Hagbarth, Hongell and Wallin, 1972b) but no systematic change in SSA (DeLius, Hagbarth, Hongell and Wallin, 1972c). Approximately at the same time as these human data were obtained other evidence of differentiation was reported from animal studies (Walther, Iriki and Simon, 1970; Iriki, Walther, Pleschka and Simon, 1971; see also Jänig, 1985).

The implication of these results is that there is no common sympathetic tone. On the contrary, the sympathetic system is highly differentiated with many subdivisions, each of which has its own characteristic outflow governed by its own reflex mechanisms. Of course this does not imply that sympathetic reactions always differ in different nerves. Instead it means that sympathetic effectors can be used like the keys of a piano: depending on the functional situation different keys are combined to produce different chords.

In contrast to the differences between SSA and MSA, simultaneous recordings of sympathetic activity from two muscle nerve branches show a remarkable similarity between neurograms from different extremities. In double recordings from two leg nerves or from one arm and one leg nerve almost all sympathetic bursts can be identified in both neurograms -- the only systematic difference is that a burst in arm nerves occurs a little earlier than the corresponding burst in a leg nerve (because of the shorter conduction distance in the arm). Occasionally, small bursts are seen only in one neurogram but such variations occur randomly and if one quantifies the activity by counting the number of mean voltage bursts the agreement between the two records is remarkably good (Sundlöf and Wallin, 1977). From a practical point of view this is an important finding since it means that a recording from a randomly chosen muscle nerve branch gives a measure of MSA which is representative for the individual. In skin nerve recordings there is a similar close parallelism between sympathetic neurograms from hands and feet; neurograms from two forearm nerves are also similar but there are differences between nerves to the forearm and the palm of the hand (Bini, Hagbarth, Hynninen and Wallin, 1980b).

This impressive parallelism between sympathetic records from different extremities suggests that supraspinal influence is similar on most spinal sympathetic motoneurones to muscle and that there are equally strong but

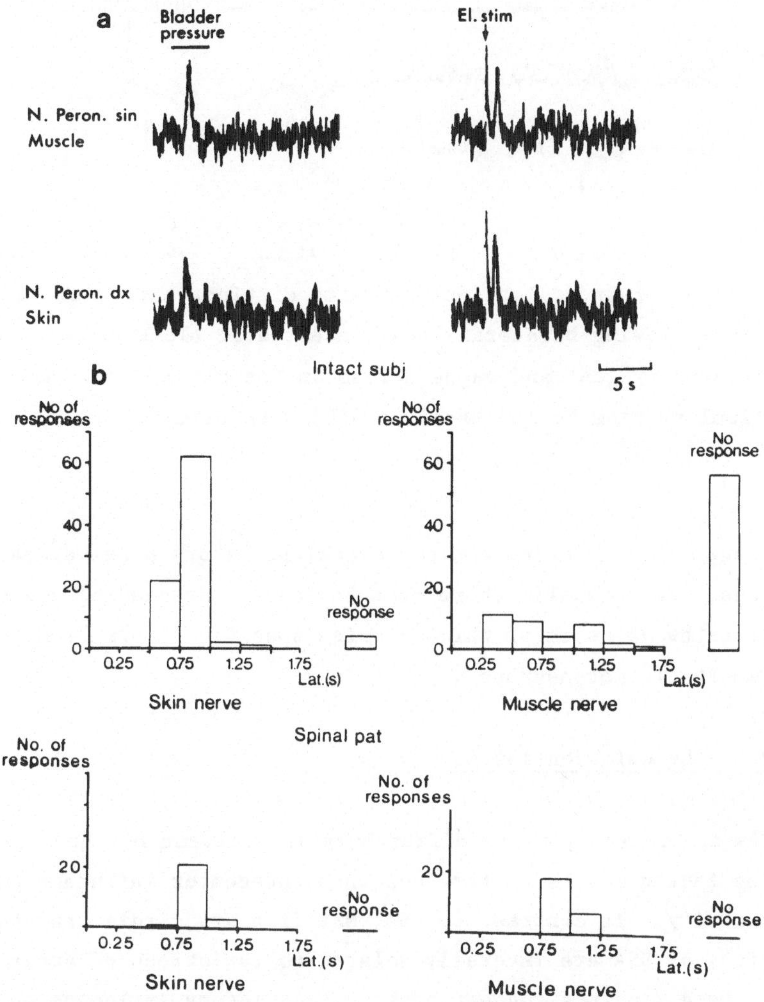

Fig. 3. A. Mean voltage neural activity recorded simultaneously from a muscle fascicle in the left peroneal nerve and a skin fascicle in the right peroneal nerve in a patient with spinal cord lesion at the T7 level. Neural discharges were induced synchronously in both fascicles by bladder pressure and electrical stimulation to the skin of the left hip.
B. Relationship between electrical skin stimuli applied to the upper part of the thigh and sympathetic discharges in nerves to skin and muscle in five intact subjects (upper) and two patients with spinal cord injury (lower). Sympathetic discharges recorded simultaneously from the two peroneal nerves. In intact subjects skin sympathetic bursts were usually (85 cases) recorded 0.5 - 1.0 sec after stimuli and only in four cases were there no bursts within 0.25 - 1.75 sec after stimuli. In muscle nerves bursts were often lacking (56 cases) and the bursts recorded showed no systematic relationship to the stimuli. In the patients all stimuli gave rise to discharges both in skin and muscle nerves. Modified from Stjernberg et al. (1986) with permission.

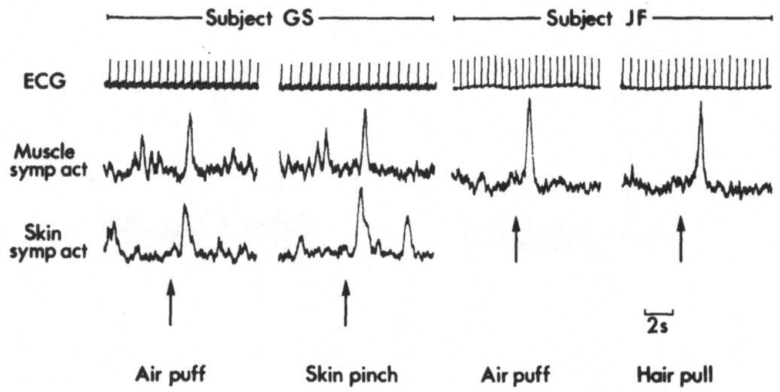

Fig. 4. Examples of arousal stimuli evoking distinct bursts of MSA (and SSA) following bilateral local anaesthetic block of glossopharyngeal and vagus nerves in the neck. Arrows indicate stimulus. From Fagius et al. (1985) with permission.

different supraspinal influences on several pools of spinal sympathetic skin neurones. Functionally there must be several descending spinal sympathetic pathways, each of which carries specific information to its own pool of sympathetic motoneurones.

The origin of the differentiation

What is the reason for the differentiation between MSA and SSA? A contributing factor may be differences in baroreceptor influence on the two types of activity. In contrast to SSA, MSA displays cardiac rhythmicity and variations of MSA are inversely related to variations of arterial blood pressure. These findings suggest a strong modulatory influence on MSA from arterial baroreceptors but little or no such influence on SSA. If differences in baroreceptor influence contribute importantly to the differentiation one would expect differentiation to be reduced if baroreceptor influence were to be eliminated. A clinical situation when this occurs is in patients with complete spinal cord lesions. When MSA was recorded in nerves leaving the spinal cord below the lesion spontaneous activity was sparse and absence of baroreceptor influence was evidenced by absence of cardiac rhythmicity (Stjernberg and Wallin, 1983; Stjernberg et al., 1986). The most interesting finding was, however, that skin stimuli below the level of the lesion gave rise to reflex discharges occurring in parallel in skin and muscle nerves. As mentioned above this does not occur in normal subjects. The results are illustrated in Fig. 3.

Another way of eliminating baroreceptor influence is to deafferent the baroreceptors, and in man temporary deafferentation can be obtained by local anaesthetic block of the vagus and glossopharyngeal nerves in the neck (Fagius, Wallin, Sundlöf, Nerhed and Engelesson, 1985). When such blocks were applied blood pressure and heart rate increased. At the same time there was a strong increase of MSA and the pulse synchrony was replaced by a 0.4 - 0.7 Hz irregular rhythm similar to that seen in SSA. SSA on the other hand, did not change much, supporting the claim that there is little or no baroreceptor influence on SSA. In this situation when all signs of baroreceptor inhibition had disappeared arousal stimuli were applied and again there were parallel responses in muscle and skin nerves (Fig. 4). Consequently both the anaesthesia and the spinal cord lesion had similar effects -- the differentiation between MSA and SSA was reduced, i.e. the results supported the hypothesis that baroreceptor influence is important for the differentiation. The results also suggest that normal baroreceptor influence, in addition to contributing to blood pressure homeostasis, also has a powerful inhibitory effect on arousal reflexes in MSA. This raises the possibility that deafferentation may result in qualitatively abnormal reflex patterns giving rise to clinical symptoms. Such an hypothesis would be in accordance with other examples of lesions in the nervous system, revealing reflex patterns which normally are concealed by inhibitory regulation. The Babinski sign is the best known example on the somatomotor side.

REFLEX CONTROL OF MUSCLE SYMPATHETIC ACTIVITY

MSA is dominated by vasoconstrictor impulses and today there is evidence of reflex influence on MSA from a number of different receptor stations.

Baroreceptor influence

The baroreceptor influence is of special interest with regard to man's ability to maintain upright posture. Starting with the so-called low pressure receptors, i.e. volume receptors in the chest, there is general agreement that they have a tonic inhibitory effect on the vascular bed of skeletal muscle. In agreement with this, unloading of the low pressure receptors by lower body negative pressure caused a static increase of MSA (Sundlöf and Wallin, 1978b).

With regard to the arterial baroreceptors the situation is more complex. Our initial observation was a close inverse correlation between variations of MSA and arterial BP. Sequences of vasoconstrictor bursts occurred during transient blood pressure reductions and disappeared when blood pressure increased (DeLius et al., 1972a). In contrast there was no correlation between a subject's mean number of sympathetic bursts and his long term blood pressure level, suggesting that MSA was more important for buffering changes of pressure than for setting the blood pressure level (Sundlöf and Wallin, 1978a). Results along the same lines were obtained when carotid baroreceptor activity was modulated by neck pressure or neck suction. Onset and offset of pressure gave clear dynamic responses but adaptation was rapid and there were only minor sustained effects (Båth, Lindblad and Wallin, 1981; Wallin and Eckberg, 1982). The dynamic character of the response may be due in part to counterregulation from the aortic arch receptors which are uninfluenced by the stimulus. However, application of neck pressure does to some extent mimic the physiological effects on arterial baroreceptors during transition from the lying to the upright posture: when the head is raised there is a sustained pressure reduction in the carotid sinus but since the aortic arch remains approximately at heart level the aortic arch pressure does not change much. Consequently, when man goes from lying to sitting or standing the arterial baroreceptors probably contribute importantly to the initial phase of the compensatory vasoconstriction that occurs (Burke, Sundlöf and Wallin, 1977). Once upright posture is achieved, however, it is likely that effects from arterial baroreceptors are small and the main stimulus for sustained muscle vasoconstriction is the reduction of blood volume in the chest which causes unloading of the low pressure receptors.

Chemoreceptor influence

The effects of general hypoxia and hypercapnia on MSA are not established but preliminary results from an ongoing study suggest that hypoxia and hypercapnia both cause increases of MSA (Blumberg, personal communication).

MSA is influenced not only from systemic chemoreceptors but effects can also be evoked from intramuscular chemoreceptors. In a recent study a 2 min isometric handgrip at 30% of maximal power was found to cause an increase of MSA in leg nerves (i.e. in nerves to inactive muscles) during the 2nd min of contraction (Mark, Victor, Nerhed and Wallin, 1985). If the circulation in the contracting forearm was arrested prior to the release of the contraction the increase of MSA persisted suggesting that the response was triggered from chemosensitive nerve endings in the contracting muscles. Consequently, if one makes sufficiently strong isometric contractions one gets a generalized increase of MSA which lasts as long as there is muscle ischaemia. Apart from being physiologically interesting the observation may have clinical importance. Patients with stress syndromes often display chronically increased muscle tension and it would not be unreasonable if there are adverse effects of long lasting increases of MSA.

Influence from cutaneous receptors

Another afferent input that may influence MSA comes from cutaneous receptors. The cold pressor test (submersion of one hand in ice water) leads to an increase of both MSA (Victor et al., unpublished) and SSA (Fagius and Blumberg, 1985). Submersion of the face in cold water evokes a "diving reflex" which also includes an increase of MSA (Fagius and Sundlöf, 1986). Interestingly enough in this situation SSA decreases, once again illustrating the marked capacity for differentiation of sympathetic outflow. The results agree well with those from previous animal studies and it was concluded that the effects on MSA during simulated diving were due to a combination of cutaneous stimulation in the face and arrested respiration but that the facial stimulation was more important.

MSA AND PLASMA NORADRENALINE

It is well known that noradrenaline is the principal postganglionic sympathetic transmitter and there is a circulating level of noradrenaline in the blood which has been thought to provide a measure of net activity in the sympathetic system. The precise relationship between the sympathetic activity and the plasma level of noradrenaline is, however, incompletely understood. We noted quite early that there were wide interindividual differences in MSA which were reproducible from day to day over many months (Sundlöf and Wallin, 1977). From the literature we knew that the situation was similar for plasma concentrations of noradrenaline: wide interindividual differences at rest but reproducible levels in the same subject on different occasions (Lake, Ziegler and Kopin, 1976). When MSA and plasma concentrations of noradrenaline were compared in the same subjects a significant positive correlation was found, i.e. subjects with high resting levels of MSA had high plasma levels of noradrenaline (Wallin,

Sundlöf, Eriksson, Dominiak, Grobecker and Lindblad, 1981). After this initial observation similar correlations have been obtained both in patients with essential hypertension (who have normal MSA and NA) (Mörlin, Wallin and Eriksson, 1983) and in patients with cardiac failure (who have elevated levels of MSA and plasma noradrenaline) (Leimbach, Wallin, Victor, Aylward, Sundlöf and Mark, 1986). At rest there is also a significant positive correlation between MSA and spill-over of noradrenaline to venous plasma (blood sampled from the femoral vein) measured with radioactive tracers (Wallin et al., unpublished observations).

These findings agree with the notion that MSA is an important determinant of venous plasma concentrations of noradrenaline. Since sympathetic outflow is differentiated it may seem surprising with a dominating influence from one particular type of sympathetic nerve. The explanation is probably multifactorial. One contributing factor is that muscle is a large organ and contributes approximately 20% of total spill-over of noradrenaline (Esler, Jennings, Leonard, Sacharias, Burke, Johns and Blombery, 1984). The site of sampling of blood is also important. In venous blood from the forearm approximately 45% of the noradrenaline in the sample has been shown to derive from the forearm (Hjemdahl, Freyschuss, Juhlin-Dannfelt and Linde, 1984). Since skin vasoconstrictor activity usually is low at comfortable ambient temperatures (Bini, Hagbarth, Hynninen and Wallin, 1980a) most of the noradrenaline will come from muscle which will be overrepresented in the sample. A third factor may be that splanchnic blood to a large extent is cleared of noradrenaline in the liver.

Lately, we have also studied the quantitative relationship between changes of MSA and changes of plasma noradrenaline concentration in forearm venous blood. Both a 30% isometric muscle contraction (Wallin et al., unpublished) and the cold pressor test (Victor et al., unpublished) induce increases of MSA, which after a delay of 1-2 minutes are followed by increases of plasma noradrenaline concentration. When comparing peak values of MSA and plasma noradrenaline (expressed in percent of control values) the increases of MSA were 3-4 times larger than corresponding increases of noradrenaline. In another study, intravenous infusions of vasoactive drugs (phenylephrine and sodium nitroprusside) were found to induce changes of MSA and plasma noradrenaline concentrations which were linearly related to each other (Eckberg et al., unpublished). Again, however, the percentage change of MSA was approximately 3 times larger than that of plasma noradrenaline. In contrast to these interventions, different types of mental stress did not change MSA and then there was also little change of the plasma concentration of noradrenaline (Wallin et al., unpublished). Thus even if plasma concentrations of noradrenaline in forearm venous blood to a large extent are determined by the strength of MSA, plasma noradrenaline is a blunt instrument for determining changes of MSA; only 20-30% of a change in MSA is reflected as a change in the noradrenaline concentration.

CONCLUDING REMARKS

Microneurographic recordings provide a powerful tool for improving our understanding of the physiology of the sympathetic nervous system. We have used the approach for about 15 years and it is only during the last couple of years that the technique has become more widespread. Since the field is vast and the technique can be used in many areas outside those touched upon here, I believe that over the next decade there will be a rapid development not only of human sympathetic physiology but also of pathophysiology and pharmacology.

ACKNOWLEDGEMENTS

Supported by Swedish Medical Research Council Grant No. B86-04X-03546-15B. I thank Mrs. Eva Guggenheim for her skillful secretarial work.

REFERENCES

Adrian, E.D., Bronk, D.W. and Philips, G. (1932) Discharges in mammalian sympathetic nerves. J. Physiol. (Lond.) 74, 115-133.

Andersson, P.O. (1983) Adrenergic, cholinergic, and VIPergic neuro-effector control. With special reference to high-frequency burst excitation patterns. Thesis, Bloms boktryckeri: Lund.

Båth, E., Lindblad, L.-E. and Wallin, B.G. (1981) Effects of dynamic and static neck suction on muscle nerve sympathetic activity, heart rate and blood pressure in man. J. Physiol. (Lond.) 311, 551-564.

Bini, G., Hagbarth, K.-E., Hynninen, P. and Wallin, B.G. (1980a) Thermoregulatory and rhythm-generating mechanisms governing the sudomotor and vasoconstrictor outflow in human cutaneous nerves. J. Physiol. (Lond.) 306, 537-552.

Bini, G., Hagbarth, K.-E., Hynninen, P. and Wallin, B.G. (1980b) Regional similarities and differences in thermoregulatory vaso- and sudomotor tone. J. Physiol. (Lond.) 306, 553-565.

Blumberg, H. and Wallin, B.G. (1986) Direct evidence of neurally mediated vasodilatation in hairy skin of the human foot. J. Physiol. (Lond.) In press.

Burke, D., Sundlöf, G. and Wallin, B.G. (1977) Postural effects on muscle nerve sympathetic activity in man. J. Physiol. (Lond.) 272, 399-414.

DeLius, W., Hagbarth, K.-E., Hongell, A. and Wallin, B.G. (1972a) General characteristics of sympathetic activity in human muscle nerves. Acta Physiol. Scand. 84, 65-81.

DeLius, W., Hagbarth, K.-E., Hongell, A. and Wallin, B.G. (1972b) Manoeuvres affecting sympathetic outflow in human muscle nerves. Acta Physiol. Scand. 84, 82-94.

DeLius, W., Hagbarth, K.-E., Hongell, A. and Wallin, B.G. (1972c) Manoeuvres affecting sympathetic outflow in human skin nerves. Acta Physiol. Scand. 84, 177-186.

Esler, M., Jennings, G., Leonard, P., Sacharias, N., Burke, F., Johns, J. and Blombery, P. (1984) Contribution of individual organs to total noradrenaline release in humans. Acta Physiol. Scand. Suppl. 527, 11-16.

Fagius, J., Wallin, B.G., Sundlöf, G., Nerhed, C. and Englesson, S. (1985) Sympathetic outflow in man after anaesthesia of the glossopharyngeal and vagus nerves. Brain, 108, 423-438.

Fagius, J. and Blumberg, H. (1985) Sympathetic outflow to the hand in patients with Raynaud's phenomenon. Cardiovasc. Res. 19, 249-253.

Fagius, J. and Sundlöf, G. (1986) The diving response in man: Effects on sympathetic activity in muscle and skin nerve fascicles. J. Physiol. (Lond.) 377, 429-443.

Folkow, B. (1960) Range of control of the cardiovascular system by the central nervous system. Physiol. Rev. 40, Suppl. 4, 93-99.

Hagbarth, K.-E. and Vallbo, Å.B. (1968) Pulse and respiratory grouping of sympathetic impulses in human muscle nerves. Acta Physiol. Scand. 74, 96-108.

Hagbarth, K.-E., Hallin, R.G., Hongell, A., Torebjörk, H.E. and Wallin, B.G. (1972) General characteristics of sympathetic activity in human skin nerves. Acta Physiol. Scand. 84, 164-176.

Hallin, R.G. and Torebjörk, H.E. (1974) Single unit sympathetic activity in human skin nerves during rest and various manoeuvres. Acta Physiol. Scand. 92, 303-317.

Hjemdahl, P., Freyschuss, U., Juhlin-Dannfelt, A. and Linde, B. (1984)

Differentiated sympathetic activation during mental stress evoked by the Stroop test. Acta Physiol. Scand. Suppl. 527, 25-29.

Iggo, A. and Vogt, M. (1960) Preganglionic sympathetic activity in normal and in reserpin-treated cats. J. Physiol. (Lond.) 150, 114-133.

Iriki, M., Walther, O.-E., Pleschka, K. and Simon, E. (1971) Regional cutaneous and visceral sympathetic activity during asphyxia in the anesthetized rabbit. Pflügers Arch. 332, 167-182.

Jänig, W. (1985) Organization of the lumbar sympathetic outflow to skeletal muscle and skin of the cat hindlimb and tail. Rev. Physiol. Biochem. Pharmacol. 102, 119-213.

Lake, C.R., Ziegler, M.G. and Kopin, I.J. (1976) Use of plasma norepinephrine for evaluation of sympathetic neuronal function in man. Life Sci. 18, 1315-1326.

Leimbach, W.N., Wallin, B.G., Victor, R.G., Aylward, P.E., Sundlöf, G. and Mark, A.L. (1986) Direct evidence from intraneural recordings for increased central sympathetic outflow in patients with heart failure. Circulation, 73, 913-919.

Mark, A.L., Victor, R.G., Nerhed, C. and Wallin, B.G. (1985) Microneurographic studies of the mechanisms of sympathetic nerve responses to static exercise in humans. Circ. Res. 57, 461-469.

Mörlin, C., Wallin, B.G. and Eriksson, B.-M. (1983) Muscle sympathetic activity and plasma noradrenaline in normotensive and hypertensive man. Acta Physiol. Scand. 119, 117-121.

Nilsson, H., Ljung, B., Sjöblom, N. and Wallin, B.G. (1985) The influence of sympathetic impulse pattern on contractile response of rat mesenteric arteries and veins. Acta Physiol. Scand. 123, 303-309.

Stjernberg, L. and Wallin, B.G. (1983) Sympathetic neural outflow in spinal man. A preliminary report. J. Auton. Nerv. Syst. 7, 313-318.

Stjernberg, L., Blumberg, H. and Wallin, B.G. (1986) Sympathetic activity in man after spinal cord injury. Outflow to muscle below the lesion. Brain. In press.

Sundlöf, G. and Wallin, B.G. (1977) The variability of muscle nerve sympathetic activity in resting recumbent man. J. Physiol. (Lond.) 272, 383-397.

Sundlöf, G. and Wallin, B.G. (1978a) Human muscle nerve sympathetic activity at rest. Relationship to blood pressure and age. J. Physiol. (Lond.) 274, 621-637.

Sundlöf, G. and Wallin, B.G. (1978b) Effect of lower body negative pressure on human muscle nerve sympathetic activity. J. Physiol. (Lond.) 278, 525-532.

Vallbo, Å.B., Hagbarth, K.-E., Torebjörk, H.E. and Wallin, B.G. (1979) Somatosensory, proprioceptive and sympathetic activity in human peripheral nerves. Physiol. Rev. 59, 919-957.

Wallin, B.G., Sundlöf, G., Eriksson, B.-M., Dominiak, P., Grobecker, H. and Lindblad, L.-E. (1981) Plasma noradrenaline correlates to sympathetic muscle nerve activity in normotensive man. Acta Physiol. Scand. 111, 69-73.

Wallin, B.G. and Eckberg, D.L. (1982) Sympathetic transients caused by abrupt alterations of carotid baroreceptor activity in man. Am. J. Physiol. 242, H185-H190.

Wallin, B.G. and Fagius, J. (1986) The sympathetic nervous system in man - aspects derived from microelectrode recordings. Trends in Neurosci., 9, 63-67.

Walther, O.-E., Iriki, M. and Simon, E. (1970) Antagonistic changes of blood flow and sympathetic activity in different vascular beds following central thermal stimulation. II. Cutaneous and visceral sympathetic activity during spinal heating and cooling in anaesthetized rabbits and cats. Pflügers Arch. 319, 162-184.

THE COMPUTATIONAL STUDY OF VISION

Ellen C. Hildreth

Artificial Intelligence Laboratory and
Center for Biological Information Processing
Massachusetts Institute of Technology
Cambridge, MA, U.S.A.

This article considers the role of computational studies in the analysis of biological vision systems. We begin with a general discussion of the computational approach to vision -- we describe some essential components of computational studies, the sorts of insights about vision that can be gained by this approach, and how computational studies can be used together with experimental studies from perceptual psychology and the neurosciences to explore how biological systems perform complex tasks like vision. To make the discussion more concrete, we explore some problems in the analysis of visual motion, in which there has been close interaction between computational and experimental studies.

To begin, vision is a deceptively simple task to perform. We open our eyes and suddenly capture many important aspects of the world -- its structure, movement, colour, texture, and so on. Hidden beneath this apparent simplicity are complex processes that take the visual information that enters the eye and transform it into this rich description of the world. The computational approach to the study of vision inquires directly into the sort of information processing required to extract important information from the changing visual image -- information such as 3-D structure and motion. There are two basic principles that underlie the computational approach. The first is the separation of the tasks performed by a vision system from the hardware that carries out these tasks. The second is the study of vision by synthesizing it on a machine.

The development of computers with increasing power and sophistication often stimulates comparisons between computers and the brain, especially since computers have been applied more and more to tasks that were once considered uniquely human, like understanding natural language. We must be careful in making this comparison, however. At the level of their hardware, computers and brains are obviously very different. The electrochemical environment of neurones, their means of transmitting information and their overall architecture are very different from that of the wires and etched crystals of semiconducting materials of which computers are made. We can, however, describe the processes that take place in these two systems at a level that is independent of this hardware -- that is a description of the tasks that they perform. In much the same way that we can describe the theory of arithmetic independent of the computing device that performs the arithmetical operations, we can describe the theory of vision independent of the hardware that carries it out -- whether it be biological hardware or a machine.

This idea of separating tasks from hardware was central to the work of David Marr (Marr, 1982; Marr and Poggio, 1977). Marr argued that there are at least three different levels at which problems in vision can be described, which he labelled the theory, algorithm, and implementation or mechanism. The theory of a task includes a description of what any visual system needs to compute in order to accomplish this task and why it is necessary to compute this information. In addition, we know from our experience that the human visual system generally derives a single, stable interpretation of what is in the scene, where it is located and how it is changing with time. For most problems that are solved in the early stages of vision, however, there is actually an infinity of possible interpretations. For example, there are many patterns of movement that could be assigned to features in the image that would be consistent with the changing intensities that reach the eye. To obtain a single interpretation, we need to make assumptions about the physical world that allow most interpretations to be ruled out, leaving one that is most plausible from a physical standpoint. The assumptions that we make do not always hold true, of course, and sometimes give rise to optical illusions.

A final component of visual computations that could be considered an aspect of the theory or the algorithm is the issue of how visual information is represented. From the initial retinal image, the visual system does not achieve an understanding of what is in the scene in a single step. Vision proceeds in stages, with each stage producing increasingly more useful descriptions of the world. The process of vision can be viewed as the construction of a series of representations of visual information, with explicit computation that transforms one representation into the next. It is not yet known exactly how biological systems represent visual information, but computational studies have suggested several intermediate representations that are useful in visual processing (see, for example, Marr, 1982; Barrow and Tenenbaum, 1978; Ballard and Brown, 1982; Horn, 1986). The choice of which representation to use is critical in a computational study, as some representations facilitate the solution to vision problems more than others. For example, we can represent an image in terms of its spatial structure or we can construct a Fourier representation of its spatial frequency components. In computer vision, the spatial representation has generally been more effective for solving problems such as motion and colour analysis, stereopsis, and even recognition.

Thus the theory of a visual task describes what is computed and why, what assumptions are needed and what representations are used. From this theoretical analysis, we can already gain useful insights about vision. For example, it might be shown that a particular assumption by itself is sufficient to guarantee a single solution to the problem, or that the use of one representation facilitates a solution more than another. We can often state clearly when a computation should succeed at solving a problem and when it is likely to fail, because it often boils down to the question of when the assumptions hold and when they do not.

At another level, one can describe the particular algorithms used to perform the computation. Algorithms themselves can be described at different levels of detail. Deriving an algorithm is a useful exercise because it forces one to be explicit about the methods proposed for solving a problem. At this stage, it is shown in detail how the assumptions are used in the computation. Finally, one can describe the particulars of how a computation is carried out in the hardware of the brain or a machine. There are no hard and fast divisions between the three levels of theory, algorithm and implementation. What is important is that an understanding of a problem in vision is not complete until it covers the full range from theory to implementation.

The second principle that underlies the computational approach is the study of vision by building a machine that performs complex visual tasks. This step forces a specification of the details of an algorithm and provides a rigorous test of a theory. If a model for how to solve a vision problem can be embodied in a computer program, these ideas can be tested by examining whether or not the computer program is successful at solving the desired problem. Usually what happens at this stage is that a new aspect of the problem is discovered that was not realized in the theoretical analysis, or some aspect of the problem that was thought to be easy to solve turns out to be very difficult. The implementation of a model also provides a powerful predictive tool for experimental studies.

How can this approach be useful in studying biological vision systems? First, the brain can be considered an information processing device, much like a computer. Brains are not literally like computers, but we can describe the functions and workings of the brain in the language of computation. We can describe the theory that underlies tasks such as vision that are performed by the brain, elucidate the methods or algorithms by which the brain accomplishes vision, and describe how visual processing is implemented in neural hardware. We can propose as one goal for the study of vision the description of visual processing in the human system, or other biological systems, at the above three levels.

The approaches of perceptual psychology and neuroscience each provide valuable and different insights into how the brain processes visual information. From a psychological perspective we can ask what vision problems are solved and how good is the human system at solving them. Often there are many choices for the assumptions, representations and algorithms used to solve a particular problem and critical perceptual experiments can be designed that indicate which choices are made by the human system. Physiological studies show what parts of the brain are involved in carrying out particular computations, what the elementary operations are that the brain uses to carry out its solution, and can provide insight into the nature of the representations that the brain uses to store visual information. Computational studies can help to guide the experimental questions addressed through perceptual and physiological studies.

THE MEASUREMENT OF MOTION

To make this discussion more concrete, we explore some problems in the analysis of visual motion. The analysis of motion serves many essential functions for the visual system. The detection of sudden movements in the scene can serve as an early warming system, possibly alerting us to potential dangers. The measurement of motion allows us to track objects with our eyes. Discontinuities in motion often occur at object boundaries and can be used to carve up the scene into distinct objects. Relative motion in the image is also used to infer the 3-D structure of object surfaces and the movement of the observer relative to the scene.

The pattern of movement of features in the image is not given to the visual system directly, but must be inferred from the changing intensities that reach the eye. The 3-D shape of objects, the locations of their boundaries and the movement of the observer can then be inferred from the pattern of image motion. In computational models the analysis of motion is typically divided into two stages: the first measures movement in the changing 2-D image and the second uses motion measurements, for example to recover the 3-D layout of the environment. It is not yet clear whether motion analysis in biological systems is performed in these two distinct stages, but this division has served to facilitate theoretical studies and to focus experimental questions for perceptual and physiological studies.

The uses of motion impose different requirements on the accuracy and completeness with which image motion must be represented. The localization of object boundaries, for example, requires the detection of sharp changes in the direction or speed of movement, but may not require a precise representation of absolute velocities everywhere. Object tracking requires knowledge of the gross translation of an object, without much concern for the detailed relative movements that may take place within the object. On the other hand, the recovery of the accurate 3-D shape of an object appears to require a more precise and complete representation of the variations in motion across the surface of the object.

To suit these many visual tasks, motion measurement may be performed in different ways in biological systems. There is evidence that in human vision, motion may be analyzed by at least two different systems, termed short-range and long-range processes by Braddick (1974, 1980). Based on many perceptual phenomena (see, for example Ternus, 1926; Anstis and

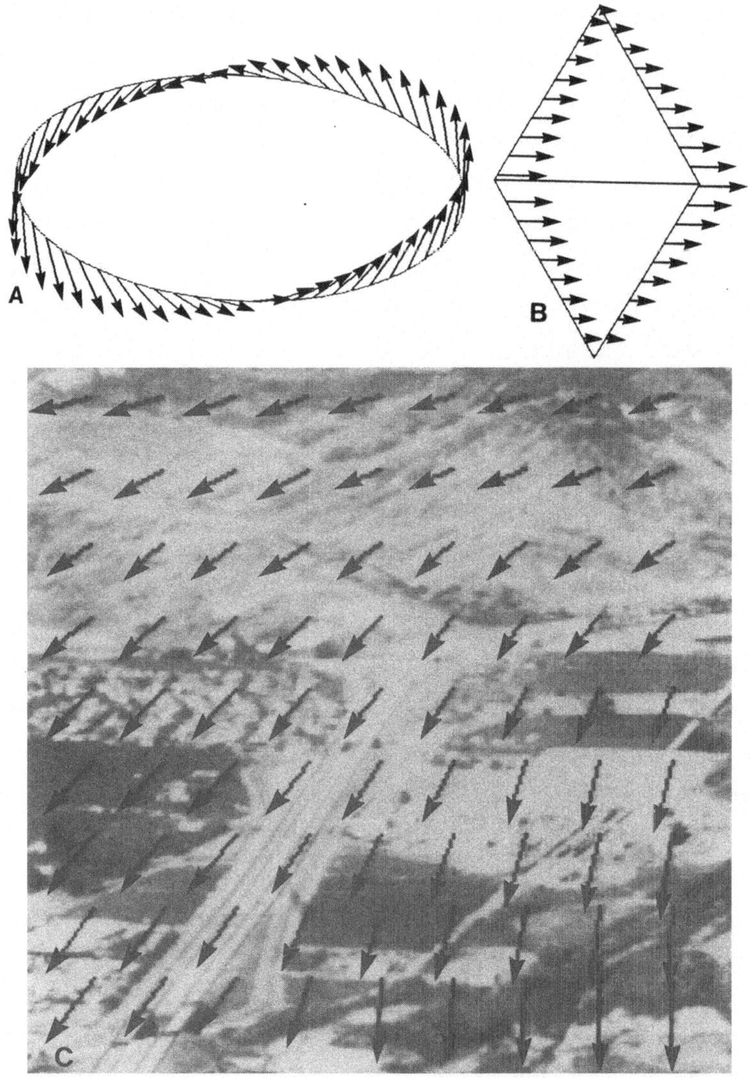

Fig. 1. Examples of velocity fields. (a) An elliptical contour, rotating in the image plane about its centre. The arrows along the contour represent the direction and speed of movement of individual points along the contour. (b) The projection of a 3-D figure undergoing rotation about a central vertical axis in space. The arrows represent the projected 2-D motions of points on the contours. (c) The velocity field generated by the motion of an airplane relative to terrain.

Rogers, 1975; Pantle and Picciano, 1976; Shepard and Judd, 1976; Petersik and Pantle, 1979; Anstis, 1980; Burt and Sperling, 1981; Green, 1983), it has been hypothesized that there exists a short-range process that analyses continuous motion or motion presented discretely but with small spatial and temporal separations between frames. A long-range process is also presumed to exist, which can infer motion over larger spatial and temporal displacements. There also appear to exist specialized mechanisms that may play a role in detecting discontinuities in motion or looming motion toward the observer (see, for example, Sterling and Wickelgren, 1969; Bridgeman, 1972; Collett, 1972; Nakayama and Loomis, 1974; Cynader and Regan, 1978, 1982; Frost, 1978; Regan and Beverley, 1979; Frost, Scilley and Wong, 1981; Miezin, McGuinness and Allman, 1982; Frost and Nakayama, 1983; Allman, Miezin and McGuinness, 1985; Nakayama, 1985).

Motion analysis in the human visual system may ultimately involve interaction between several processes, some fast but rough, others slow and more accurate, and still others specialized for specific tasks like detecting object boundaries. This article focuses on two aspects of the measurement of motion. The first is the computation of an instantaneous velocity field or optical flow field from the changing retinal image - this computation might underlie a short-range motion measurement process. We then discuss two aspects of long-range motion correspondence.

Consider first the computation of a velocity field. Fig. 1 shows some examples of velocity fields. In Fig. 1a is an elliptical contour, rotating in the image around its centre. The arrows represent the direction and speed of motion of individual points on the contour. Fig. 1b shows the projection of a 3-D contour rotating about a vertical axis in space. The arrows here represent the projected 2-D motions of points in the image. It can be seen that in the true velocity field, the direction and speed of motion changes around the contour. This variation is an important cue to the 3-D shape of the object. Fig. 1c is a natural image - one of a sequence of aerial photographs. The arrows superimposed on the image indicate the movement of image features as the plane flies along. This field of motions is also referred to as the "optical flow field" by Gibson (1950).

How do we go about computing a velocity field such as this? The first important observation to make is that in theory, there is not a single solution to this problem. There are many possible patterns of motion that

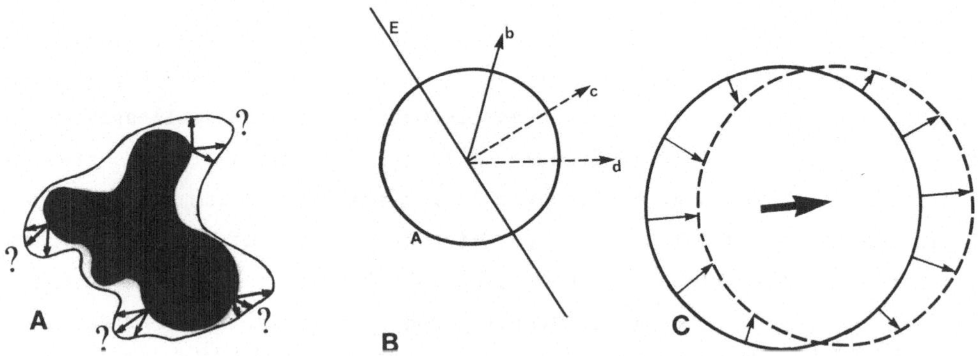

Fig. 2. The aperture problem in motion measurement. (a) An amoeboid shape that deforms as it moves. The motions of points around the contour are ambiguous. (b) An operation that views the moving edge E through the local aperture A can compute only the component of motion c in the direction perpendicular to the orientation of the edge. The true motion of the edge is ambiguous. (c) The circle undergoes pure translation to the right; the arrows represent the perpendicular components of velocity that can be measured from the changing image.

are consistent with the changing image intensities. If we consider a textured image such as the aerial photograph, we might be deceived into thinking that the measurement of motion ought to be simple - we just take a patch from one image, move to the next image in the sequence and look for a patch that has the same pattern of intensities. How then do we measure the movement of the amoeboid shape shown in Fig. 2a, which deforms as it moves? When we view a distorting object of this sort, we also see a well-defined motion. In theory, however, the motions of individual points on the object are entirely ambiguous. To identify a single pattern of motion, we need to impose additional constraint on the motion measurement computation that will narrow down the possibilities to one interpretation that is most plausible from a physical standpoint. This problem does not arise only in the analysis of unusual deforming motions; it is a problem that is faced all the time, everywhere in the changing image. Even when objects undergo pure translation, we see the simple translation not because this is the only pattern of motion consistent with the changing image, but because, given our assumptions about how physical objects behave under motion, the simple translation is the most plausible interpretation.

A second important observation to be made about the motion measurement computation is that in both biological and computer vision systems, the first operations for detecting motion examine only a small part of the changing image. As a consequence, these operations generally provide only partial information regarding the 2-D pattern of movement in the image, due to a problem that is often referred to as the "aperture problem" (Wallach, 1976; Fennema and Thompson, 1979; Burt and Sperling, 1981; Horn and Schunck, 1981; Marr and Ullman, 1981; Adelson and Movshon, 1982). Suppose the edge E in Fig. 2b moves across the image and its movement is observed through a limited window as defined by the circular aperture A. All that can be measured here is the movement of the edge in the direction c perpendicular to its orientation. The component of motion along the orientation of the edge is invisible through this aperture. It is not possible therefore to distinguish between motion in the three directions indicated by b, c and d in Fig. 2b; it is only possible to measure the component of motion in the perpendicular direction.

As a consequence of the aperture problem, motion measurement must take place in two stages. The first stage computes the components of motion in the direction perpendicular to the orientation of image features and the second stage combines these local motion measurements to compute the full 2-D pattern in motion. In Fig. 2c is a circle translating to the right.

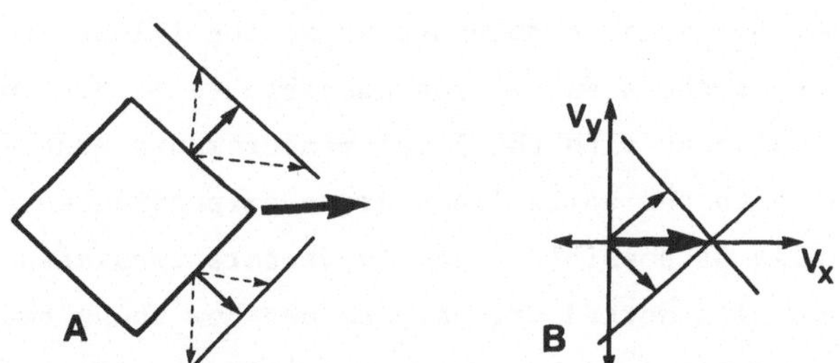

Fig. 3. The assumption of pure translation. (a) A square is translating to the right, and two perpendicular components of motion are shown with the solid arrows along two sides of the square. The true motions of these two points must project to the solid lines drawn perpendicular to the two components. Examples are indicated with dotted arrows. (b) The two motion constraints are transferred to a common origin in a velocity space. The intersection of the two constraint lines indicates the single velocity vector that is consistent with both measurements.

The arrows along the contour represent the perpendicular components of motion that can be measured directly from the image. Each measurement provides some constraint on the motion of the circle and must be combined to determine its overall motion. The aperture problem is illustrated here with edges and contours, but it is a general problem faced by any motion detection mechanism that examines only a limited area of the image. Biological motion detectors with limited receptive fields give rise to the aperture problem. There are features like corners that can be localized in 2-D and tracked unambiguously across the image, but in general, the first measurements of motion provide only partial information and must be integrated to compute the full pattern of motion in the image.

As we noted earlier, the combination of local motion measurements is difficult because in theory there are infinitely many patterns of motion consistent with the changing image. Additional constraint is needed to identify a single pattern of motion. There are many constraints that could be used to solve this problem. We discuss two that represent extremes in the range of possible constraints that have been proposed in the literature.

At one end of the scale is the assumption of pure translation; it is assumed that locally, velocities are constant and a single velocity vector is computed for each patch in the image. In theory, the use of this constraint is quite simple. In Fig. 3a is a square translating to the right, with two perpendicular components of motion illustrated by the solid arrows on two sides of the square. The possible velocities of the overall figure that are consistent with the local measurements must extend from the points on the square to the solid lines drawn perpendicular to the two components. Examples of possible velocity vectors consistent with each component are shown with dotted arrows. To see how these two measurements interact, they are transferred to a common origin in a "velocity space", in which the X and Y axes represent the X and Y components of velocity. As shown in Fig. 3b, there is only one overall motion vector that is consistent with both of these components, given by the intersection of the two solid "constraint" lines. This intersection point corresponds to translation to the right. In practice, using the assumption of pure translation is more difficult, because there is error in the image measurements, velocities may not be constant locally, and there may not be an adequate range of orientations of features in a particular area of the image to obtain a good estimate of the direction and speed of translation.

There are, however, many algorithms used in computer vision, as well as biological models of motion analysis, that embody this basic assumption (for example, Lappin and Bell, 1976; Pantle and Picciano, 1976; Fennema and Thompson, 1979; Anstis, 1980; Marr and Ullman, 1981; Thompson and Barnard, 1981; Adelson and Movshon, 1982; Lawton, 1983).

The second type of additional constraint to be considered here is often referred to as a "smoothness" constraint (Horn and Schunck, 1981; Hildreth, 1984; Nagel, 1984; Nagel and Enkelmann, 1984; Anandan and Weiss, 1985). This constraint is more general and rests on the assumption that physical surfaces tend to be smooth, and as a consequence when they move, nearby points tend to move with similar velocities. There exist discontinuities in motion at object boundaries, but most of the image is the projection of relatively smooth surfaces. It is natural to assume that image velocities will vary smoothly over most of the visual field. A unique pattern of movement can be obtained by computing a velocity field that is consistent with the changing image and has the least amount of variation possible. In other words, a pattern of motion is computed for which nearby points in the image move with velocities that are as similar as possible. One can be more precise about how variation in the velocity field is measured and can formulate specific algorithms for computing the smoothest velocity field (Horn and Schunck, 1981; Hildreth, 1984; Nagel, 1984; Nagel and Enkelmann, 1984; Anandan and Weiss, 1985).

Using the smoothness assumption has a number of attributes from a computational perspective. First, it allows the analysis of more general classes of motion. Surfaces can be rigid or nonrigid, undergoing any movement in space, and a projected velocity field of least variation can always be computed. Second, this assumption can be embodied in the velocity field computation in a way that guarantees a unique solution (Hildreth, 1984). For some classes of motion, it can be shown that the computation yields the physically correct velocity field (Hildreth, 1984). Finally, the computation can be implemented using simple, parallel algorithms, a property that is useful from a biological perspective.

The assumption of smoothness is especially useful for measuring motion for the purpose of recovering 3-D structure. For this task, the variations in velocity across the surface of an object must be preserved in the representation of image motion. An algorithm that assumes only pure translation might provide a fast, rough computation of direction and speed of movement that could be useful for object tracking or the detection of discontinuities in the velocity field.

There is another important issue that arises in the design of computational models. In principle there are many different stages in the analysis of the image at which the measurement of motion could first take place. For example, motion could be inferred from changes in the raw image intensities or changes in the filtered intensities. The detection of motion also could be focused at the locations of features in the image, such as sharp intensity changes or edges. When the measurements of local components of motion are combined, they could be combined over 2-D areas of the image or along features such as edge contours.

How can computational studies shed light on the measurement of motion in biological systems? There are at least four questions that arise from what has been discussed so far. Experimental studies from psychophysics and physiology have begun to explore each of these questions. The first is the basic question of whether the overall analysis of motion takes place in two stages, in which motion is first measured in the changing 2-D image and these measurements are then used to infer 3-D structure and motion. The second question considers whether motion measurement itself takes place in two stages, where the local perpendicular components of motion are first measured and then combined to solve the aperture problem. Third, we can ask whether the local motion measurements are integrated over areas or along contours. Finally we can ask what assumptions the human visual system uses to compute a unique pattern of motion from the changing image.

Consider the second question, regarding whether there is a two-stage motion measurement process. To study this question, Adelson and Movshon (1982) and Movshon, Adelson, Gizzi and Newsome (1985) worked with visual displays consisting of superimposed sinewave gratings of different orientations, shown schematically in Fig. 4a. If the two oblique gratings are moved in the directions perpendicular to their orientation and superimposed, the combination will generally be seen as a checkerboard pattern moving rigidly to the right. By exploring the ability of the human system to combine these two components into a single rigid pattern under different conditions, Movshon et al. (1985) were able to find psychophysical evidence for a two-stage motion measurement process.

Movshon et al. (1985) also found direct physiological evidence for two

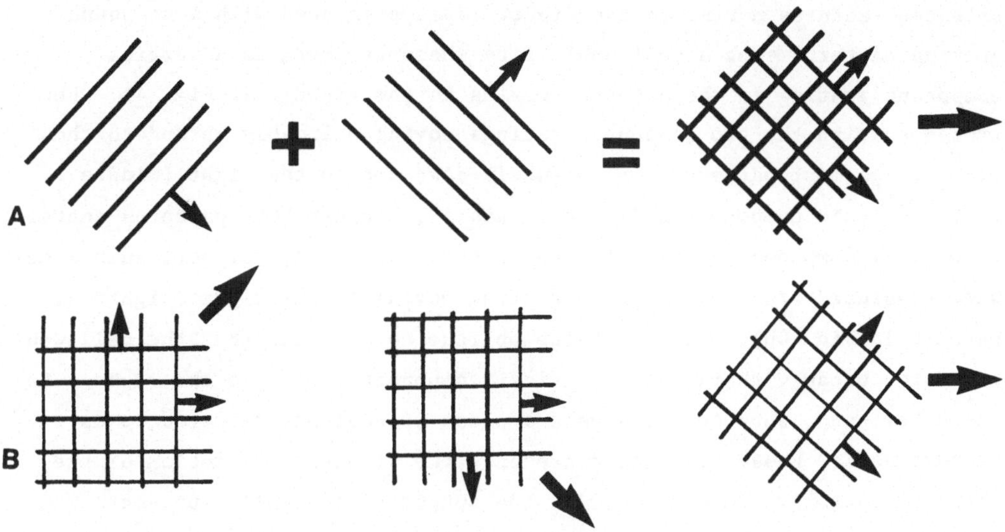

Fig. 4. The experiments of Movshon et al. (1985) (a) The visual stimuli used by Adelson and Movshon consisted of superimposed sinewave gratings, each moving in the direction perpendicular to their orientation. (b) The experiments of Movshon et al. used compound grating patterns moving in different directions. A "component cell" that prefers vertically oriented components moving to the right would respond to the leftmost and central compound patterns, but not to the rightmost pattern. A "pattern cell" that prefers motion to the right would only respond to the rightmost pattern, whose overall motion is to the right.

stages and a possible locus for these stages in the visual pathway. Two visual areas in the monkey that contain an abundance of motion-sensitive neurones are cortical areas V1 and MT (with regard to the properties of MT neurones, see, for example, Maunsell and Van Essen, 1983; Van Essen and Maunsell, 1983; Allman, Miezin and McGuiness, 1985). Lesions in area MT have been shown to result in specific behavioural deficits at tasks such as smooth pursuit eye movements (Newsome, Wurtz, Dursteler and Mikami, 1985) and the recovery of 3-D structure from motion (Seigel and Andersen, 1986), which require an accurate measurement of image motion. Movshon et al. (1985) explored the nature of the motion measurement taking place in these two areas using superimposed sinewave grating patterns. The basic idea behind their experiment is the following. Suppose a class of neurones extracted only the components of motion perpendicular to the orientation of image features. Consider a subset of this class that prefers vertically oriented features moving to the right. When presented with a compound grating pattern, such a cell would fire whenever there is a vertical component grating in the pattern, moving to the right. In Fig. 4b, when presented with the two compound gratings moving obliquely up and to the right (rightmost pattern), or obliquely down and to the right (middle pattern), this component cell would respond, because both patterns contain a vertical component that is moving to the right. If, however, such a cell were presented with the compound grating moving to the right (rightmost pattern in Fig. 4b), which contains oblique components, then the cell would not fire, because there is no vertical component moving to the right. On the other hand, suppose there were a class of neurones referred to as pattern cells, whose response represents the direction of motion of the combined pattern. In other words, the output of the cell represents the result of solving the aperture problem. Consider a subset of this class that prefers motion to the right. Then such a cell would not fire for the first two compound gratings shown on the left and in the middle of Fig. 4b, because their overall motion is in the two oblique directions. It would, however, respond to the last compound pattern that has an overall motion to the right. In the physiological studies, Movshon et al. (1985) found that cells in area V1 all had the behaviour of component cells - they only responded when there was a component in the pattern that had the preferred orientation and direction of motion for the cell. In area MT, many of the cells also behaved like component cells, but there was a subpopulation of cells in some layers that behaved like pattern cells. That is, they were responsive to the overall direction of translation of the combined pattern, independent of the components that made up the pattern. These neurones may serve to combine motion components to compute the real 2-D direction of velocity of a moving pattern. This physiological study provides strong evidence for a two-stage motion measurement computation in the primate visual system.

The third question mentioned above was the issue of whether motion measurements are integrated along contours or over areas of the image. This question was addressed in a perceptual study by Nakayama and Silverman (1984). They began with a simple distorted line as shown in Fig. 5a, and moved it up and down. The central diagonal section of the line appears to move in the oblique direction in this case, so that the entire figure appears nonrigid. The figure can be made to appear to move rigidly up and down by introducing additional features in the image that are unambiguously moving up and down. Nakayama and Silverman introduced both breaks on the

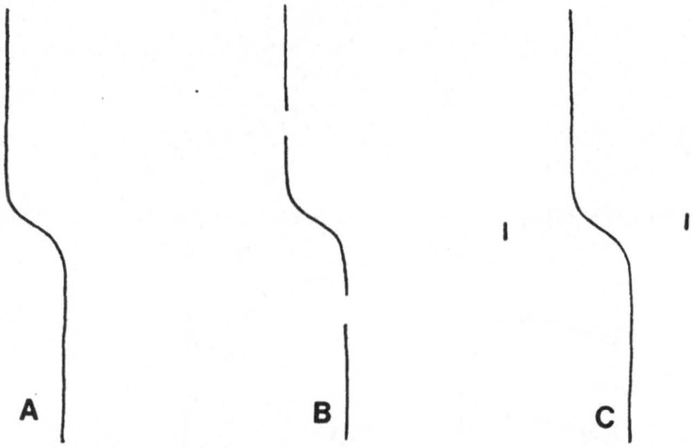

Fig. 5. The experiment by Nakayama and Silverman (1984). (a) A distorted line, translating up and down. (b) Breaks on the contour also move up and down. (c) Segments off the contour move up and down.

contour as shown in Fig. 5b, and short segments off the contour as shown in Fig. 5c. It was found that both the breaks on the line and the segments off the line could cause the central part of the figure to appear to move up and down, but the features on the contour had a much stronger effect, in that their distance from the diagonal part of the figure could be very large. The segments off the contour had to be close to the contour to exert any influence on the perception of its motion. This phenomenon suggests that the integration of local motion measurements along contours may play a stronger role in the human visual system. This phenomenon is also supported by other perceptual demonstrations (for example, Hildreth, 1984). This general observation has implications for physiology - it suggests, for example, that we might expect stronger interactions between motion-sensitive cells along their preferred orientations, rather than perpendicular to their orientation.

The fourth question was, what additional assumptions are we making in order to compute a unique pattern of motion? There is evidence from early perceptual studies that the human visual system perceives patterns of motion consistent with those predicted by a computation that embodies the smoothness assumption. The algorithm for computing the smoothest velocity proposed by Hildreth (1984) has the property that in theory, it yields the correct solution when objects are undergoing pure translation, or when there is a general translation and rotation of a rigid 3-D object whose edges are essentially straight. For the case of smooth curves undergoing rotation, the computation sometimes yields incorrect solutions, but the

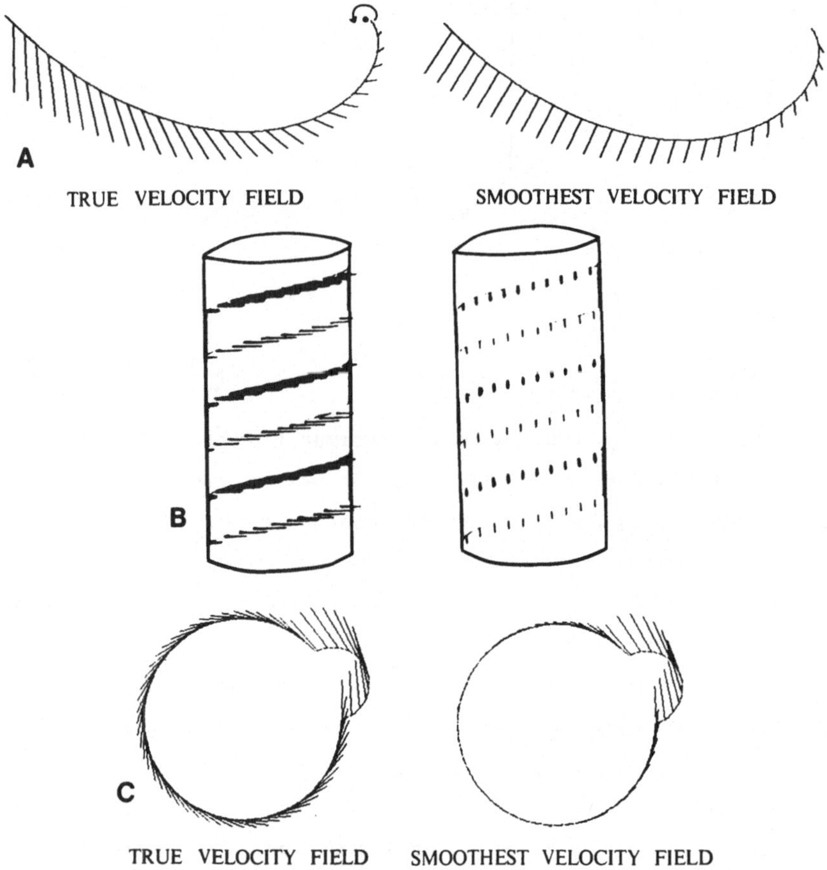

Fig. 6. Examples of the smoothest velocity field. (a) On the right is the true velocity field (represented by the short segments along the contours) for a spiral rotating about its centre. On the left is the smoothest velocity field consistent with the changing image, which has a large radial component of motion. (b) The true velocity field for the rotating barberpole is shown on the left, and the smoothest velocity field on the right. (c) On the left is the true velocity field for a deformed circle, rotating about its centre. The smoothest velocity field is shown on the right.

human system also appears to derive an incorrect perception of motion in these situations. Three simple examples are shown in Fig. 6. In Fig. 6a is a rotating spiral. A spiral appears to expand and contract as it rotates. This suggests that there is a strong radial component in our perceived motion, while the true motion is pure rotation, as shown on the left in Fig. 6a (the short line segments represent the movement of individual points along the contours). The smoothest pattern of motion that is consistent with the rotating spiral, shown on the right in Fig. 6a, also has a strong radial component, much like our perception. A second example is the rotating barberpole. The true motion is strictly horizontal, but we perceive a downward motion of the stripes. The smoothest pattern of motion in this case is also vertical, consistent with our perception. In the 1950's, Wallach, Weisz and Adams (1956) conducted perceptual experiments in which simple geometric figures were placed on a rotating turntable, and observers were asked to describe the perceived motion. It was found that simple smooth curves generally appeared to deform continuously as they rotated. In the case of a deformed circle such as that shown in Fig. 6c, rotating about its centre, the circle appears to remain stationary, while the bump travels around its perimeter. This is again consistent with the smoothest pattern of motion, in which there is significant velocity around the bump, but almost no movement as we move away from the bump. There are many other examples of perceptual illusions in motion that are consistent with a computation that embodies the smoothness assumption (Hildreth, 1984). New illusions can also be generated quite easily, because the precise conditions under which this algorithm will give an incorrect solution are known.

The use of the translation versus smoothness assumptions has implications for how the motion measurement computation might be carried out in neural hardware. The main distinction is that to implement the computation of the smoothest pattern of motion, communication is required between nearby elements that are involved in the solution to the aperture problem, in order to find an overall pattern of motion for which nearby velocity vectors are as similar as possible. A mechanism that implements the more restricted assumption of pure translation does not require this communication. Movshon et al. (1985) suggested one possible mechanism by which pattern cells in area MT could combine outputs from component cells to determine the direction of translation of patches of the image. Poggio and Koch (1985) have proposed a hypothetical neural implementation of the computation of a smoothest velocity field, illustrated in Fig. 7. They began by designing an analogue electrical network that could perform this computation and then transformed this analogue circuit into a neural circuit using known biophysical mechanisms. This model was not intended as a specific model of the motion analysis that might take place in area MT, but rather was intended to show that in principle, it is quite simple to embody this type of computation in neural hardware.

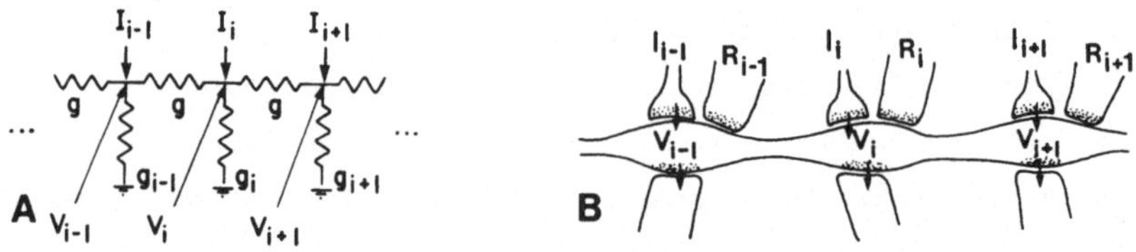

Fig. 7. Analog models of the velocity field computation. (**A**) A simple resistive network that computes the smoothest velocity field. The conductances g and g_i, and the currents I_i represent properties of a moving contour that are measured directly from the image. The 2-D velocity field along the contour is represented implicitly by the combination of these inputs and the resulting voltages V_i. (**B**) A hypothetical neural implementation of the circuit shown in (**A**). Synaptic mediated currents I_i, and additional inputs R_i (possibly a $GABA_A$ type of synapse) represent properties of a moving contour. The resulting voltages V_i, sampled by dendro-dendritic synapses, together with the input currents, represent local velocities along the contour.

The last question that we address with regard to the velocity field computation is the question of whether it is appropriate to consider the overall analysis of motion as being divided into two parts, in which movement in the 2-D image is measured first, and then the 3-D motion and structure of objects in the scene is recovered. Most computational models assume distinct stages, although some make explicit different information in the first stage (see, for example, Waxman and Ullman, 1985; Waxman and Wohn, 1985). Recent work by Bruss and Horn (1983), Subbarao and Waxman (1985), and Negahdaripour and Horn (1985) has attempted to compute 3-D structure directly from the changing image intensities. Alternatives to this two-stage analysis of motion are therefore being considered in computational studies.

With regard to the human visual system, it remains an open question whether motion analysis takes place in two distinct stages. There are actually two separate questions here: first, if only the motion cue is present, do we compute 2-D motion first and then recover 3-D motion and structure? Second, if other 3-D cues are available, for example from binocular stereo, at what stage are these additional 3-D cues incorporated into the motion computation? There is not yet a clear answer to this question, but we mention here two or three aspects of the human visual system that are relevant. First, the time course is different for measuring 2-D motion and recovering 3-D structure. Studies from McKee, Nakayama and others (see, for example, McKee and Welch, 1985; McKee, Silverman and Nakayama, 1986) suggest that the time frame for measuring image velocities is at most 200 msec or so, and usually much less. Andersen and Siegel (1986) have recently conducted a study with both monkeys and humans as observers, in which they were given three tasks to perform. Two required the detection of 2-D organizations in a motion field established with moving random dots - the organizations to be detected were a global 2-D rotation or expansion of the motion field; the third task required the detection of 3-D structure in a display in which points were moving as though they were projected from the surface of a rotating cylinder. Andersen and Siegel then varied the percentage of dots undergoing the correct global motion in the displays and measured the reaction time for detecting the different organizations. The 2-D tasks had reaction times around 600 msec, while the 3-D detection task required around 900 msec. In recent studies by Hildreth, Inada, Grzywacz and Adelson (1986), it appears that at least a second or so is required for the human system to build up an accurate perception of 3-D structure from the relative movements of a few isolated points. These observations suggest that the computation of 3-D structure and motion has a more extended time frame than the simple computation of 2-D image motions.

With regard to the second question, Adelson (1984) has shown that stereo disparity is taken into account in the solution to the aperture problem. If two motion components are presented at very different depths, they are not combined. Physiological studies show that MT neurones also tend to be selective for stereo disparity. On the other hand, the structure-from-motion computation can proceed in the presence of conflicting stereo cues. Prazdny recently constructed a demonstration consisting of a random-dot stereogram with a central patch of points that are moving as though they were projected from the surface of a rotating cylinder in space. The background was stationary. The stereo pair was constructed with the central moving patch shifted uniformly in the horizontal direction. The motion cue then indicated that there is a rotating 3-D cylinder in the centre of the pattern, while the stereo indicated that there are two flat planes. People perceive the cylinder in this case. Thus stereo may be used early on in the analysis of motion, but it does not constrain motion analysis in such a strict way that it prevents the structure-from-motion process from proceeding on its own. These are some observations that bear on the general issue of whether we perform motion analysis in two stages, but there is no solid answer to the question yet.

Models for computing a velocity field could only underlie a short-range motion process, because they require that motion in the image be roughly continuous. The perception of motion by the human visual system does not require that objects move continuously across the visual field, however. Motion can be inferred when features are presented discretely at positions separated up to several degrees of visual angle and with long time intervals between presentations. The long-range motion phenomena illustrate the ability of the human system to derive a correspondence between elements in the changing image, over considerable distances and temporal intervals. Under these conditions there is no continuous motion of elements that can be measured directly with short-range motion detection mechanisms. We need to perform a correspondence computation, in which image elements are matched from one moment to the next that correspond to the same physical feature in motion. There are at least three important issues that arise regarding this computation. First, which features in the image are matched and what properties of these features are used to establish a match? Second, what rules are used to compute a unique correspondence of features? Similar to the velocity field computation, there are in theory many possible matchings between features in two images and additional constraint must be imposed to compute a single correspondence. Finally, with regard to the human visual system, we can ask what is the role of the long-range process and how does it interact with the short-range process?

We focus here on two issues that have been raised in computational studies, and perceptual demonstrations that address these issues for the human system. The first regards whether the correspondence is initially established in two dimensions or three. The second deals with a particular assumption used in a model proposed by Ullman (1979), referred to as the "independence" hypothesis.

Early perceptual studies suggested that the distance between image features and their potential matches at the next moment plays a role in the matching of features (for example, Ternus, 1926; Kolers, 1972; Burt and Sperling, 1981). In general, if there are a few isolated features in motion, a given feature will prefer to match its nearest neighbour in the subsequent frame. We also prefer to have all points in adjacent frames

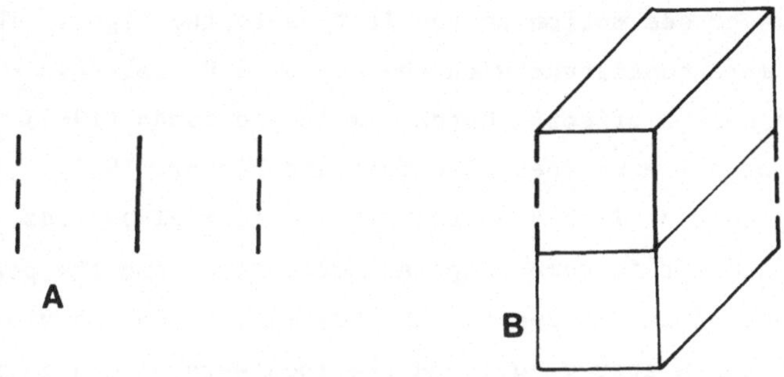

Fig. 8. The effect of the 3-D perspective cue on motion correspondence. (a) A simple competing motion display, in which the first frame consists of a single vertical bar of light, and the second frame consists of two bars of light, equally spaced to the right and left. (b) The pattern in (a) is embedded in a perspective drawing of a 3-D figure.

match something, and these two constraints can interact in a way that violates the nearest-neighbour rule.

Ullman (1979) proposed a model for establishing a correspondence between isolated features that used three constraints: first, every point in two adjacent frames should have a match in the other; second, the computation minimizes the total distance travelled by all of the elements together; and third, each point is assumed to move independently of other points. There are models that use other constraints as well, but we examine here only two of the above constraints.

For any model that incorporates a constraint on the distance travelled by moving features, the question arises, do we use 2-D or 3-D distances? So far, perceptual studies seem to indicate that we use 2-D distances. One piece of evidence can be drawn from Ullman's (1979) study. Suppose a competing motion display is constructed in which a single bar appears in the first frame, and two bars appear in the second, displaced equal distances to the left and right, as shown in Fig. 8a. Half of the time, observers report motion to the left, and half of the time they see motion to the right. Suppose this display is now embedded in a 3-D perspective drawing as shown in Fig. 8b. The two bars in the second frame are still displaced equal amounts to the left and right, but the perceived 3-D distance to the right is much larger. If again the middle bar is presented first, and then the two flanking bars, observers still perceive motion and are equally like to see motion to the left as to the right. The perceived motion here is more consistent with the use of 2-D distances - the 3-D perspective cue has no effect. Mutch, Smith and Yonas (1983) introduced a larger variety of 3-D cues (see also Tarr and Pinker, 1985). They used a physical arrangement of lights on the tips of plexiglass rods. The lights were turned on and off in some temporal succession, and the positions of the rods were varied in 3-D space. The physical setup was viewed binocularly, giving a stereo cue, and the rods were placed in a visible grid that provided a perspective cue regarding the background plane. Even in this situation where many 3-D cues are available, the perceived movement of the lights was based on 2-D retinal distances. If for example, a central light was turned on first and then turned off and the two other lights were turned on, observers would see motion to the light whose retinal distance from the central light is smallest, regardless of the 3-D distances present. Perceptual studies therefore suggest that motion correspondence is initially established in two dimensions rather than three.

A second aspect of Ullman's model that has been explored in perceptual studies is the independence hypothesis. Do we assume that features in the image tend to move independently of one another, or are there more complex spatial interactions between the correspondences established for nearby features? We present two examples of more complex interactions that can occur. The first is from Ramachandran and Anstis (1985). They created a motion stimulus in which the first frame consists of two points of light at the upper left and lower right corners of a square, and the second frame contains lights at the opposite two corners, as shown in Fig. 9a. If the two frames are alternated back and forth, the motion of the points is ambiguous. Sometimes the points will appear to bounce up and down, and other times they will appear to bounce left and right. This local pattern was then embedded in a large array, as shown in Fig. 9b. Each subpattern could now, in principle, be perceived as moving up and down, or left and right. If we incorporated an independence hypothesis, we would expect the entire grid to have subpatterns moving either way, but perceptually, the grid tends to synchronize. Typically, all of the subpatterns appear to move up and down, or they all appear to move left and right. The correspondence computed in one region of the image influences the correspondence derived in surrounding regions in a nontrivial way.

The second example is based on a demonstration by Hildreth (1984).

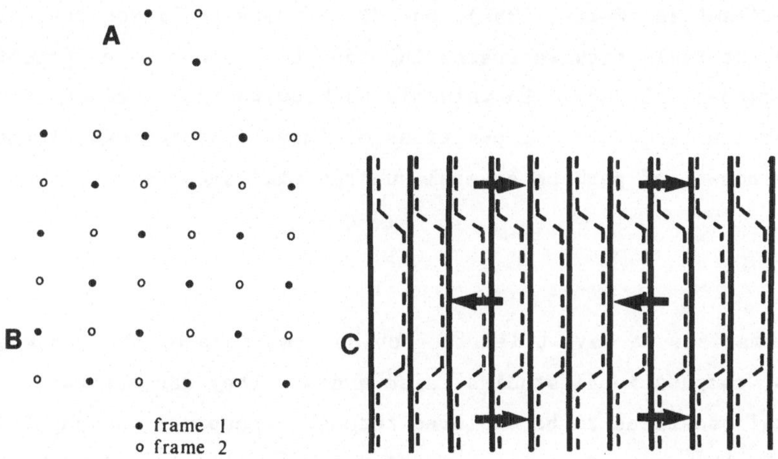

Fig. 9. Violations of the independence hypothesis. (a) An apparent motion stimulus, in which the first frame consists of two points of light at the upper left and lower right corners of a square and the second frame consists of the opposite two points of light. (b) The local pattern in (a) is embedded in a large array of similar patterns. (c) The solid lines represent the bars of light presented in the first frame, and the dashed lines represent the bars of light presented in the second frame.

There are again two frames that are alternated back and forth. One frame contains vertical bars of light and the other contains trapezoid-shaped contours, as shown in Fig. 9c. If there is only a short time between frames, a splitting motion is perceived - the top and bottom parts of the display shift in one direction while the middle shifts in the opposite direction. If there is a long time interval between the frames, however, on the order of 200 msec or so, the entire vertical contour is seen as jumping in the same direction, to the entire trapezoidal contour. That is, elements that are connected in one frame are seen as moving to connected contours in the second. This suggests an interaction between the correspondences computed along connected features in the image. Currently, there are no models that can account for these more complex spatial interactions.

With regard to physiology, a class of neurones has not yet been found whose behaviour reflects the long-range motion measurement abilities of the human visual system. MT neurones show directionally selective responses for larger spatial displacements between spots of light than V1 neurones

(Newsome, Mikami and Wurtz, 1982), but MT neurones still require relatively small time intervals between frames in order to respond in a directionally selective manner. It would be valuable to know whether a class of neurones exists that can integrate the positions of image features over larger temporal windows, of perhaps several hundred milliseconds.

SUMMARY

To summarize, we have tried to provide some idea of the types of issues that computational studies raise and how they can motivate experimental questions to be explored through perceptual and physiological studies. In the area of motion analysis, computational studies have helped to define precisely, what the goals of motion analysis are. They have provided some insight into the nature of problems in motion - for example, that initial local motion measurements provide only partial information about image motion and have to be integrated in some way, and that additional assumptions are needed to identify a unique pattern of motion. These studies have also provided insight into the nature of possible solutions. For example, they have elucidated additional assumptions that we could use and have shown how these assumptions can be embodied in the motion measurement computation. They have suggested possible algorithms for deriving solutions. Implementations of these algorithms can be applied to natural imagery to see whether they give rise to reasonable solutions, and also provide a powerful predictive tool that can be used to design critical tests of whether the computational models exhibit behaviour similar to that of the human visual system.

It is often the case that the first attempts at analyzing a problem from a computational perspective provide some insight into the nature of the problem, but have little resemblance to the behaviour of the human system. Subsequent studies then take into account information about the biological system more and more, and by working back and forth between computational and experimental studies, we hope eventually to converge on models of visual processing that are both consistent with the behaviour of biological systems, and really solve useful problems in vision.

ACKNOWLEDGEMENTS

This article describes research done within the Artificial Intelligence Laboratory and the Center for Biological Information

Processing (Whitaker College) at the Massachusetts Institute of Technology. Support for the A.I. Laboratory's artificial intelligence research is provided in part by the Advanced Research Projects Agency of the Department of Defense under Office of Naval Research contract N00014-80-C-0505. Support for this research is also provided by a grant from the Office of Naval Research, Engineering Psychology Division.

REFERENCES

Adelson, E.H. (1984) Binocular disparity and the computation of two-dimensional motion. J. Opt. Soc. Am. A 1, 1266.

Adelson, E.H. and Movshon, J.A. (1982) Phenomenal coherence of moving visual patterns. Nature 300, 523-525.

Allman, J., Miezin, F. and McGuinness, E. (1985) Direction- and velocity-specific responses from beyond the classical receptive field in the middle temporal area (MT). Perception 14, 105-126.

Anandan, P. and Weiss, R. (1985) Introducing a smoothness constraint in a matching approach for the computation of optical flow fields. Proc. IEEE Workshop on Computer Vision: Representation and Control, Bellaire, MI, October, pp 186-194.

Anderson, R.A. and Siegel, R.M. (1986) Two- and three-dimensional structure from motion sensitivity in monkeys and humans. Soc. Neurosci. Abstracts, in press.

Anstis, S.M. (1980) The perception of apparent motion. Phil. Trans. Roy. Soc. B: 290, 153-168.

Anstis, S.M., Rogers, B.J. (1975) Illusory reversal of visual depth and movement during changes of contrast. Vision Res. 15, 957-961.

Ballard, D.H. and Brown, C.M. (1982) Computer Vision. Prentice-Hall: Engelwood Cliffs.

Barron, J. (1984) A survey of approaches for determining optic flow, environmental layout and egomotion. Univ. Toronto Tech. Rep. Res. Biol. Comp. Vision RBCV-TR-84-5.

Barrow, H.G. and Tenenbaum, J.M. (1978) Recovering intrinsic scene characteristics from images. In: Computer Vision Systems, Hanson, A.R. and Riseman, E.M., eds., Academic Press: New York.

Braddick, O.J. (1974) A short-range process in apparent motion. Vision Res. 14, 519-527.

Braddick, O.J. (1980) Low-level and high-level processes in apparent motion. Phil. Trans. Roy. Soc. B: 290, 137-151.

Bridgeman, B. (1972) Visual receptive fields sensitive to absolute and relative motion during tracking. Science 178, 1106-1108.

Bruss, A. and Horn, B.K.P. (1983) Passive navigation. Comput. Vision Graph. Image Proc. 21, 3-20.

Burt, P. and Sperling, G. (1981) Time, distance, and feature trade-offs in visual apparent motion. Psych. Rev. 88, 171-195.

Collett, T. (1972) Visual neurons in the anterior optic tract of the privet hawk moth. J. Comp. Physiol. 78, 396-433.

Cynader, M. and Regan, D. (1978) Neurons in the cat parastriate cortex sensitive to the direction of motion in three-dimensional space. J. Physiol. (Lond.) 274, 549-569.

Cynader, M. and Regan, D. (1982) Neurons in cat visual cortex tuned to the direction of motion in depth: effect of positional disparity. Vision Res. 22, 967-982.

Fennema, C.L. and Thompson, W.B. (1979) Velocity determination in scenes containing several moving objects. Comput. Graph. Image Proc. 9, 301-315.

Frost, B.J. (1978) Moving background patterns after directionally specific responses of pigeon tectal neurons. Brain Res. 151, 599-603.

Frost, B.J. and Nakayama, K. (1983) Single visual neurons code opposing motion independent of direction. Science 220, 744-745.

Frost, B.J, Scilley, P.L. and Wong, S.C.P. (1981) Moving background patterns reveal double opponency of directionally specific pigeon tectal neurons. Exp. Brain Res. 43, 173-185.

Gibson, J.J. (1950) The Perception of the Visual World. Houghton Mifflin: Boston.

Green, M. (1983) Inhibition and facilitation of apparent motion by real motion. Vision Res. 23, 861-865.

Hildreth, E.C. (1984) The Measurement of Visual Motion. MIT Press: Cambridge.

Hildreth, E.C., Inada, V.K., Grzywacz and N.M., Adelson, E.H. (1986) The perceptual buildup of 3-D structure from motion. MIT Artificial Intelligence Laboratory Memo.

Horn, B.K.P. (1986) Robot Vision Cambridge: MIT Press.

Horn, B.K.P. and Schunck, B.G. (1981) Determining optical flow. Artif. Intell. 17: 185-203.

Kolers, P.A. (1972) Aspects of Motion Perception. Pergamon: New York.

Lappin, J.S. and Bell, H.H. (1976) The detection of coherence in moving random dot patterns. Vision Res. 16, 161-168.

Lawton, D.T. (1983) Processing translational motion sequences. Comput. Vision Graph. Image Proc. 22, 116-144.

Marr, D. (1982) Vision. Freeman: San Francisco.

Marr, D. and Poggio, T. (1977) From understanding computation to understanding neural citcuitry. Neurosci. Res. Prog. Bull. 15, 470-488.

Marr, D. and Ullman, S. (1981) Directional selectivity and its use in early visual processing. Proc. Roy. Soc. B: 211, 151-180.

Maunsell, J.H.R. and Van Essen, D.C. (1983) Functional properties of neurons in middle temporal visual area of the macaque monkey. I. Selectivity for stimulus direction, speed and orientation. J. Neurophysiol. 49, 1127-1147.

McKee, S.P., Silverman, G.H. and Nakayama, K. (1986) Precise velocity discrimination despite random variations in temporal frequency and contrast. Vision Res., in press.

McKee, S.P. and Welch, L. (1985) Sequential recruitment in the discrimination of velocity. J. Opt. Soc. Am. A 2, 243-251.

Miezin, F., McGuinness, E. and Allman, J. (1982) Antagonistic direction specific mechanisms in area MT in the owl monkey. Soc. Neurosci. Abstracts 8, 681.

Movshon, J.A., Adelson, E.H., Gizzi, M.S. and Newsome, W.T. (1985) The analysis of moving visual patterns. In: Pattern Recognition Mechanisms, Chagas, C., Gattas, R. and Gross, C.G., eds., Vatican Press: Rome.

Mutch, K., Smith, I.M. and Yonas, A. (1983) The effect of two-dimensional and three-dimensional distance on apparent motion. Perception 12, 305-312.

Nagel, H.-H. (1984) Recent advances in image sequence analysis. Premier Colloque Image - Traitment, Synthèse, Technologie et Applications, Biarritz, pp 545-558.

Nakayama, K. (1985) Biological motion processing: a review. Vision Res. 25, 625-660.

Nakayama, K. and Loomis, J.M. (1974) Optical velocity patterns, velocity-sensitive neurons, and space perception: a hypothesis. Perception 3, 63-80.

Nakayama, K. and Silverman, G.H. (1984) Propagation of velocity information along moving contours. J. Opt. Soc. Am. A 1, 1266.

Negahdaripour, S. and Horn, B.K.P. (1985) Direct passive navigation. MIT Artifical Intelligence Laboratory Memo 821.

Newsome, W.T., Mikami, A. and Wurtz, R.H. (1982) Direction selective responses to sequentially flashed stimuli in extrastriate area MT in the awake macaque monkey. Soc. Neurosci. Abstracts 8, 812.

Newsome, W.T., Wurtz, R.H., Dursteler, M.R. and Mikami, A. (1985) Deficits in visual motion processing following ibotenic acid lesions of the middle temporal visual area of the macaque monkey. J. Neurosci. 5, 825-840.

Pantle, A.J. and Picciano, L. (1976) A multistable display: evidence for two separate motion systems in human vision. Science 193, 500-502.

Petersik, J.T. and Pantle, A. (1979) Factors controlling the competing sensations produced by a bistable stroboscopic motion display. Vision Res. 19, 143-154.

Poggio, T. and Koch, C. (1985) Ill-posed problems in early vision: from computational theory to analog networks. Proc. Roy. Soc. B: 226, 303-323.

Ramachandran, V.S. and Anstis, S.M. (1985) Perceptual organization in multistable apparent motion. Perception 14, 135-143.

Regan, D. and Beverley, K.I. (1979) Visually guided locomotion: Psychophysical evidence for neural mechanisms sensitive to flow patterns. Science 205, 311-313.

Shepard, R.N. and Judd, S.A. (1976) Perceptual illusion of rotation of three-dimensional objects. Science 191, 952-954.

Siegel, R.M. and Andersen, R.A. (1986) Motion perceptual deficits following ibotenic acid lesions of teh middle temporal area (MT) in the behaving Rhesus monkey. Soc. Neurosci. Abstracts, in press.

Sterling, P. and Wickelgren, B.G. (1969) Visual receptive fields in the superior colliculus of the cat. J. Neurophysiol. 32, 1-15.

Subbarao, M. and Waxman, A.M. (1985) On the uniqueness of image flow solutions for planar surfaces in motion. Workshop on Computer Vision: Representation and Control, IEEE Computer Soc. Press: Bellaire, pp 129-140.

Tarr, M.J. and Pinker, S.P. (1985) Nearest neighbors in apparent motion: Two or three dimensions? Proc. Annual Meeting Psychonomics Soc., Boston, p.19.

Ternus, J. (1926) Experimentelle Untersuchung über phänomenale Identität. Psychologische Forschung 7, 81-136. Translated in A Source Book of Gestalt Psychology, (1967) W.D. Ellis, ed., Humanities Press: New York.

Thompson, W.B. and Barnard, S.T. (1981) Lower-level estimation and interpretation of visual motion. IEEE Computer, August, pp. 20-28.

Ullman, S. (1979) The Interpretation of Visual Motion. MIT Press: Cambridge.

Van Essen, D.C. and Maunsell, J.H.R. (1983) Hierarchical organization and functional streams in the visual cortex. Trends in Neurosci. 6, 370-375.

Wallach, H. (1976) On perceived identity: 1. The direction of motion of straight lines. In On Perception, Wallach, H., ed. Quadrangle: New York.

Wallach, H., Weisz, A. and Adams, P.A. (1956) Circles and derived figures in rotation. Am. J. Psych. 69, 48-59.

Waxman, A.M. and Ullman, S. (1985) Surface structure and 3-D motion from image flow: a kinematic analysis. Int. J. Robotics Res. 4, 72-94.

Waxman, A.M. and Wohn, K. (1985) Contour evolution, neighborhood deformation and global image flow: planar surfaces in motion. Int. J. Robotics Res. 4, 95-108.

SOME CONCEPTS DERIVING FROM THE NEURAL CIRCUIT FOR A HORMONE-DRIVEN MAMMALIAN REPRODUCTIVE BEHAVIOUR

Donald W. Pfaff and Charles V. Mobbs

Laboratory of Neurobiology and Behavior
The Rockefeller University
1230 York Avenue
New York, NY 10021 U.S.A.

One purpose for analyzing the neural and endocrine mechanisms for an individual behaviour in considerable cellular and molecular detail is to provide the empirical platform for increasingly precise investigations of nerve cell function. Thus, with respect to neuroendocrine mechanisms of interest for reproductive behaviour, we use cDNA/RNA in situ hybridization as a measure of gene expression in hypothalamic cells, for looking at levels of ribosomal RNA (Jones, Chikaraishi, Harrington, McEwen and Pfaff, 1986), LHRH messenger RNA (Shivers, Harlan, Hejtmancik, Conn and Pfaff, 1986; Rothfeld, et al., 1986), and the message for proenkephalin (Romano et al., 1986). Proteins synthesized in ventral medial hypothalamic neurones and transported to the midbrain may be of direct relevance for the control of female reproductive behaviour (Mobbs, Harlan and Pfaff, 1985, 1986), for example through influences on the electrical excitability of midbrain central gray neurones (Sakuma and Pfaff, 1980).

A second purpose for such a coherent set of investigations is to yield general formulations, and see how these fit with concepts of how ensembles of nerve cells function as a system. Can they address long-standing conceptual questions about how brain governs behaviour? In this chapter we deal primarily with this latter aim. The most fundamental questions in neurobiology are to explain why a given behaviour occurs, and, at that point in time, why other behaviours do not. We have solved the first half of this problem at a level of cellular detail greater than for any other mammalian behaviour. Below, we will briefly summarize (Section I) what has been shown (and presented elsewhere in much greater detail) of the neural mechanisms for the female reproductive behaviour, lordosis, and the effects of estrogen and progesterone on the hypothalamic cells that control the overall behavioural circuit. Then, we will discuss (Sections II and III) mathematical and conceptual questions brought into the focus allowed by the solid neural and neuroendocrine base of information provided.

I. STEROID HORMONE BINDING, AND NEURAL CIRCUITRY FOR LORDOSIS BEHAVIOUR

Using steroid hormone autoradiography, precise locations of neurones with steroid sex hormone receptors have been mapped. Estrogen-concentrating neurones, androgen-concentrating neurones and, in a few cases, progestin-concentrating neurones have been studied across a wide range of vertebrate species. The original work (Pfaff, 1968) and the greatest amount of autoradiographic work has been done with estrogen-binding neurones in mammals. However, certain findings about the neuroanatomical distribution of estrogen-, androgen- and progestin-concentrating cells remain true across all the vertebrate species studied, indicating an orderly, lawful development of such hormone concentrating cells in the vertebrate brain (references in Morrell and Pfaff, 1978; Pfaff, 1980). First, in representative species from all major vertebrate classes, neurones specifically concentrating steroid sex hormones can always be detected autoradiographically. Second, in all species, these steroid hormone concentrating cells can be found in the medial preoptic area, in cell groups of the basomedial (tuberal) hypothalamus, in specific limbic forebrain structures such as the medial amygdala and lateral septum, and in a specific area of the mesencephalon deep to the tectum (Fig. 1). Thirdly, papers on the autoradiographic findings in each species (references in Morrell and Pfaff, 1978; Pfaff, 1980) include evidence from the endocrine and behavioural physiology of that species that nerve cell groups which bind estrogenic or androgenic hormones actually participate in the control of hormone modulated functions. These physiological findings usually emphasize the participation of such cell groups in the regulation of gonadotropin release or mating behaviour.

Using our knowledge of estrogen and progestin binding cells, and focussing on a relatively simple, stereotyped, hormone-driven reproductive behaviour has not only allowed a coherent series of investigations of steroid hormone actions on nerve cells, but also has permitted discovery of the first circuit known for a mammalian behaviour. Lordosis behaviour, the

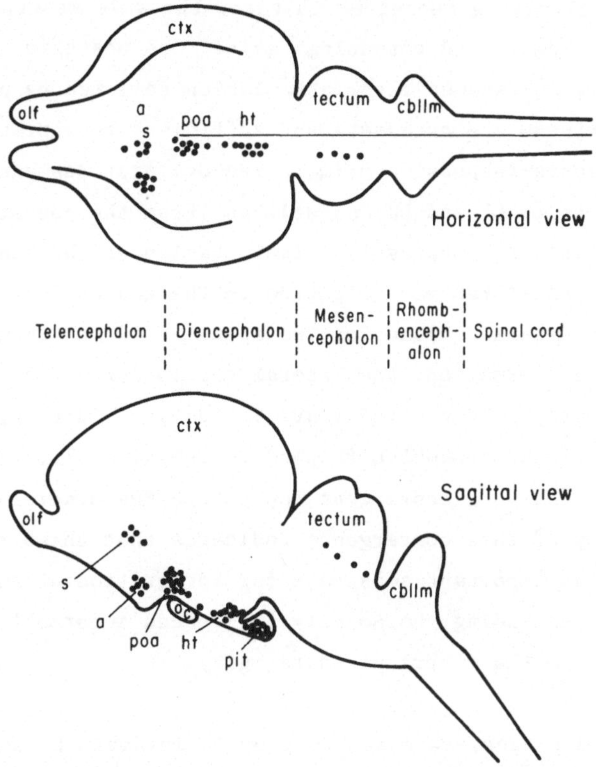

Fig. 1. Locations of estrogen and androgen binding neurones which are constant across all vertebrate species studied, represented on a drawing of a generalized vertebrate brain. Vertebrate brain schematic is presented looking from the top at a horizontal view, and looking from the left side (sagittal view). Clusters of black dots indicate cell groups which are always found in our lab by steroid autoradiography to contain estrogen and androgen binding neurones. Prominent are hormone binding cells in the preoptic area (POA), the tuberal cell groups of the hypothalamus (HT), and in structures of the phylogenetically ancient limbic system such as the medial amygdala (a) and the lateral septum (s). (From Morrell and Pfaff, 1978).

primary reproductive behaviour of female rodents, is strongly estrogen and progestin dependent, and is required for fertilization. We are able to follow the flow of neural activity through the nerve cell groups responsible for lordosis, all the way from application of stimuli by the male to muscular contraction by the female (Pfaff, 1980; Pfaff and Schwartz-Giblin, 1986). This proves that it is possible to explain the neural circuitry responsible for initiating an entire mammalian behaviour.

During natural mating behaviour in rats, the male mounts by grasping the flanks of the female and thrusting against the posterior rump, tail base and perineum. Cutaneous flank stimulation followed by pressure on the tail base and perineum are necessary and sufficient for lordosis to occur. The relevant pressure-responsive primary sensory neurones enter the spinal cord over dorsal roots L1 and L2, as well as (from the posterior rump and perineum) L5, L6 and S1. Supraspinal facilitation of the behaviour is required. Ascending fibres participating in the supraspinal loop travel through the anterolateral columns of the spinal cord and terminate in the medullary reticular formation, the lateral vestibular nucleus, and in the midbrain central gray. In the medullary reticular formation, virtually all the reticulospinal neurones which respond to behaviourally adequate sensory stimulation also receive a convergent input from the midbrain central gray -- the specificity of this convergence indicates that these reticulospinal neurones provide an important substrate for integration of somatosensory information with descending, hormonally-influenced information from the hypothalamus through the midbrain central gray.

Estradiol and progesterone act to promote lordosis by local effects on neurones in the ventromedial nucleus of the hypothalamus. Estrogen can increase the electrical activity of slowly firing neurones in this region, in part by increasing responsiveness to muscarinic cholinergic agonists. Estrogen also increases the expression of the gene for ribosomal RNA, yields ultrastructural changes consistent with new protein synthesis, and increases the appearance of certain proteins, notably one with molecular weight about 70,000 (Mobbs and Pfaff, 1985, 1986). Control over lordosis by these medial hypothalamic neurones is exerted through a tonic output to the midbrain. Axons from ventromedial hypothalamic neurones reach the midbrain central gray via a periventricular as well as a sweeping lateral route, with the contribution of the lateral-running axons being more important. It appears that it is through a combination of electrical and secretory changes in ventromedial hypothalamic neurones that the electrical excitability of behaviourally relevant dorsal midbrain neurones is raised (Pfaff, 1983).

Midbrain neurones facilitate lordosis over a much faster time course than do those of the medial hypothalamus. Some of these midbrain central gray cells send descending axons to the medullary reticular formation, and electrical stimulation of the central gray greatly potentiates the effects of reticulospinal neurones on the deep back muscles important for lordosis.

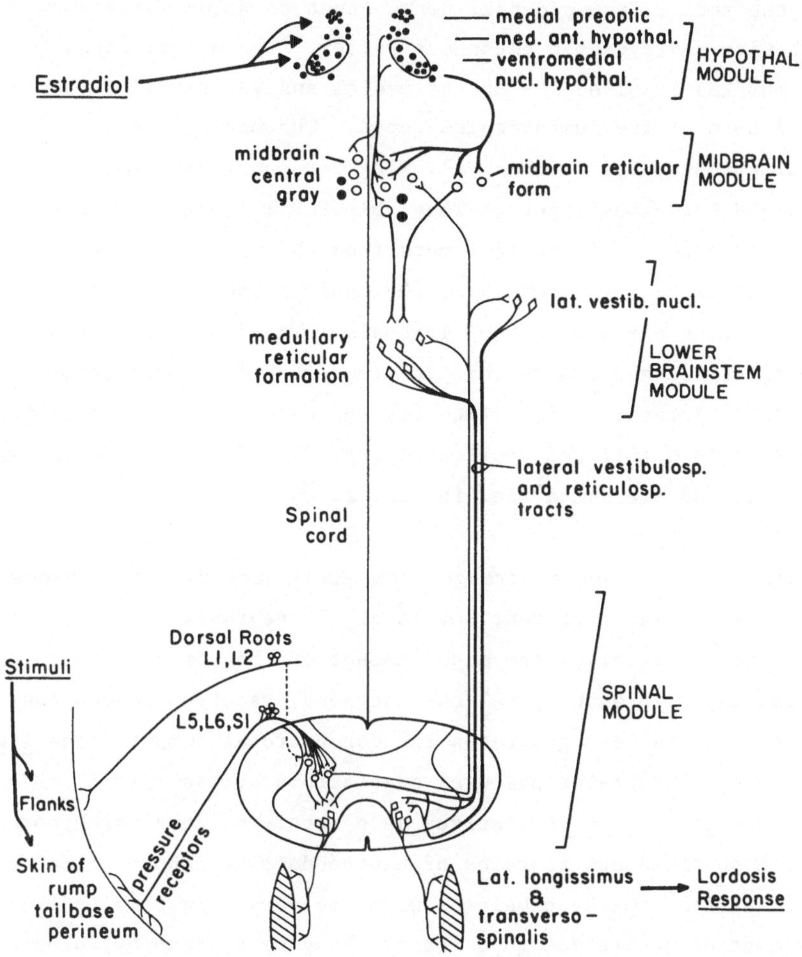

Fig. 2. Neural circuitry for estrogen-progestin dependent lordosis, the primary reproductive behavioural response in the female rodent. For clarity of presentation, mechanisms are represented on only one side; all stimuli, hormone actions and responses are bilaterally symmetrical. (From Pfaff and Schwartz-Giblin, 1986).

In the lower brainstem, the two tracts required for lordosis are the lateral vestibulospinal tract and the medullary reticulospinal tract, and these two groups of cells synergize in their effects on the behaviourally relevant deep back muscles. The properties of these two descending systems fit perfectly the requirements for control of lordosis as a whole (Pfaff, 1980; Pfaff and Schwartz-Giblin, 1986). Lateral reticulospinal and vestibulospinal tracts prepare lumbar spinal circuits for lordosis initiation. They have monosynaptic connections to the motoneurones for deep back muscles in the rat, but may also operate through an interneurone.

Timing of the reflex from pudendal nerve input to motor nerve output is consistent with a disynaptic cutaneous reflex. The motoneurones for the deep back muscles involved lie on the medial and ventro-medial borders of the ventral horn of the lumbar spinal cord. The monosynaptic reflex of these axial motoneurones is weak, allowing for powerful effects of descending and cutaneous inputs. Thus, bilateral inputs from the skin of the flanks followed by bilateral inputs from the pudendal nerve, whose effects are amplified by lateral vestibulospinal and reticulospinal actions, determine the strength of lordosis. The deep back muscles, lateral longissimus and transversospinalis, execute the vertebral dorsiflexion of lordosis. The circuitry and mechanisms thus described have been presented in detail and replicated (Pfaff, 1980; Pfaff and Schwartz-Giblin, 1986), and are summarized in Fig. 2.

Mechanisms for lordosis arrange themselves according to obvious modules represented at different levels of the neuraxis (Fig. 2). The spinal cord module receives the major impact of the somatosensory input, filters that input in each spinal cord segment, receives descending facilitating signals, and generates the motoneuronal output. The lower brainstem module integrates postural adaptations across spinal cord segments. The midbrain module serves as a receiving zone for hypothalamic and preoptic peptides and proteins of neuroendocrine import. It translates signals coming from the hypothalamus with the very slow time course typical of neuroendocrine mechanisms into faster changing electrophysiological signals typical of the rest of the nervous system. Hypothalamic input facilitates midbrain neuronal responses to synaptic input and, in turn, hierarchically the midbrain module facilitates reticulospinal cells in the lower brainstem module. The hypothalamic module, the most rostral component which is required for activating lordosis, adds the endocrine control component to this behavioural mechanism. Steroid hormone binding, and the effects of estrogens and progesterone on electrical, transcriptional and protein synthetic mechanisms here, primarily account for the steroid hormone dependency of the neural circuit and behaviour as a whole.

One important consequence of defining a complete behavioural circuit is that it provides a base for more precise thinking about how nerve cells control mammalian behaviour. Hypotheses about neural circuits and integrated behaviours need a formal language for their accurate expression, but what language is suitable? One approach used a simple form of

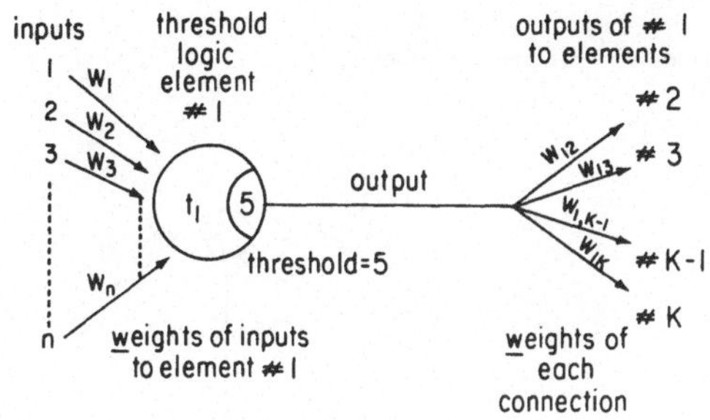

Fig. 3. Drawing of a threshold logic element used in the threshold logic form of network theory (Dertouzos, 1965). Here, element number 1 is pictured with inputs W_1-W_n. The weights of the inputs are shown, and the threshold illustrated = 5. The weights of the outputs of this threshold logic element to other elements may have different values, as shown. (From Cohen and Pfaff, 1985).

mathematical logic called "threshold logic" to express formally some of our knowledge of the neural circuit for lordosis (Cohen and Pfaff, 1985). Threshold logic describes the performance of circuits of threshold elements, each element performing a function more powerful than a conventional diode or transistor gate (Dertouzos, 1965). Complicated system functions can usually be achieved with fewer components in threshold logic than can circuits composed of large numbers of conventional gates. An attractive aspect is the visual similarity of threshold elements to neurones (Fig. 3). Each threshold element is a multiple-input, single-output device, without memory, which has been useful in some branches of information processing theory. Each input to a threshold element is associated with a real number called its weight, which can easily be envisioned as the strength of a synaptic connection. The output of each threshold element is zero unless the weighted sum of the inputs equals or exceeds a numerical threshold. The analogy to the nerve cell is clear. Taking the empirically discovered neural circuit for lordosis (Fig. 2) and representing nerve cells and the appropriate connections between them with threshold logic elements as shown in Fig. 3, one comes to the threshold logic circuit shown in Fig. 4. The weights of the connections between threshold logic elements and the numerical threshold for each element were chosen to reflect the electrophysiological results which were the basis for the lordosis circuit (Pfaff, 1980). Construction of the truth tables which reflect the binary performance of the circuit (Cohen and Pfaff, 1985) faithfully showed the domination of the behavioural output by the requirement for a combination of adequate sensory plus adequate hormonal

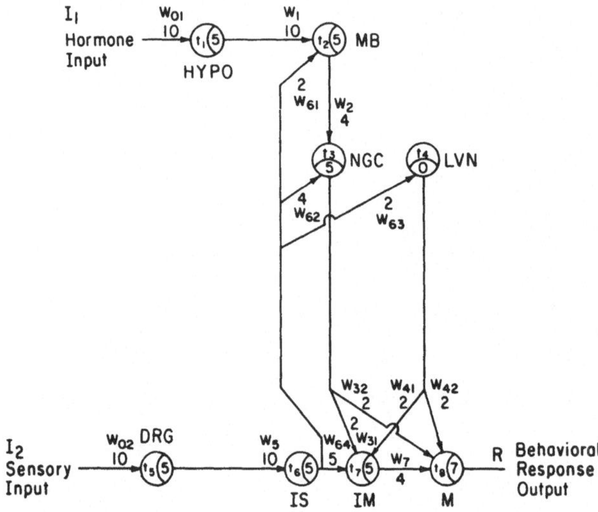

Fig. 4. Neural circuitry for lordosis represented in a form to submit to the calculations of threshold logic. Each input and element is named according to the abbreviations shown, and weights and thresholds assigned to reflect electrophysiological and behavioural results. (From Cohen and Pfaff, 1985).

input, and also demonstrated the central integrative role of the motoneurone in a behavioural circuit. One also was struck, however, by the rigidity of a binary representation, and so it also seemed wise to construct equations in which the output of each threshold logic element in the circuit equals the sum of the products of each of the element's inputs and its respective weight minus that element's threshold (Cohen and Pfaff, 1985). For the behaviourally relevant circuit this yields an equation for each element, so that the performance of the entire network can be modeled by a set of simultaneous equations. Combining these simultaneous equations yields a general equation for the entire network. With the thresholds and weights supplied in Fig. 3, this formula can be reproduced to an output value (with quantification in arbitrary units) as a function of Input 1 (hormone input) and Input 2 (sensory input). Graphs representing this simplified equation are shown in Fig. 5. When the output of the network (R) is plotted as a function of the value of the hormonal input, the low slopes, true for any value of sensory input, are surprising. In contrast, when the output is plotted as a function of the magnitude of the sensory input, the slopes are much higher. Despite the many advantages of this model, according to this result sensory input drives the behaviour with much greater quantitative power than does the hormonal facilitation. This

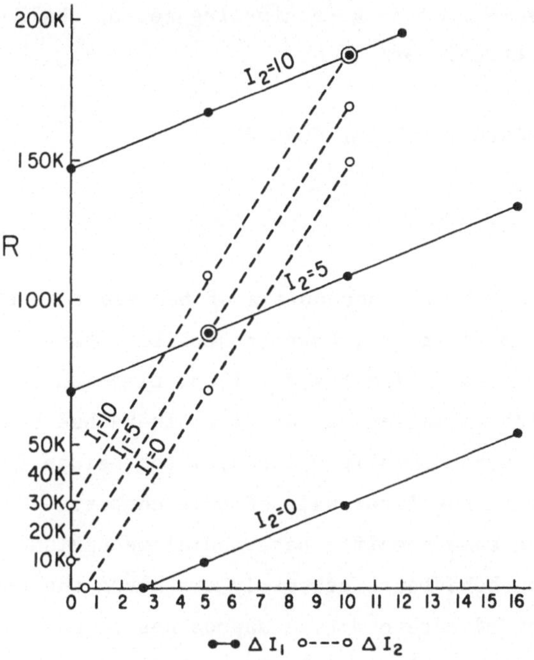

Fig. 5. Magnitude of behavioural response output (R), quantified in arbitrary units, as a function of changes in the magnitude of hormonal input (I_1) and sensory input (I_2). When I_1 is varied in this graph, I_2 is the parameter given values of 0, 5 or 10 (solid lines). When I_2 is varied, I_1 is the parameter given values of 0, 5 or 10 (dashed lines). Surprisingly, with the circuitry values shown in Fig. 4, response output was much more sensitive to changes in sensory input than hormonal input. The possible implications of this calculation for our conceptions of hormone-dependent hypothalamic output mechanisms are discussed in the text. (From Cohen and Pfaff, 1985).

contrast between the quantitative power of hormonal and sensory inputs does not fit the facts, since the reproductive behaviour of female rodents depends absolutely on physiological estrogen levels, supplemented by progesterone. We suspect that the facilitating role of hormone-dependent hypothalamic output is underestimated in this model -- the synthesis of hypothalamic proteins and peptides under control of steroid hormones probably has physiological importance beyond what is represented in this threshold logic model, including possible trophic actions (Pfaff and Cohen, 1986). The exact means through which hypothalamic axons, carrying newly synthesized peptides as well as electrical signals, facilitate midbrain neuronal control of lordosis -- and the formal mathematical representation of these mechanisms -- provide a fascinating set of challenges for new empirical and theoretical work.

II. CONCRETE APPROACHES TO GLOBAL CONCEPTS

Motivation

For research on cellular mechanisms of behaviour in the mammalian brain, concepts of "motivation", however sensible, have not been in fashion for at least two decades. With the detailed, time-consuming experiments required to deal with mammalian nerve cells, there has been scant attention paid to more global brain-behaviour concepts that arose from biological and behavioural studies in the first half of this century. Now, with a solid base of knowledge on some specific behavioural mechanisms -- hormone controlled mammalian behaviours certainly are among the best examples -- we have the possibility of more concrete approaches to the broad questions of the mechanisms by which motivational signals direct behaviour.

The logical necessity of the concept of "motivation" derives from the following type of reasoning (Pfaff, 1982). Given a mammalian organism, studied in a specific context, suppose that it is presented with a well defined stimulus and the well defined response expected does not occur. A short time later (much too soon for changes due to maturation) the same organism in the same context (with no reinforcement or learning) at the same time of day is presented with the same stimulus and the response does occur. What explains the existence of a response on the second trial? A change in the internal state of the organism, called motivation, must be inferred. Thus, as an intervening variable, the concept of motivation identifies a logically required cause of behavioural change.

This spare, logical treatment of motivation is quite distinct from the rich elaborations of the term that might be adduced when discussing human psychological states, and this careful definition is important, because it permits a mechanistic explanation.

By the reasoning above, the occurrence of lordosis reflects a motivational state (Pfaff, 1982, page 287). An ovariectomized female rat not treated with estrogen or progesterone, though given large numbers of applications of behaviourally adequate somatosensory input during mounts by stud male rats or during reflex tests with skin pressure applied by an experimenter, virtually never performs lordosis. As little as one day later, clearly without changes due to maturation, following a schedule of estrogen and progesterone treatment the same females tested in the same context can respond to exactly the same somatosensory stimuli with strong and frequent lordoses. This well established set of observations requires a variable in the behavioural equation, "sexual motivation", for an explanation of the behavioural change in an input-output manner.

Thus, the cellular mechanisms for estrogen action on lordosis circuitry, briefly summarized above, are in fact mechanisms of a motivational influence. Abstracting and reorganizing the full description of lordosis circuitry (Pfaff, 1980; Pfaff and Schwartz-Giblin, 1986), one can characterize the main physiological mechanisms of this motivational effect of estrogen as follows. Estrogen circulating in the blood is accumulated by cells in and adjacent to the ventrolateral portion of the ventromedial nucleus of the hypothalamus. Receptors in the nucleus of these nerve cells concentrate the hormone. As a result estrogen increases both the electrical excitability of slowly firing nerve cells, and certain biosynthetic capacities. Through the combined action of electrical signals reaching the midbrain central gray and peptides reaching the terminals of these hypothalamic neurones in the midbrain central gray through normal axoplasmic flow, the ventromedial hypothalamic cells prepare midbrain and lower brainstem circuitry for lordosis to occur, given adequate somatosensory input, according to the mechanisms previously described (Pfaff and Schwartz-Giblin, 1986).

Arousal

Arousal as a behavioural concept refers to the obvious cluster of behavioural reactions that we associate with an alert animal. Electrophysiologically, during the 1940's and 1950's, this type of behavioural state was investigated extensively because of correlations with the cortical EEG. Relaxed or drowsy animals had cerebral cortical surface activity dominated by slow waves, while alert animals had an EEG dominated with fast ("beta") wave activity. The anatomical complexity of the brainstem reticular formation which controls EEG electrophysiology, and the apparent designation of reticular circuits as "nonspecific" were discouraging, and questions of arousal received a smaller volume of experimental work in subsequent years.

A new approach to the electrophysiological mechanisms by which reticular neurones affect behaviour was to record single unit activity in unrestrained animals, discern the behavioural responses most closely correlated with significant increases or decreases in the discharge rate of a particular neurone, and deduce the most parsimonious formulation for the manner of that cell's participation (Siegel, 1979). From that type of work, it looked as though many brainstem reticular cells could most easily be conceived as related primarily to the excitation of particular groups of muscles. This is consistent with indications from our electrical recordings in axial muscles which, in any case, are correlated with the behavioural changes usually associated with "arousal".

We found that changes in the spontaneous electromyographic activity of the deep back muscle lateral longissimus were highly correlated with the occurrence of changes in the cortical EEG, in urethane anaesthetized rats (Sullivan, Schwartz-Giblin and Pfaff, 1986). Increased muscle electrical activity was always accompanied by EEG desynchronization (fast wave activity), while decreased muscle activity always was accompanied by EEG synchronization (slow wave activity). Moreover, pudendal nerve stimulation (representing cutaneous input from the skin region involved in triggering lordosis) was often able to desynchronize slow wave EEG within five seconds. In 88% of the cases in which stimulation of the pudendal nerve desynchronized EEG, the stimulus also evoked the electrical activity of motor units in the lateral longissimus. In dramatic contrast, if the slow wave EEG was not desynchronized, motor units were never evoked. In terms of the order of electrical changes, the EEG desynchronization always preceded the onset of motor unit activity.

We conceive of this pattern of electrophysiological changes as being due to a spino-bulbo-spinal reflex with reticulospinal cells at the top of the loop, and that these cells are responsible both for the changes in arousal and the facilitation of a specific concrete behaviour: deep back muscle dorsiflexion in response to a pudendal nerve input. Individual gigantocellular neurones in the rat medullary reticular formation have been shown to project both caudally to the spinal cord and rostrally to the diencephalon (Scheibel and Scheibel, 1958). Moreover, this reticular neuronal group has single units which have been proven to respond to pressure applied to the perineal skin (signalled by the pudendal nerve) (Kow and Pfaff, 1982). In terms of a concrete behaviour, repeated pudendal nerve stimulation clearly activates deep back muscle motoneuronal responses partly by the spino-bulbo-spinal reflex in which earlier pudendal stimuli will facilitate responses to later pudendal stimuli. The connection to overall arousal depends on the ascending projections of the same reticular neurones -- activating more rostral brainstem reticular formation and cortical EEG -- as those which facilitate back muscle activity.

III. SETTING IN A MORE GENERAL NEUROSCIENTIFIC THEORY

It is possible that the facts bearing on the development and the adult operation of the neural circuits for hormone-driven mammalian behaviours fit into a group selection theory of higher brain function (Edelman, 1978). Such a neuroscientific theory states that the brain selects sensory-motor information "through the temporally coordinated interactions of collections or repertoires of functionally equivalent units consisting of small groups of neurones" (Edelman, 1978, page 52). The brain may also process information and stored signals in a phasic way, which allows reentrant signalling to amplify the neural forces leading to specific behavioural states. On the continuum between selective and instructive models of brain function, the purely abstract extremes cannot exist. There is a formative role for sensory inputs, so the pure selective model cannot be 100% correct. Likewise, there are always limits placed by the capacity of the neuronal substrate, so the pure instructive case cannot exist. Nevertheless, the facts of reproductive neuroendocrinology may well favour a selective model in which steroid hormones "select" specific cell groups for active control, even in the adult, and, more obviously, during brain development.

Edelman (1978) has applied the group selection theoretical approach to issues of perception, memory and consciousness, all functions which depend heavily on the cerebral cortex -- "Appolonian" aspects of brain function. The areas of neurobiology treated by the research summarized in this chapter are more "Dionysian" -- the cellular explanation of instinctive mammalian behaviours which lead to concepts like motivation and arousal. What applies to one group of functions may apply to the other.

Hormone-dependent Plasticity

To what extent might the study of estrogen effects and lordosis provide any information on higher brain functions? Estrogen alterations of brain function may provide an advantageous model system for studying neuronal plasticity (Mobbs and Pfaff, 1987). Estrogen-responsive neurones regulating lordosis may be viewed as functioning like a switch which "enables" or completes the lordosis circuit. Without estrogen, the connections between flank afferents and back muscle motoneurones are only weakly functional. In the presence of estrogen the efficacy of these connections is greatly enhanced. The time-course of estrogen's effects suggest that new anatomical connections are unlikely to be involved. Thus estrogen appears to modulate the strength of pre-existing neuronal connections, which suggests that the regulation of lordosis may be considered as an expression of neuronal plasticity (relying on William James' use of the term "plasticity" to denote any regulated change of behaviour). Currently most studies of neuronal plasticity focus on anatomical plasticity during development and in response to injury, and the modulation of synaptic efficacy during learning, but clearly the regulation of functional connectivity is a more general problem in neurobiology.

Neuronal plasticity may therefore be viewed in the context of the general question: how much information in biological systems arises from instruction from the environment, and how much information depends on selection? From diversity, all selectionist processes select and amplify an appropriate unit at the expense of inappropriate units. The immune system maintains a pre-existing repertoire of immunoglobulins, and appropriate antibodies are "amplified" upon challenge. A selectionist view of neuronal functions, for example the formation of memories in response to environmental input, suggests that a superabundance of anatomical connections already exists largely independent of environmental influences, and that environmental input selects and amplifies appropriate connections. A contrasting instructional view would be that environmental influences directly cause new connections to develop, and that most or all anatomical connections which regulate responses to environmental signals originally arose in response to environmental influences. Thus studies guided by the selectionist view would tend to focus on modulation of pre-existing connections, whereas studying the formation of new connections would be suggested by the instructional view. The study of estrogen-dependent lordosis, as a model for the regulation of neuronal functional connectivity, may draw upon data and hypotheses informed by the selectionist view.

What generally argues in favor of the selectionist view of neuronal function? First, mechanisms of evolution and immune response may share enough properties with the nervous system, particularly with respect to thermodynamic constraints, at least to rationalize an analogy. Second, studies of the development of synaptic connections strongly suggest that such development takes place largely by selection from an abundance of connections, and that similar processes could be modulated by environmental signals. Third, studies in humans and in birds suggest that the acquisition of speech and birdsong takes place by selecting specific subsets of grammars within a pre-existing template. Fourth, recent studies of the readiness potential suggest that volitional control arises from a conscious selection of a single action from many potential actions generated by lower brain processes. Finally, the neuroanatomy of higher processing centres suggests the kind of pre-existing superabundance of connections which is required by selectionist theories, and has led to the formulation of the group selection theory of higher neuronal function (Edelman, 1978).

To illustrate possible commonalities between hormone-dependent lordosis and higher function, we will compare our postulated mechanisms of lordosis with the mechanisms of learning and consciousness postulated by Edelman's group selection theory. In highly simplified form, the group selection theory supposes that the units of association in the cortex are groups of neurones, with connections within and between neurones in each group. Each group can resonate with signals generated outside the group, either from the environment or by another group. Furthermore, groups are similar to but different from each other, such that for any given signal, several groups can resonate, and any given group can resonate to any of a range of signals. The key to memory is that any given environmental signal would cause a number of groups to resonate together. Then, by an unspecified mechanism, the association of these resonant groups would be stabilized, presumably by stengthening the previously weak connections. Stimulating any subset would then tend to stimulate the entire associated set of groups by recalling the resonance, thus recreating the initial signal.

The most direct commonality between lordosis and the group selection theory is the requirement for strengthening a previously weak synaptic connection. Less obviously, both processes require signal discrimination which relies on associated neurones operating as an ensemble rather than, for example, a simple monosynaptic reflex. Thus elucidation of our simpler neuroendocrine mechanisms would likely lead to a better understanding of those aspects of higher function which involve similar underlying mechanisms: ensemble discrimination and modulation of neuronal connectivity.

In the adult case, then, steroid hormones, estrogens and progestins, can be conceived as selecting groups of individual sex hormone concentrating neurones which, by their elevated electrical activity and peptide secretion select, in turn, midbrain and brainstem groups of neural circuits which facilitate the relevant motoneurones for reproductive behaviour. Developmental aspects of neuroendocrinology are theoretically more easily approached by Edelman's concept of "secondary repertoire" (Edelman, 1978, page 65): neuronal groups whose synaptic function has been altered by selection during previous experience. Clearly, elevated levels of androgens in the blood during the neonatal period in the rat lead to a permanent commitment of certain hypothalamic cell groups to masculine-typical behaviour, and select against feminine aspects of neuroendocrine function, both with respect to ovulation and behaviour.

Comparison to Elementary Theoretical Requirements

The experimental facts deriving from the large number of studies now performed on estrogen and progestin effects on brain and the control of reproductive behaviour comport well with the theoretical requirements for a theory based on selection (Edelman, 1978, page 92). For example, among the predictions of this theory, it is clear that groups of hormone-dependent cells, not individual cells, govern reproductive behaviour circuitry; steroid concentrating cells even in specific anatomical regions have not been shown to have highly individualized hormone binding characteristics (Morrell and Pfaff, 1986). Steroid binding cell groups are indeed multitudinously represented, and hormone binding neuronal groups project to each other densely, allowing for all the functional convergence required by selection theory (Pfaff and Keiner, 1973; Cottingham and Pfaff, 1986). Pontifical neurones, representing single cell decision units, have not been found. During brain development, through a process of controlled cell death, circuits adequate for directing masculine or feminine reproductive function are determined, but there is no indication that this is controlled on a synapse-by-synapse basis.

Furthermore, the conditions for falsification of a theory based on selection of neuronal groups have not been found to obtain. There is no indication that single cells are capable alone of handling the signalling involved in this neuroendocrine behavioural circuit. Removal of single cells or very small numbers of cells typically is not effective in blocking the effects of steroid hormones on lordosis: large lesions at various points in the lordosis circuit are required (Kow and Pfaff, 1982; Zemlan, Kow and Pfaff, 1983). Since we have been able to conceive the different contributions of modules in the overall lordosis neural circuit, we have no indication that the cell groups treated are merely redundant. There is no problem accounting for the range or specificity of stimulus-behaviour relations with the neuroendocrine mechanisms as stated (Pfaff, 1980). The very characteristic of a stereotyped, "instinctive" reproductive behaviour is a low range of implied stimulus-response relations. Finally, the rich intrinsic connectivity in the hypothalamic cell groups related to lordosis (Nishizuka and Pfaff, 1983) and its plasticity according to hormonal influences in the adult (Matsumoto and Arai, 1979; Chung, Pfaff and Cohen, 1986a,b) give no indication of a level of determinacy which would deny a theory based on selection. In summary, the results developing from the study of the best analyzed mammalian behaviour mechanism -- steroid hormone-driven lordosis -- fit quite well, so far, with a global approach to brain function based on the notion of the brain as a selective system.

REFERENCES

Chung, S., Pfaff, D. and Cohen, R. (1986a) Neuroscience, submitted.

Chung, S., Pfaff, D. and Cohen, R. (1986b) Neuroscience, submitted.

Cohen, R.S. and Pfaff, D.W. (1985) Cell biological and math-logical theories of the neural circuit for steroid-dependent female reproductive behaviour. Integr. Psychiatry 3, 262-279.

Cottingham, S. and Pfaff, D.W. (1986) Interconnectedness of sex steroid-binding neurons: theoretical implications. In: Current Topics of Neuroendocrinology, Vol. 7, in press.

Dertouzos, M.L. (1965) Threshold Logic: A Synthesis Approach. MIT Press: Cambridge

Edelman, G.M. and Mountcastle, V.B. (1978) The Mindful Brain: Cortical Organization and the Group-Selective Theory of Higher Brain Function. MIT Press: Cambridge

Jones, K.J., Chikaraishi, D.M., Harrington, C.A., McEwen, B.S. and Pfaff, D.W. (1986) Estradiol (E_2)-induced changes in rRNA levels in rat hypothalamic neurons detected by in situ hybridization. Mol. Brain Res. submitted.

Kow, L.-M., Montgomery, M. and Pfaff, D.W. (1977) Effects of spinal cord transections on lordosis reflex in female rats. Brain Res. 123, 75-88.

Kow, L.-M. and Pfaff, D.W. (1982) Responses of medullary reticulospinal and other reticular neurons to somatosensory and brainstem stimulation in anesthetized or freely-moving ovariectomized rats with or without estrogen treatment. Exp. Brain Res. 47, 191-202.

Matsumoto, A. and Arai, Y. (1979) Synaptogenic effect of estrogen on the hypothalamic arcuate nucleus of the adult female rat. Cell Tissue Res. 198, 427-433.

Mobbs, C.V., Harlan, R.E. and Pfaff, D.W. (1985) An estradiol-induced protein synthesized in the ventral medial hypothalamus (VMN) and transported to the midbrain central gray (MCG). Soc. Neurosci. Abstracts, 11, 1271.

Mobbs, C.V., Harlan, R.E. and Pfaff, D.W. (1986) An estradiol-induced protein in the hypothalamus. J. Neurosci. submitted.

Mobbs, C.V. and Pfaff, D.W. (1987) Estrogen regulated neuronal plasticity. In Strauss, G. and Pfaff, D. Molecular Neurobiology: Endocrine Approaches. Academic Press: New York, in press.

Morrell, J.I. and Pfaff, D.W. (1978) A neuroendocrine approach to brain function: Localization of sex steroid concentrating cells in vertebrate brains. Amer. Zool. 18, 447-460.

Morrell, J.I., Krieger, M.S. and Pfaff, D.W. (1986) Quantitative autoradiographic analysis of estradiol retention by cells in the preoptic area, hypothalamus and amygdala. Exp. Brain Res. 62, 343-354.

Nishizuka, M. and Pfaff, D.W. (1983) Pattern of synapses on ventromedial hypothalamic neurons in the female rat: An electron microscopic study. Soc. Neurosci. Abstracts, 9, 317.

Pfaff, D.W. (1968) Uptake of estradiol-17β-H^3 in the female rat brain. An autoradiographic study. Endocrinology 82, 1149-1155.

Pfaff, D.W. (1980) Estrogens and Brain Function: Neural Analysis of a Hormone-Controlled Mammalian Reproductive Behavior. Springer: New York.

Pfaff, D.W. (Editor) (1982) The Physiological Mechanisms of Motivation. Springer: Heidelberg.

Pfaff, D.W. (1983) Impact of estrogens on hypothalamic nerve cells: Ultrastructural, chemical, and electrical effects. Recent Progress in Hormone Research 39, 127-179.

Pfaff, D.W. and Cohen, R.S. (1986) Estrogen has trophic effects on hypothalamic neurons, and on the regions in which they synapse. In: Leung et al., Plenum: New York, in press.

Pfaff, D.W. and Keiner, M. (1973) Atlas of estradiol-concentrating cells in the central nervous system of the female rat. J. Comp. Neurol. 151, 121-158.

Pfaff, D.W. and Schwartz-Giblin, S. (1986) Physiological mechanisms of female reproductive behaviour. In Textbook of Physiology. Knobil, E. and Neill, J., eds. Raven Press: New York, in press.

Romano et al. (1986) Soc. Neurosci. Abstracts, in press.

Rothfeld et al. (1986) Soc. Neurosci. Abstracts, in press.

Sakuma, Y. and Pfaff, D.W. (1980) Excitability of female rat central gray cells with medullary projections: Changes produced by hypothalamic stimulation and estrogen treatment. J. Neurophysiol. 44, 1012-1023.

Scheibel, M.E. and Scheibel, A.B. (1958) Structural substrates for integrative patterns in the brainstem reticular core. In Jasper, H.H., Proctor, L.D., Knighton, R.S., Noshay, W.C. and Costello, R.T., eds., Reticular Formation of the Brain, Little Brown: Boston, pp 31-55.

Shivers, B.D., Harlan, R.E., Hejtmancik, J.F., Conn, P.M. and Pfaff, D.W. (1986) Localization of cells containing LHRH-like mRNA in rat forebrain using in situ hybridization. Endrocrinology 118, 883-885.

Siegel, J.M. (1979) Behavioral Functions of the Reticular Formation. Brain Res. Reviews 1, 69-105.

Sullivan, J.M., Schwartz-Giblin, S. and Pfaff, D.W. (1986) Correlations between EEG state and spontaneous and evoked axial muscle EMG. Brain Res. 368, 197-200.

Zemlan, F.P., Kow, L.-M. and Pfaff, D.W. (1983) Effect of interruption of bulbospinal pathways on lordosis, posture, and locomotion. Exp. Neurol. 81, 177-194.

THE NEURAL DETERMINATION OF SKELETAL MUSCLE FIBRE CHARACTERISTICS

T.P. Feng

Shanghai Institute of Physiology
Academia Sinica
Shanghai, China

ABSTRACT

This lecture presents the results of a series of experiments in search of evidence for a role of some specific trophic or chemical influence in addition to impulse activity in the neural determination of skeletal muscle fibre characteristics. An indirect approach to the problem through experiments on muscle fibres doubly innervated by both slow and fast nerves was first attempted. This led to the finding that the Z-band width of the fast EDL muscle fibres can be transformed within two weeks after cross-innervation by the slow soleus nerve. Using the Z-band as the test characteristic, cross-innervation experiments were performed under the condition of muscular inactivity produced by chronic TTX nerve block and by spinal cord transection. The experiments agree in showing that an inactive soleus nerve can still exert a transforming effect on the Z-band width of the EDL muscle fibres it cross-innervates, thus proving that impulse activity cannot be the sole neural determinant of skeletal muscle fibre type at least for one particular type-specific characteristic. In the light of this new development some of the interesting and relevant results given by doubly innervated muscle fibres are discussed.

Mature mammalian skeletal muscle fibres are distinguishable into several types. There are several schemes of classification of muscle fibres (Brooke and Kaiser, 1970; Peter, Barnard, Edgerton, Gillespie and Stampel, 1972; Burke, Levine, Tsairis and Zhajac, 1973), but for the purpose of this lecture, we need only adopt the simplified classification of muscle fibres into two broad types: slow-twitch or type I fibres, and fast-twitch or type II fibres. Normally each type of muscle fibre has its distinct set of correlated physiological, histochemical, biochemical and ultrastructural characteristics. It is now a well-established fact that the motoneurone exerts a determining influence on the type-specific properties of the muscle fibres it innervates. For instance, after cross-reinnervation between a fast-twitch and a slow-twitch muscle, their type-specific characteristics tend to undergo reciprocal changes (Close, 1969; Edgerton and Cremer, 1981); and all the muscle fibres innervated by the same motoneurone are of the same type (Nemeth, Solanki, Gordon, Hamm, Reinking and Stuart, 1986). But how does the motoneurone determine the muscle fibre characteristics? This is a question which was clearly brought forth and carefully considered by Buller, Eccles and Eccles in their pioneer work in 1960 and has been the subject of numerous studies since then. Two possibilities were conceived from the outset. One is that each type of motoneurone may determine the corresponding type of muscle fibres through its characteristic pattern of impulse activity producing a corresponding pattern of muscle activity. We may for brevity refer to this as the activity hypothesis. The other possibility is that it may do so through some specific trophic or chemical influence which can act independently of nerve impulses. This may be referred to as the chemical hypothesis. The activity hypothesis has received support from chronic stimulation experiments. It has been repeatedly shown that the superimposition of a slow pattern (10 Hz) chronic stimulation on a fast muscle through its nerve can make the fast muscle acquire the characteristic properties of a slow muscle (Salmons and Sreter, 1976). It has also been shown with direct stimulation of denervated muscle that intermittent high frequency (100 Hz) stimulation makes the normally slow muscle fast, while continuous low frequency (10 Hz) stimulation makes it slower than normal, indicating that the pattern of muscle activity per se, in the absence of any nerve influence, is capable of determining the contractile properties of skeletal muscle (Lømo, Westgaard and Engebretsen, 1980). In comparison the chemical hypothesis is so far in a much weaker position experimentally. Experiments using cats involving cross-reinnervation of slow and fast muscles with their nerves silenced by spinal cord isolation were made at the beginning of this line of research by Buller et al. in 1960. They used contraction speed as the test characteristic. The result was negative, i.e. under the condition of spinal cord isolation the usual reciprocal transformation of slow and fast muscles after cross-reinnervation no longer occurred. They did not regard this negative result as contradicting a specific chemical hypothesis of the neural influence on muscle speed and suggested four possible ways by which the negative result could be reconciled to such an hypothesis. Recently Eldridge and Mommaerts (1980) repeated the cat cross-reinnervation experiments with the motoneurones silenced by ventral horn isolation or spinal cord isolation, using in addition to the contraction speed myosin ATPase and LDH activities as test characteristics. They did observe some crossing effects in some experiments but the results were inconsistent, and it was difficult to draw any definitive conclusion.

In recent years, as a part of our interest in nerve-muscle trophic relations, we have made a search for experimental evidence for the chemical hypothesis, starting with a different approach. We did not doubt the importance of impulse activity in the neural determination of skeletal muscle fibre type; the question we asked was whether impulse activity is the sole neural determinant. The direct approach to this question requires the performance of cross-innervation experiments using inactive nerves. At first, wanting a practical method other than spinal cord isolation already tried by others, which could maintain the nerve inactive for the usual long duration of cross-innervation experiments, we adopted an indirect approach,

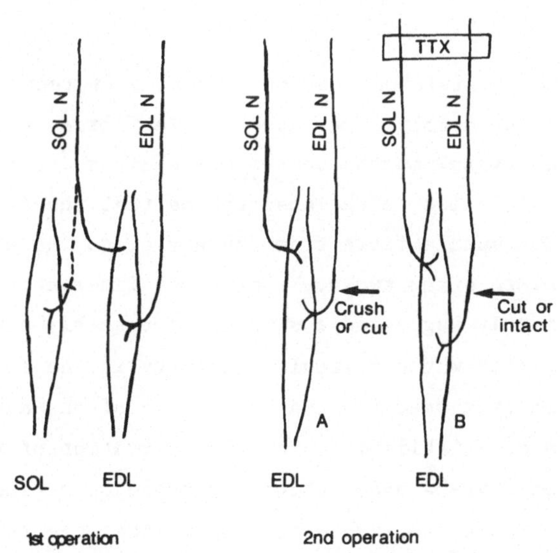

Fig. 1. Two-stage operation to get either simply cross-innervated or doubly innervated muscle fibres under normal condition (2nd operation A) or under TTX block of the nerves (2nd operation B). Usually 3 weeks were allowed to elapse between the two operations.

studying what we called doubly innervated muscle fibres (Feng, Huang, Zhu, Zhou and Qu, 1982; Feng, Zhou and Zhu, 1980, 1982a) (Fig. 1). By a two-stage operation it is possible to give the slow-twitch SOL muscle fibres additional ectopic innervation from the fast EDL nerve, or the fast-twitch EDL muscle fibres additional ectopic innervation from the slow SOL nerve. In either case the muscle fibres are innervated by both slow and fast nerves forming separate endplates, usually 3-5 mm apart. Contraction in twitch fibres being propagated, in such doubly innervated fibres, any effects due to activity would be equalized along the whole fibre. On the other hand, if the different determining influences of the nerves on the type-specific properties of the muscle fibre involve different specific trophic substances released by the nerves, we might find in such doubly innervated muscle fibres local differences in the two endplate regions in respect of some type-specific characteristics. This seemed to be a possibility worthwhile looking into. We therefore examined myosin ATPase histochemically and Z-band width along the length of the doubly innervated muscle fibres. No local differences in the two endplate regions in either of these two characteristics could be seen. Our experiments on the doubly innervated muscle fibres, while failing to fulfil our original objective, nevertheless yielded interesting and relevant results some of which we shall discuss later. For the moment the important point is that they paved the way to a direct approach to the question under study, to which we will now turn.

In pursuing the possible existence of local differences in the two endplate regions in the doubly innervated muscle fibres, we varied the time of observation from several months to two weeks after the initiation of the cross-innervation. This led to the observation that the transformation of the Z-band of the EDL muscle fibre from the narrow to the wide type was already nearly complete after two weeks of cross innervation by the SOL nerve. This immediately suggested a simple but possibly crucial experiment relating to the question whether impulse activity is the sole neural determinant for some type-specific characteristic of the muscle fibre. The experiment consists of establishing cross-reinnervation of the EDL muscle by the SOL nerve for 2 weeks under chronic tetrodotoxin block, i.e. with a completely inactive SOL nerve, and observing whether the Z-band transformation in the EDL fibres still occurs. This experiment was first done with cross-innervation established in a two-stage operation as depicted in Fig. 1, using second operation B (Feng, Zhou and Zhu 1982b). For establishing chronic TTX block, the method of inserting a TTX-filled

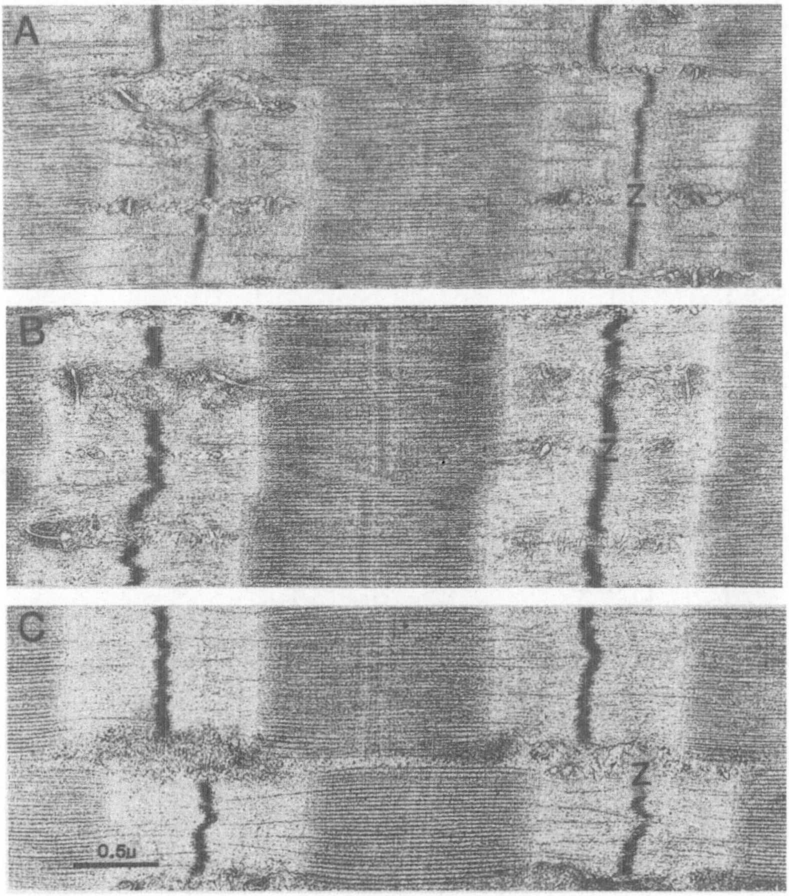

Fig. 2. EDL fibre, showing transformation of its Z-band width after cross-innervation by TTX-blocked SOL nerve. A: its own nerve blocked by TTX for 15 days; B: simply cross-innervated by TTX blocked SOL nerve for 15 days; C: its own nerve intact and additionally cross-innervated by SOL nerve for 15 days, both nerves under TTX block.

capillary under the epineurium of the sciatic nerve (Bray, Hubbard and Mills, 1979) was adopted. For simple cross-innervation, the EDL nerve was cut; for double innervation, it was left intact. The result is shown in Figs. 2, 3 and 4. It is seen that the TTX blocked SOL nerve is still capable of transforming the Z-band of the EDL muscle fibres, though not as completely as a normally active SOL nerve. It is of some special interest to note that in EDL muscle fibres doubly innervated by both TTX-blocked SOL and EDL nerves, or simply cross-reinnervated by TTX-blocked SOL nerve, the degree of the Z-band width transformation attained is very similar. Apparently the inactive EDL nerve seems to exert no trophic effect on the Z-band.

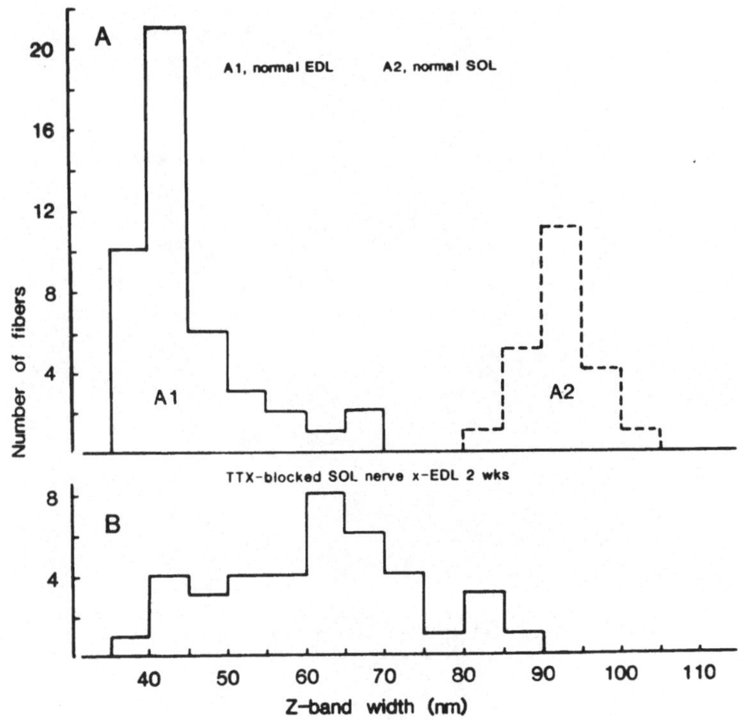

Fig. 3. Z-band width histograms: normal EDL, normal SOL, and EDL cross innervated by TTX-blocked SOL nerve for 2 weeks.

In the above experiment, the cross-innervation was effected through ectopic neuromuscular junctions, and there was a preliminary period of 3 weeks for the foreign nerve to grow and ramify among the muscle fibres. Lest these circumstances might involve some unknown complication, the experiment was repeated with a different method of establishing cross-innervation, namely, the more conventional end-to-end anastomosis between the proximal stump of the SOL nerve with the distal stump of the EDL nerve. The TTX block was started 10 days after the nerve cross-union operation and maintained for 3 weeks, preliminary experiments having shown that within 2 weeks after the nerve-crossing operation there was no response of the EDL to stimulation of the SOL nerve. In this case, cross-innervation presumably took place mostly at the original endplates and there should be only very limited preliminary ramification of the foreign nerve fibres among the muscle fibres. In the previous experiment, cross-innervated muscle fibres, whether simply cross-innervated or doubly innervated, could be identified by finding the ectopic endplate in addition to the original endplate with the help of ChE staining and dissected out for electron microscopy. In the present experiment after ChE staining single muscle

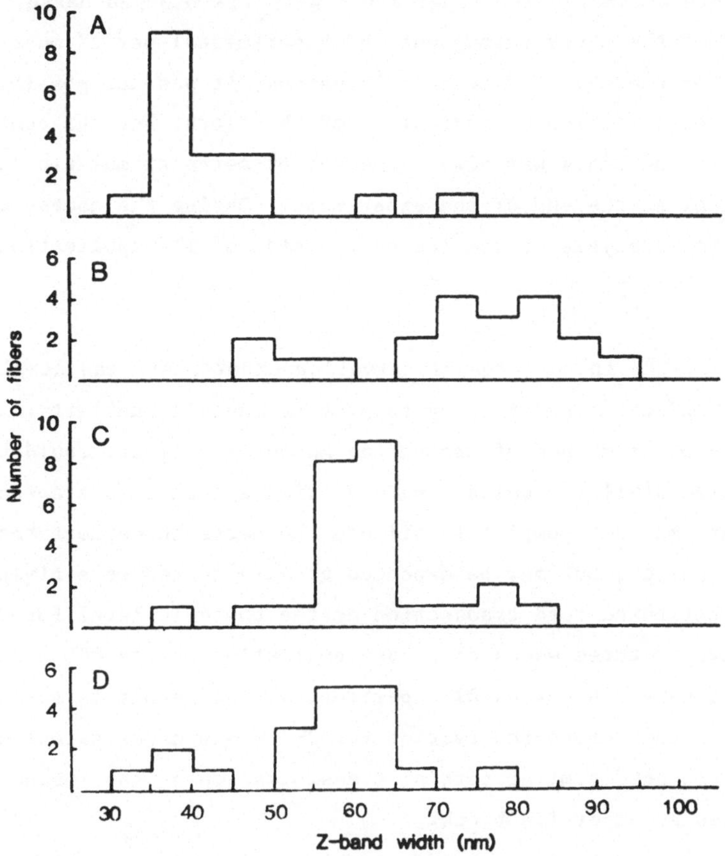

Fig. 4. Z-band width histograms of EDL fibres. A: its own nerve blocked by TTX for 15 days; B: cross-innervated by normal SOL nerve for 15 days; C: cross-innervated by TTX blocked SOL nerve for 15 - 25 days; D: doubly innervated by both SOL and EDL nerves for 15 - 25 days under TTX block.

fibres were taken at random and EM sections of each fibre were made through the endplate which may be denervated or reinnervated. Only those fibres with endplates containing nerve terminals and therefore reinnervatred were used for Z-band measurement. This was a laborious procedure. Useful observations had to accumulate from several animals to make up a good sample. The cumulative result of observations (Fig. 5) indicates the occurrence of a partial transformation of the Z-band width of the EDL muscle fibres by inactive SOL nerve when the cross-innervation was established through an end-to-end anastomosis of the nerve stumps, thus lending support to the conclusion of the previous TTX experiment.

In the cross-innervation experiments with TTX blocked nerve, the completeness of the block throughout the experimental period must be assured. Unfortunately, for technical reasons, it was not possible to carry out direct, continuous monitoring of the block, but the continuous maintenance of the block was always checked by nerve stimulation above and below the block at the end of the experiment. During the course of the experiment the paralysis of the leg on the side of TTX application was observed twice a day.

The use of TTX in our cross-innervation experiments was simply to block nerve conduction and thereby to prevent muscular activity. In principle any other method of preventing muscular activity could be used and should give similar results. We had tried spinal cord transection. This procedure may not completely silence the nerve throughout the whole experimental period, but may be expected greatly to reduce activity. First we simply substituted cord transection at the thoracic level for TTX block during the two to three weeks of cross-innervation of the EDL muscle by the implanted SOL nerve in our usual experiments. The result is shown in Fig. 6. It is seen that cross-innervation for 2-3 weeks under spinal cord transection produced similar partial transformation of the Z-band width of the EDL fibres as under TTX block.

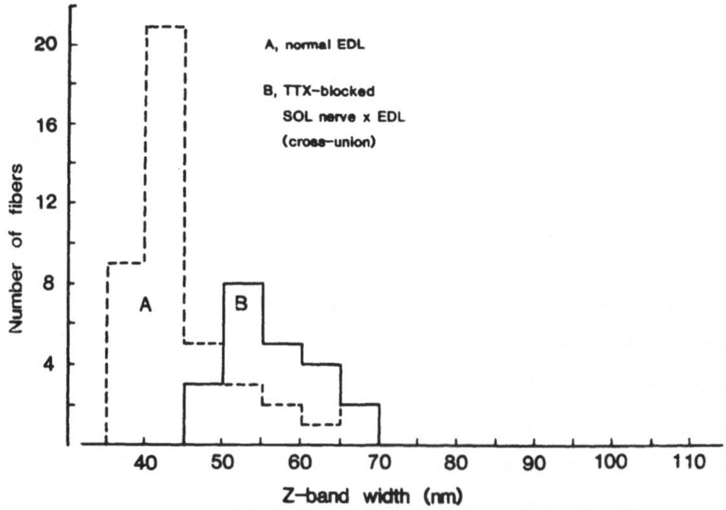

Fig. 5. Z-band width histograms: A: normal EDL; B: EDL cross-innervated by TTX-blocked SOL nerve for 3 weeks, cross-innervation being established by end-to-end anastomosis of the nerve stumps.

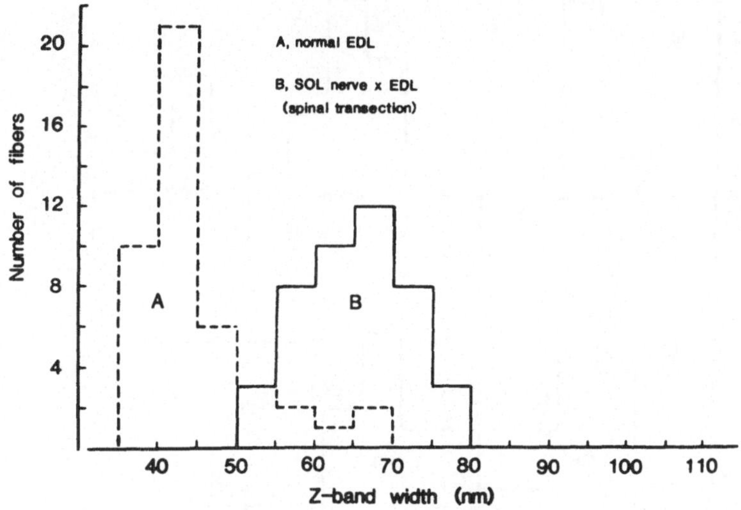

Fig. 6. Z-band width histograms: A: normal EDL; B: EDL cross-innervated by implanted SOL nerve under spinal cord transection at thoracic level for 2-3 weeks.

With cord transection substituted for TTX block to produce muscular inactivity, we were no longer limited to a duration of cross-innervation as short as 2-3 weeks. We therefore tried to extend the above result by performing end-to-end cross-union of the proximal stump of the SOL nerve with the distal stump of the EDL nerve and allowing enough time for a large proportion of the muscle fibres to become cross-reinnervated, usually two months. With nearly complete cross-innervation, muscle fibres could be taken at random for EM study. Cord transection was usually done one week after the nerve-crossing operation. As a preliminary we first examined what effect simple cord transection might have on the Z-band width of the SOL and EDL muscle fibres. As shown in Fig. 7, after cord transection for 50 days, the Z-band widths of the SOL muscle fibres showed a considerable narrowing while those of the EDL fibres remained unchanged. This is perhaps in keeping with the observations reported by others (Hoh, Kevan, Dunlop and Kim, 1980) and confirmed by ourselves that after cord transection the contraction of the SOL became faster while that of the EDL remained unchanged. In our experiments after 50-60 days cord transection the contraction time of the SOL often became as short as that of the EDL.[1]

---

1. I have just been informed by Prof. Wilfried Mommaerts that Dr. Lynn Eldridge has found that in her spinal cord isolation cats the soleus muscle only became fast when allowed to shorten and remained slow when kept

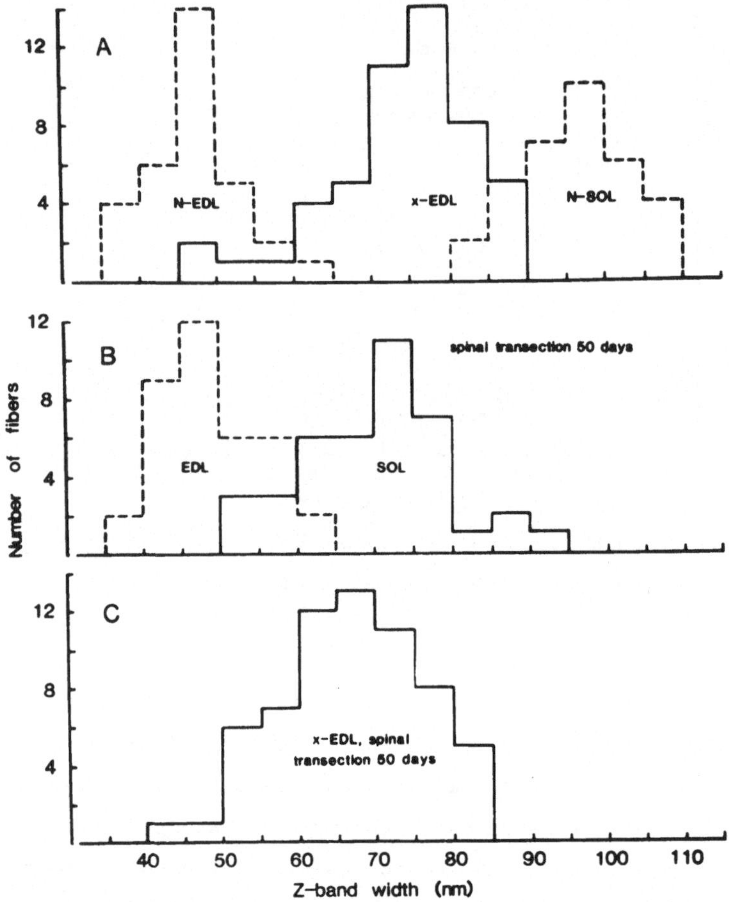

Fig. 7. Z-band width histograms: A: normal EDL, normal SOL and EDL cross-innervated by normal SOL nerve; B: EDL and SOL after spinal cord transection for 50 days; C: EDL cross-innervated by SOL nerve under spinal transection for 50 days.

It should be noted, however, that in these cases, the Z-band width of the SOL muscle fibres, though narrowed, remained distinctly different from that of the EDL fibres. After 50 days cross-innervation of the EDL fibres by the SOL nerve with the spinal cord transected, their contraction time showed no change. This is to be expected from the fact just mentioned that after simple cord transection for 50 days the SOL could become as fast as the EDL and is also in agreement with previous reports. After 50 days

---

at normal length. We are examining cord-transected rats to see whether keeping the ankle normally flexed or leaving it extended will make any difference to the soleus muscle becoming fast.

cross-innervation by the SOL nerve under cord transection, however, the EDL muscle fibres still underwent a striking transformation in their Z-band width, as shown in Fig. 7c. This together with the result for 2-3 weeks cross-innervation under cord transection described above, provides additional evidence that the slow nerve, even when inactive, can still exert a transforming influence on the Z-band of the fast muscle fibres it cross-innervates.

All the above results were obtained using one particular type-specific characteristic of the muscle fibre, the Z-band width. We had sought to extend our study to other type-specific characteristics, but one circumstance strictly limited our effort. For any particular type-specific characteristic to be used as a test criterion in the study of its possible transformation after cross-innervation by an inactive foreign nerve, in such short-term experiments involving TTX block as described above, two requirements must be met: (1) it must be capable of being transformed in a relatively short time by cross-innervation; (2) it must remain essentially unchanged as a result of muscular inactivity alone. We were fortunate in having chosen Z-band width as the test characteristic in our experiments. So far we have not been able to find another equally suitable characteristic. The use of cord transection for preventing muscular activity in cross-innervation experiments has potentialities not yet fully exploited and may be expected to extend the range of type-specific characteristics whose possible transformation by nerve-crossing under the condition of muscular inactivity can be studied.

It should also be noted that with Z-band width, our experiments were limited to its transformation in fast muscle fibres by slow nerve. This is because the narrow to wide transformation of the Z-band is much easier than the reverse transformation. This may be a more inherent limitation which reflects the general rule that the fast to slow transformation of muscle fibres by cross innervation occurs more readily than the reverse (Jolesz and Sreter, 1981).

We may now return to our experiments on doubly innervated muscle fibres with which we began, to give a striking illustration of the assymmetry of the cross-innervation effects. Using myosin ATPase histochemistry or Z-band width as the test characteristics, the doubly innervated EDL muscle fibres became completely transformed into SOL-like fibres, but the doubly innervated SOL fibres remained unchanged as SOL

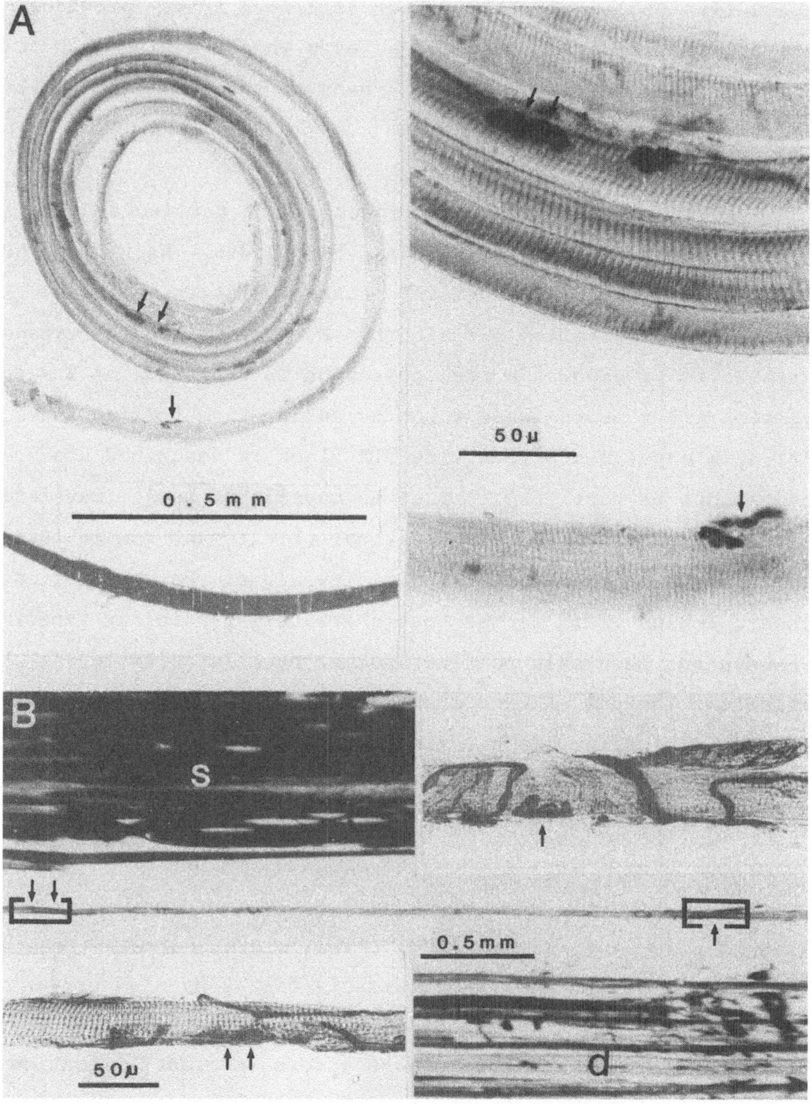

Fig. 8. A: SOL fibre and B: EDL fibre, doubly innervated for 120 and 199 days, respectively. Combined Myosin-ATPase (pH 10.4) preincubation and AChE staining. Single arrow, self-reinnervated endplate; double arrow, foreign endplate; endplate regions for both fibres also separately shown enlarged. s and d in B are longitudinal sections from self-reinnervated and doubly innervated parts of the EDL muscle, showing contrast in staining. Note complete transformation of EDL fibre and no change of SOL fibre after double innervation.

fibres (Feng, Huang et al., 1982; Feng et al., 1980, 1982a) (Fig. 8 and 9). Apparently when a muscle fibre receives innervation from a slow nerve and a fast nerve simultaneously the slow nerve may dominate over the fast nerve completely. This however, depends on the characteristic studied. The statement just made refers to myosin ATPase histochemistry and Z-band width. Judged by contraction time, however, both EDL and SOL muscle fibres, when doubly innervated, showed a contraction speed intermediate between those of normal SOL and normal EDL (Feng et al. 1980, 1982a). Judged by myotendinous junctional ACh sensitivity which is normally present in SOL but absent from EDL, double innervation abolished this sensitivity in the SOL muscle fibres but left the EDL fibres unchanged without this sensitivity (Zhu and Li, 1986). In this case, then, the EDL nerve influence apparently dominated over the SOL nerve influence.

In the early stage of our research, when we found in doubly innervated muscle fibres the predominance of the influence of the SOL nerve over that of the EDL nerve on Z band width and myosin ATPase histochemistry, we had supposed, considering at that time only the influence of the activity factor, that when a slow pattern and a fast pattern of activity coexist, the influence of the former predominates over that of the latter. Now that we know that inactive SOL nerve can still exert a transforming influence on the Z-band width of the EDL muscle fibre, and that furthermore this influence is equally manifest even in doubly innervated EDL muscle fibres in which the original EDL nerve, also inactive, is intact, the inadequacy of that interpretation becomes evident. It now appears that the SOL nerve has, besides its impulse activity pattern, some additional trophic factor which the EDL nerve lacks.

It has been pointed out that the original objective of our experiments on doubly innervated muscle fibres was to test the possibility of finding local differences in the two endplate regions in respect of some type-specific characteristics. Such differences, if found, would constitute strong evidence for different nerves releasing different specific trophic substances. Although we cannot claim that our search has been exhaustive, we no longer entertain much hope of finding such differences. At present we are completely ignorant about the mode of transmission of the trophic message from nerve to muscle and along the muscle fibre, and we have no good reason for predicting local differences in any type-specific characteristic in the two endplate regions of doubly innervated muscle fibres.

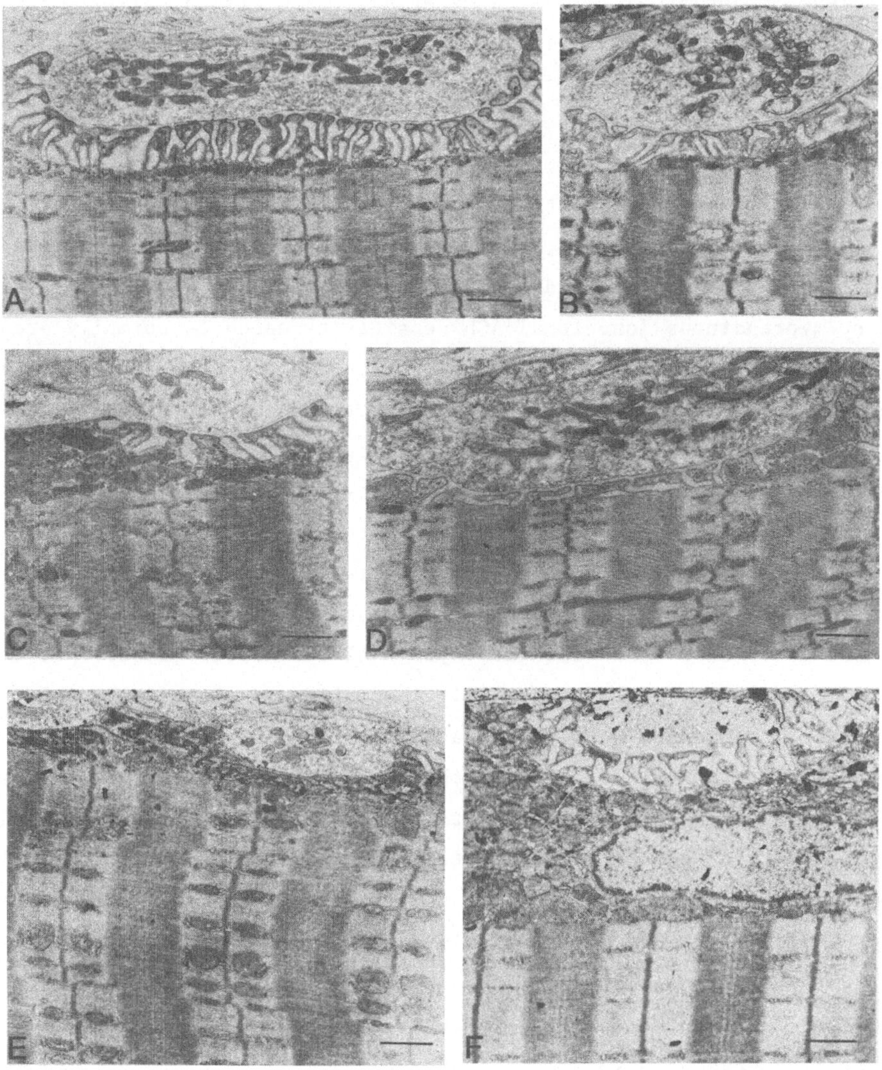

Fig. 9. Z-band width of normal EDL fibre (A), normal SOL fibre (B), the two endplate regions of doubly innervated EDL fibre (C, D) and doubly innervated SOL fibre (E, F). Note (1) Z-band width of EDL fibre completely transformed, that of SOL fibre unchanged, and (2) no local Z-band differences in the two endplate regions in the doubly innervated fibres. Calibration bar = 1 μm.

The two ways by which the motoneurone may exert a determining influence on the type-specific properties of skeletal muscle fibres, as depicted by the activity hypothesis and the chemical hypothesis, are of course not mutually exclusive. In experimental analysis, the two ways have first to be demonstrated separately. Now that this has been done, we have to picture the two ways as two factors in the normal neural control of the skeletal muscle fibre type. Here it must first be emphasized that the extent of this control and the relative importance of the two factors in this control appear to be different for slow-twitch and fast-twitch motor units in the first place and for different muscle fibre characteristics in the second. There is considerable evidence both from the study of muscle development and from experiments in adult animals to show that there is much less neural control on fast-twitch muscle than on slow-twitch muscle. After spinal cord transection in neonatal animals, the fast-twitch muscle develops in the usual way while the slow-twitch muscle fails to develop its characteristic slow contractile properties and becomes like a fast muscle. In adult animals, after cord transection, the contractile properties of the fast-twitch muscle remain unchanged, while those of the slow-twitch muscle undergo a slow-to-fast transformation (see recent review by Kuno (1984)). Kuno has suggested that only neural discharges with low frequencies such as given by the motoneurones of slow-twitch motor units, are effective in the control of the contractile properties of muscle under physiological conditions. Thus the two muscle types differ greatly in their dependence on activity for the development and maintenance of their respective contractile properties. As to the specific trophic factor, all the evidence we have put forward for its presence in the neural control of muscle fibre type is derived from the action of the motoneurones of slow-twitch motor units on the Z-band. In general red slow muscle fibres have wide Z-bands while white fast fibres have narrow Z-bands, but there is no simple relation between Z-band width and contraction speed (Gauthier, 1979). It is impressive as well as puzzling that the Z-band, an integral structural component of the myofibril, can so readily undergo transformation in width under neural influence, and that furthermore it is sensitive to the control by both the activity and non-activity factors of this influence. It is now time to try to look into the molecular events underlying the Z-band transformation. Characteristic differences between fast-twitch and slow-twitch muscle fibres in some Z-band protein components are to be expected from the work of Suzuki et al. (Suzuki, Goll, Singh, Allen, Robson and Stromer, 1976; Suzuki, Gall, Stromer, Singh and Temple, 1973) and of Schachat, Canine, Briggs and Reedy (1985). We have begun a biochemical study of the Z-band changes under various conditions, especially after the cross-innervation of fast muscle by slow nerve.

While different muscle fibre characteristics can be made to vary independently under experimental manipulation, in a motor unit the various characteristics of the muscle fibres are normally well coordinated and properly matched to the functional properties of the motoneurone. While the two factors of the neural control of muscle fibre type are experimentally separable, in the normal functioning of the neuromuscular system they may be expected to interact and to be complementary in their effects. The impulse activity of any motoneurone varies over a wide range in normal life, yet the muscle fibres of the corresponding motor unit normally assume a stable and distinct type. This discrepancy makes it difficult to conceive any simple relation between motoneurone activity pattern and muscle fibre type, and makes one feel the need of the operation of some complementary factor which may well be provided by trophic substance(s). As neural control of slow-twitch muscle is more manifest with respect to both factors, studies on how the two factors may interact and work together in a complementary manner, may first be focussed on slow-twitch motor units.

ACKNOWLDEGEMENTS

This lecture is based on the collaborative work with a number of colleagues including Huang Shikai, Li Ruhong, Rong Xinwei, Tian Lianming, Zhou Changfu and Zhu Peihong, with the technical assistance of Chen Keying, Li Ying, Qu Fujin and Yin Mei in the Shanghai Institute of Physiology, Shanghai, China.

REFERENCES

Bray, J.J., Hubbard, J.I. and Mills, R.G. (1979) The trophic influence of tetrodotoxin-inactive nerves on normal and reinnervated rat skeletal muscles, J. Physiol. (Lond.) 297, 479-491

Brooke, M.H. and Kaiser, K.K. (1970) Muscle fiber types: how many and what kind? Arch. Neurol., 23, 369-379

Buller, A.J., Eccles, J.C. and Eccles, R.M. (1960) Interactions between motoneurones and muscles in respect to the characteristic speeds of their responses, J. Physiol. (Lond.) 150, 417-439

Burke, R.E., Levine, D.N., Tsairis, P. and Zajac, F.E. (1973) Physiological types and histochemical profiles in motor units of the gastrocnemius, J. Physiol. (Lond.) 234, 723-748

Close, R. (1969) Dynamic properties of fast and slow skeletal muscle of the rat after nerve cross union, J. Physiol. (Lond.) 204, 331-346

Edgerton, V.R. and Cremer, S. (1981) Motor unit plasticity and possible mechanisms, In: Progress in Clinical Neurophysiology, Vol. 9, Motor Unit Types, Recruitment and Plasticity in Health and Disease, Desmedt, J.E., ed., Karger: Basel, pp 220-240.

Eldridge, L. and Mommaerts, W. (1980) Ability of electrically silent nerves to specify fast and slow muscle characteristics. In: Plasticity of Muscle, Pette, D., ed., de Gryter: Berlin, pp 325-337

Feng, T.P., Huang, S.K., Zhu, P.H., Zhou, C.F. and Qu, F.J. (1982) Ultrastructural type of rat skeletal muscle fibers doubly innervated by fast and slow muscle nerves, Acta Physiol. Sinica 34, 35-41

Feng, T.P., Zhou, C.F. and Zhu, P.H. (1980) Histochemical type and contractile property of rat soleus muscle fibers doubly innervated by fast and slow muscle nerves, Acta Physiol. Sinica 32, 122-139

Feng, T.P., Zhou, C.F. and Zhu, P.H. (1982a) Histochemical type and contractile property of rat extensor digitorum longus muscle fibers doubly innervated by fast and slow muscle nerves, Acta Physiol. Sinica 34, 26-34

Feng, T.P., Zhou, C.F. and Zhu, P.H. (1982b) Transformation of ultrastructural type of fast-twitch muscle fibers after cross-innervation by tetrodotoxin-blocked slow muscle nerve, Scientia Sinica 25, 953-960

Gauthier, G.F. (1979) Ultrastructural identification of muscle fiber types by immunocytochemistry, J. Cell Biol. 82, 391-400

Hoh, J.F.Y., Kevan, B.T.S., Dunlop, C. and Kim, B.H. (1980) Effect of nerve cross-union and cordotomy on myosin isoenzymes in fast-twitch and slow-twitch muscles of the rat. In: Plasticity of Muscle, Pette, D., ed., de Gryter: Berlin, pp 339-351

Jolesz, F. and Sreter, F.A. (1981) Development, innervation, and activity-pattern induced changes in skeletal muscle, Ann. Rev. Physiol. 43, 531-552

Kuno, M. (1984) A hypothesis for neural control of the speed of muscle contraction in the mammal, Adv. Biophys. 17, 69-95

Lømo, T., Westgaard, L.H. and Engebretsen, L. (1980) Different stimulation patterns affect contractile properties of denervated rat soleus muscles. In: Plasticity of Muscle, Pette, D., ed., de Gryter: Berlin, pp 297-309

Nemeth, P.M., Solanki, L., Gordon, I.A., Hamm, T.M., Reinking, R.M. and Stuart, D.G. (1986) Uniformity of metabolic enzymes within individual motor units, J. Neurosci. 5, 892-898

Peter, J.B., Barnard, R.J., Edgerton, V.L., Gillespie, C.A. and Stampel, K.E. (1972) Metabolic profiles of three fiber types of skeletal muscle in guinea pigs and rabbits, Biochemistry, 11, 2627-2633

Salmons, S. and Sreter, F.A. (1976) Significance of impulse activity in the transformation of skeletal muscle type, Nature 263, 30-34

Schachat, F.H., Canine, A.C., Briggs, M.M. and Reedy, M.C. (1985) The presence of two skeletal muscle α-actinins correlates with troponin-tropomyosin expression and Z-line width, J. Cell Biol. 101, 1001-1008

Suzuki, A., Goll, D.E., Singh, I., Allen, R.E., Robson, R.M. and Stromer, M.H. (1976) Some properties of purified skeletal muscle α-actinin, J. Biol. Chem. 251, 6860-6870

Suzuki, A., Goll, D.E., Stromer, M.H., Singh, I. and Temple, J. (1973) α-actinin from red and white porcine muscle, Biochim. Biophys. Acta 295, 188-207

Zhu, P.H. and Li, K.X. (1986) There is no acetylcholine sensitivity at the myotendinous junction of fast- or slow-muscle fibers with combined fast- and slow-muscle nerve supplies in rats, Acta Physiol. Sinica 38, 107-115

MUSCLE MECHANICS

D.C.S. White

Department of Biology
University of York
York YO1 5DD, United Kingdom

INTRODUCTION

Textbooks indicate that muscle shortens via the sliding filament model: the thick and thin filaments slide relative to each other, but keep a constant length (A.F. Huxley and Niedergerke, 1954; H.E. Huxley and Hanson, 1954). The idea that the force for the sliding motion is caused by cross-bridges (Hanson and H.E. Huxley, 1955; A.F. Huxley, 1957) is often included as part of the sliding filament model.

It is generally still held that for most muscles the filaments maintain virtually constant length (i.e. that the backbone of the filaments are not responsible for the active generation of force, and they are non-compliant). How force tending to produce relative sliding of the filaments is developed has been the main thrust of recent research. A.F. Huxley (1974) summarises the evidence that the cross bridges are responsible, and that they act as independent force generators acting cyclically. Although there have been some spirited attempts to show that the key experimental experiments on intact fibres (Gordon, Huxley and Julian, 1966) were wrongly interpreted (ter Keurs, 1978), reinvestigation confirms the original results (e.g. Altringham and Botinelli, 1985); in this paper I shall assume that cross-bridges are responsible for the active mechanism of contraction.

White and Thorson (1973) and Woledge, Curtin and Homsher (1985) give comprehensive reviews of aspects of some of the work covered in this paper.

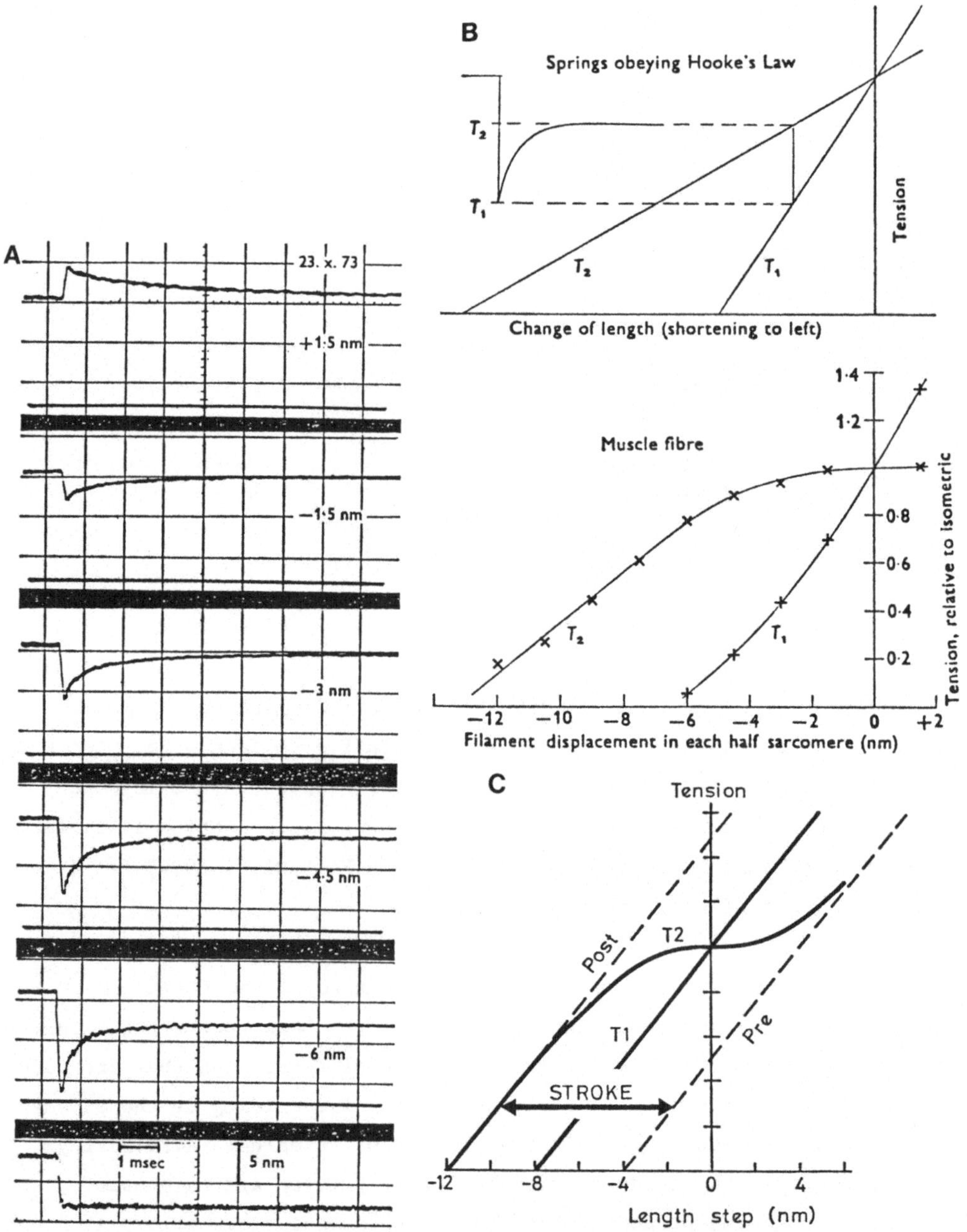

Fig. 1. A. Response of isometrically contracting frog muscle to applied step changes of length of the magnitude shown.
B. Top: method of analysis of the transients. T_1 is the extreme tension reached at the end of the applied length step; T_2 the tension reached after the relaxation process. Bottom: T_1 and T_2 curves plotted from the data of part A. Parts A and B taken from A.F. Huxley (1974).
C. Theoretical T_1 and T_2 curves of the model of Huxley and Simmons (1971). The dashed lines represent the mechanical properties of the cross-bridges in the states preceding and following the working stroke of the muscle. The magnitude of the working stroke is the horizontal separation of these lines.

WHAT IS THE MAGNITUDE OF THE CROSS-BRIDGE STROKE?

It is commonly imagined that a conformational change in the cross-bridge is transmitted mechanically to cause the thin filament to move past the thick filament. This was the basis of A.F. Huxley's original model (1957), and is tacitly assumed by most papers. The demonstration that the cross bridge could take up different angles to the filament axis (Reedy, Holmes and Tregear, 1965) gave a structural basis to this notion, suggesting that the active stroke was such an angle change, and that the subfragment-1 (S1) head was responsible for active contraction. This was clearly formulated by H.E. Huxley (1969).

In this case, the movement from one cross-bridge operation would not be expected to be greater than the length of the head, say not more than about 15 nm. The mechanical transient experiments of A.F. Huxley and Simmons (1971) and Ford, A.F. Huxley and Simmons (1977, 1981, 1986) are consistent with this magnitude. In these experiments rapid (step) length changes were applied to an isometrically active muscle fibre, and the resulting tension transients measured (Fig. 1A). Huxley and Simmons (1971) analysed the very rapid tension transients as described in Fig. 1B. They interpreted the very rapid tension changes in terms of transitions between the states on either side of the working stroke. In the model they propose to account for their results, the magnitude of the working stroke is the horizontal separation of the asymptotes of the T_2 curves at high stretches and high releases as described in Fig. 1C. This is 8 nm using the values given in their paper. More recent data (Ford, et al., 1977) show that a higher value, about double this figure, is required.

Considerable effort has been put into finding structural changes in the muscle associated with the working stroke; these experiments, including X-ray diffraction, orientation of e.p.r. probes and electron microscopy, have had trouble in identifying any change in conformation at all. Huxley and Kress (1985) have summarised this evidence, and suggest that from such studies a working stroke of 4 nm is a likely maximum. On the basis of this, and taking note of the biochemists' concept of weakly and strongly bound bridges discussed below, they propose a modification to the concepts of Huxley and Simmons (Fig. 2). They suggest that the initial attachment might be with weakly attached cross-bridges able to bind over a fairly extensive region (I), but unable to develop force by going to the strongly bound state unless the position of the actin to which they were bound put

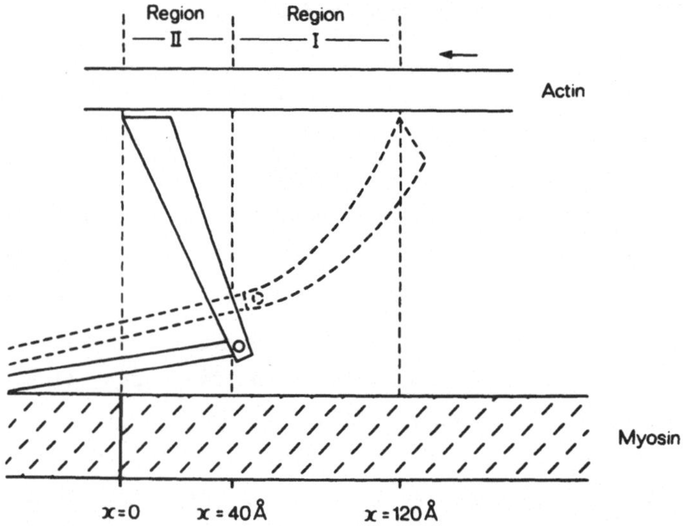

Fig. 2. H.E. Huxley and Kress (1985) concept of cross-bridge action. Cross bridges can attach over a range of 12 nm. However, in region I the bridges are weakly attached and develop no force. Possible attachment is at a single site. Only when relative filament sliding brings the head into region II does the binding become strong (e.g. with the attachment of one or more further sites between the myosin and actin) with the development of force.

them into a more restricted region (II). If the muscle is shortened rapidly, the effect of the sliding will be to cause bridges that were in region I to be moved into region II, and thus able to develop force, thereby generating the T_2 curve of Fig. 1B. The consequence of these ideas is that a cross-bridge will develop about three times more tension than necessary for the proposed mechanism of Huxley and Simmons, but each will give out the same amount of energy, since the working stroke is about three times less. In fully active muscle there will be fewer bridges developing active tension on the H.E. Huxley and Kress model than on the A.F. Huxley and Simmons model.

However, one technique devised to measure the working stroke gives a much larger figure. Yanagida, Arata and Oosawa (1985) measured the maximum amount of movement of a thin past a thick filament per ATP molecule hydrolyzed. They used a preparation in which the Z-lines were digested, leaving isolated, single sarcomeres with the thin filaments separated from each other. They observed the rate of unloaded shortening of these sarcomeres, and also measured their rate of hydrolysis of ATP. The surprising result was that a movement of about 60 nm was obtained for each ATP hydrolyzed, implying a working stroke far in excess of what could be achieved by mechanical means, and consequently a different mechanism.

In these experiments the digestion of the Z-lines means that the I filaments were isolated mechanically from each other. The work done in moving the I filaments is extremely low, since the force on the I filaments is close to zero; thus if other, weak, forces are present, not clearly understood at present (maybe electrostatic interactions, see e.g. Elliott, Rome and Spencer, 1970, or surface tension), these could be strong enough to contribute to filament movement in these experiments without contributing significantly to observable muscle tension. A necessary effect of the addition of ATP would be to unlock the rigour bridges, allowing sliding to take place.

If Yanagida's results do require a potential movement of 60 nm for each ATP hydrolyzed in normal muscle, then total reformulation of the dominant ideas about contraction is required.

WHERE IS THE CROSS-BRIDGE COMPLIANCE?

Yanagida's results, referred to in the previous paragraph, make us reconsider the evidence that cross-bridges have a mechanical role. One argument that they do concerns the different properties of relaxed and rigour muscle, and the corresponding properties of active muscle. The obvious mechanical difference between relaxed and rigour muscle is that relaxed muscle is highly compliant at rest length, but that rigour muscle is extremely stiff. Structurally the main difference between these two states is that the cross-bridges are detached from the thin filaments in relaxed muscle, but attached in rigour muscle (Reedy, Holmes and Tregear, 1965). The stiffness of active muscle as determined by very rapid length changes is comparable to that of rigour muscle (about 80% in frog fibres, Goldman and Simmons, 1977). This stiffness is thus attributed to the continuity across the sarcomere produced by thin and thick filaments linked by cross-bridges. Ford et al. (1981) determined that the major part of the compliance of the sarcomere was within the cross-bridges by measuring the stiffness of the sarcomere at different sarcomere lengths, when the extent of overlap between the filaments was different.

The cross-bridge has two heads (S1), linked to the thick filament by a long tail (S2). Which of these structures is the more compliant? The answer to this is unknown at the moment. Kimura and Tawada (1984) linked the cross-bridge tail to the backbone of the thick filament, and found no increase in stiffness of the muscle, as would have been expected with a compliant S2, implying that the compliance is a property of the heads. However, no change in conformation of the heads has been detected in studies in which a probe has been attached to the head (Cooke, 1981). Of course there is a region between the parts of the tail linked to the backbone in Kimura and Tawada's studies and the region of the cross-bridge to which the probe is attached in the probe studies. This includes the hinge region between the heads and the tail. However, this region is short, and it is doubtful if even full extension of the polypeptide chain in this region would be capable of accounting for the full compliance observed.

CONVERSION OF CHEMICAL TO MECHANICAL ENERGY

The immediate form of chemical energy used to drive muscle is the high concentration of ATP relative to ADP in the cell. An understanding of the way that the chemical potential of ATP in the cell is converted to mechanical energy requires knowledge of the chemical events occurring during the cross-bridge cycle. This problem has been extensively studied for the proteins myosin (and its subfragments) and actin in solution, and this subject has been reviewed several times recently (Eisenberg and Hill, 1985; Sleep and Smith, 1981; Taylor, 1979). There is now moderately good agreement about the sequence of reactions involved in the hydrolysis of ATP by actomyosin, and about the equilibrium constants between these states and the rate constants of the transitions between them. This is summarised in Fig. 3.

The affinity of myosin for actin is highly dependent upon the other ligands bound to the head (Geeves, Goody and Gutfreund, 1984). In the presence of bound ATP or its products ADP and Pi, the affinity is relatively low. The actin is said to be <u>weakly</u> bound. These weakly-bound states are further characterised by having high rate constants for actin dissociation and association. This low affinity is greatly increased at very low ionic strength. In the absence of nucleotide, or in the presence of bound ADP, the affinity of myosin for actin is high, and the bridges are said to be <u>strongly</u> bound. Since actin contains at least two

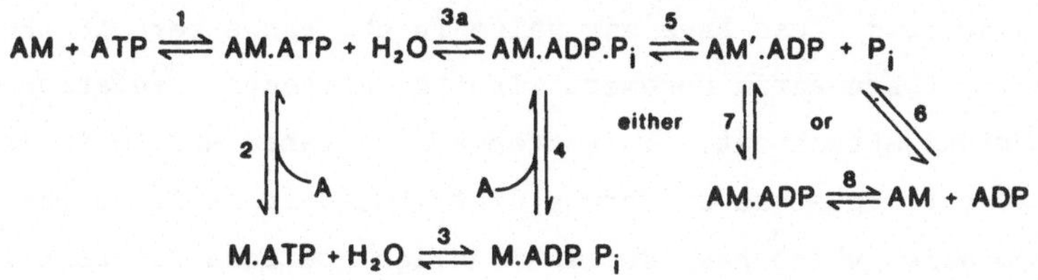

Fig. 3. Actomyosin ATPase mechanism taken from Hibberd and Trentham (1986), based on data derived largely by Eisenberg, Sleep and Taylor and their colleagues. AM denotes actomyosin.

myosin-binding sites, perhaps the weakly-bound state is binding to just one site, and the strongly-bound state is binding to all sites (as implied by Fig. 3).

Do the same reactions found for the isolated proteins in solution take place in intact fibres? There are some important differences between the situation of myosin interacting with actin in solution and in an intact fibre.

a. In solution there are no mechanical constraints on the molecules; thus conformational changes can occur without hindrance. In the intact fibre the arrangement of the molecules into filaments, which are themselves linked together via the Z and M-lines, means that a cross-bridge which is attached to a thin filament is hindered if it attempts a conformational change. Such a cross-bridge will exert a force, tending to make the filaments move. Only if the net sum of all the forces from cross-bridges exceeds the load on the fibre will there in fact be movement.

b. The filament organisation within the fibre means that cross-bridges are constrained in their orientation to potential actin sites on the thin filament. This has two consequences. Firstly, the actin-myosin association reactions, which are second-order reactions in solution and presumably governed by collision processes, are more properly thought of as being first-order reactions in the intact fibre lattice. Secondly, the proximity of the nearest available site is variable. It is conventional to describe this proximity in terms of the relative longitudinal displacement of the thick and thin filaments needed to allow the cross-bridge to attach to the thin filament with no distortion. The magnitude of this required movement we shall call distortion. It is usually given the symbol x; the value of x for a particular bridge can have any value in the range zero to the repeat distance between actin monomers with the correct orientation for cross-bridge attachment. In vertebrate striated muscle it is thought that cross-bridges are uniformly distributed in x; in insect fibrillar flight muscle, which has identical helical periods in its thick and thin filaments, the distribution is non-uniform, but for any sarcomere length will show a marked peak at a particular value of x.

c. The difference in free energy between adjacent states must include the mechanical energy required to strain the molecules when making the transition. The consequence of this is that the rate constants for the reactions between states with different conformations will be different in solution and in the intact fibre. Furthermore, these reactions, and this means the rate constants, will be distortion dependent. These concepts are discussed in detail by Hill (1974).

d. Diffusion of, in particular, ATP, ADP and Pi presents a problem experimentally in the intact fibre which is not apparent with experiments on the isolated proteins in solution.

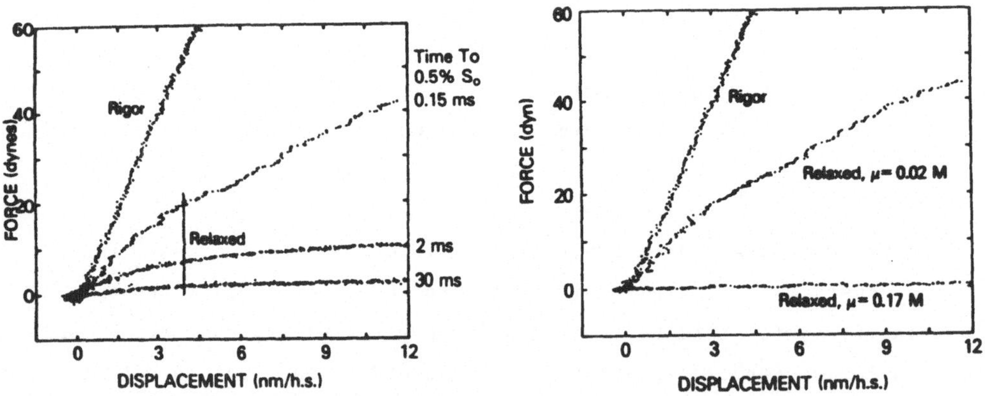

Fig. 4. Force versus displacement during rapidly applied stretches to rabbit psoas fibres. A. Effect of changing speed of stretch on the response of relaxed fibres. B. Effect of changing ionic strength on the response of relaxed fibres. Rigour responses are shown in each case. From Schoenbert et al. (1984).

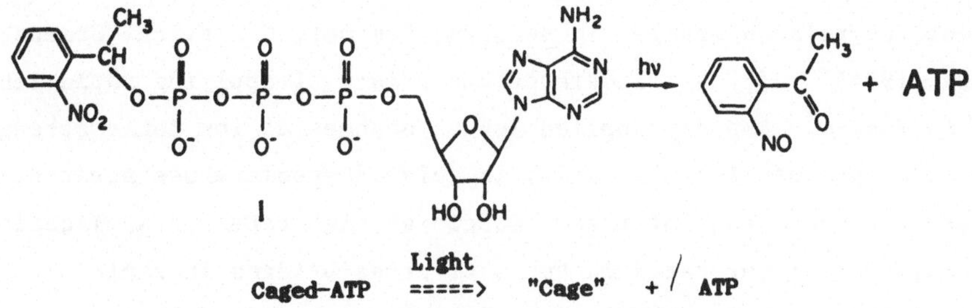

Fig. 5. Reaction of caged-ATP (P^3-1-(2 nitro)phenylethyladenosine 5'-triphosphate) to light.

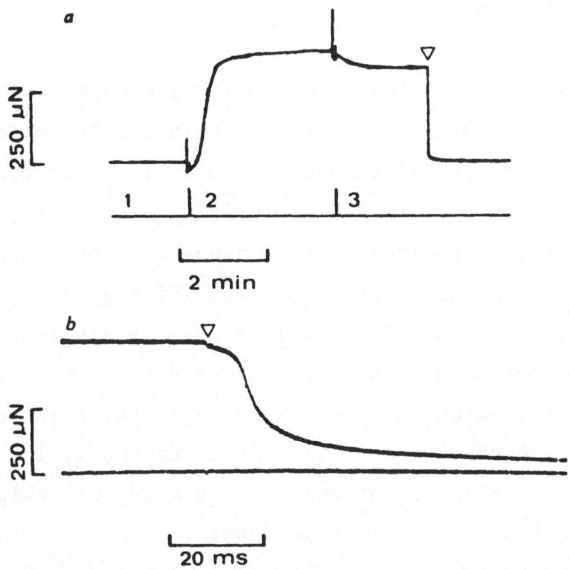

Fig. 6. Protocol and typical traces for the caged-ATP experiments. Top: slow timebase. A skinned rabbit psoas fibre is initially in relaxing solution (1). At 2 it is transferred to rigour solution (containing no ATP), and at 3 to a rigour solution containing caged-ATP in the absence of Ca^{2+}. A brief intense flash of light is given at the open triangle. Bottom: the response to the photolysis of caged-ATP at a faster timebase to show the time course of the relaxation. From Goldman et al. (1982).

WEAKLY-BOUND STATES

The presence of weakly-bound states has been inferred from mechanical experiments on relaxed muscle (Schoenberg, Brenner, Chalovich, Greene and Eisenberg, 1984). Fig. 4 illustrates the effect of applying rapid length changes to muscle. Rapidly applied length changes at low ionic strength show the presence of viscoelasticity in relaxed muscle whose apparent stiffness approaches that of rigour muscle at high rates of application. These results are interpreted in terms of cross-bridges in rapid equilibrium between a detached and an attached state. If the change of length is applied rapidly enough the bridges do not have time to detach, and show high stiffness. During slower length changes the bridges have time to detach (and perhaps "slip" to an adjacent actin monomer), thereby not contributing to the stiffness.

RAPID-REACTION TECHNIQUES

What techniques are available for studying the reactions between identifiable states of the actomyosin? In solution the most successful techniques have been stopped-flow and quenched flow. In both of these techniques, the reactants are mixed as rapidly as possible, and the reaction followed either optically or by quenching the reaction and analysing the current state of the reaction. Mixing can be achieved within 1-3 msec. This type of experiment is less readily achievable with fibres, partly due to the problem of diffusion referred to above. Thus changing the concentration instantaneously of a ligand in the solution surrounding a fibre does not change the concentration instantaneously in the centre of the fibre; there is a significant delay, of a magnitude which prevents the observation of the faster reactions in muscle.

The synthesis of so-called caged compounds gets around the problem of diffusion. These are photolabile compounds which, on photolysis, yield a biologically interesting molecule, but which are inert before photolysis. Caged-ATP (Fig. 5) was the first such compound to be used with muscle, by Goldman, Trentham and their colleagues. Hibberd and Trentham (1986) have recently published an excellent review of this work. In a typical experiment (Fig. 6) a rabbit psoas muscle fibre is relaxed with MgATP, and then transferred to a rigour solution (i.e. containing no nucleotide). Caged-ATP is added; there is little change in tension. At the point indicated a high-intensity brief flash of light is given. ATP is rapidly

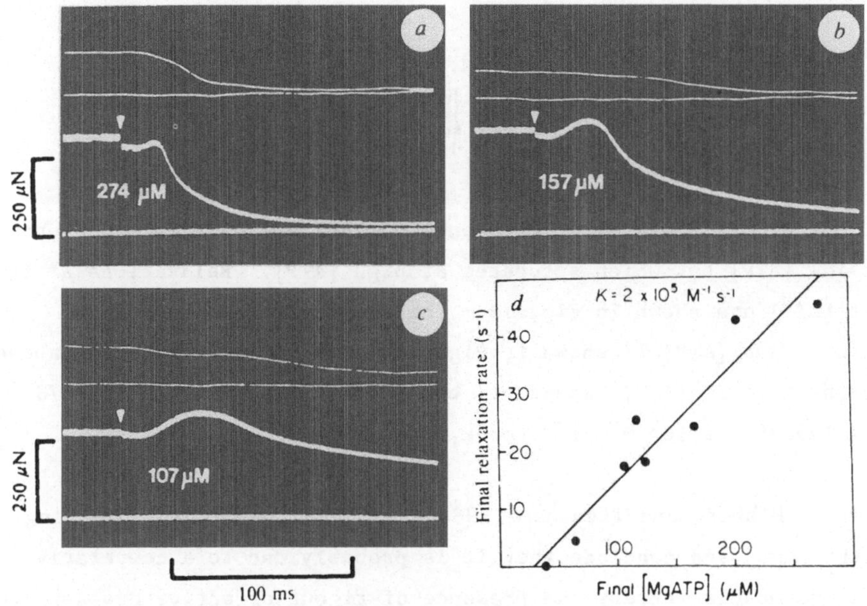

Fig. 7. Effect of changing the final [ATP] concentration on the relaxation in a caged-ATP experiment of the type outlined in Fig. 6. Part d shows the rate constant of the final, exponential, phase of the relaxation plotted against the final [Mg-ATP]. From Goldman et al. (1982).

released from the caged-ATP (with a rate constant in excess of 100 s^{-1}). The experiment illustrated in Fig. 6 was performed in the absence of Ca^{2+}; the fibre relaxes to zero tension with a fairly complicated timecourse. There is a final relaxation which is well described as an exponential process, but this is preceded by a "hump".

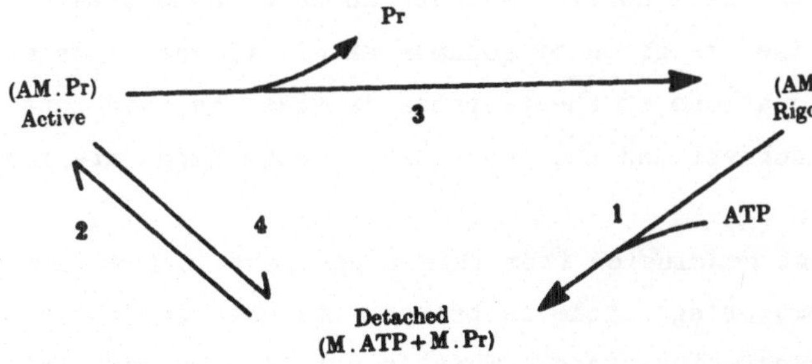

Fig. 8. Model proposed by Goldman et al. (1984b) to account for the caged-ATP results shown in Figs. 6, 7 and 9.

The simplest hypothesis to explain this experiment is

$$\begin{array}{ccc} & (1) & \\ AM + ATP & \rightleftharpoons & AM.ATP \\ & & \downarrow\uparrow (2) \\ & & A + M.ATP \end{array}$$

In this case we might expect a dependence on the rate of relaxation upon [ATP] at low [ATP] but which saturates at high [ATP]. Relaxations at three different [ATP] are shown in Fig. 7a - 7c, and the rate of the final relaxation versus [ATP] is shown in Fig. 7d. As predicted from the above equation the rate is [ATP]-dependent; the slope of the line in Fig. 7d gives a value of $2 \times 10^5\ M^{-1}.s^{-1}$ for k_1.

Goldman, Hibberd and Trentham (1984a) discuss the mechanism giving rise to the hump, and conclude that it is probably due to a cooperative effect in the muscle whereby the presence of rigour or active crossbridges allows detached cross-bridges to attach and develop tension even in the absence of Ca^{2+}. They propose that the working cross-bridge cycle can be adequately modelled by the scheme illustrated in Fig. 8.

In the presence of Ca^{2+} the response of the fibre to the sudden change in [ATP] following photolysis is as shown in Fig. 9. The tension generally dips, then rises to the level appropriate for the active fibre. That the tension is now being produced by actively cycling cross-bridges can be seen from the high value of what is known as the <u>quadrature stiffness</u>. This is showing the extent to which the phase 2 response of Huxley and Simmons (1971) is present. In rigour muscle there is very little, but in active muscle considerable, phase 2 relaxation.

The response in the presence of Ca^{2+} can also be accounted for by the scheme of Fig. 8. The time course and extent of the dip impose severe constraints on the rate constants that can be used to obtain a satisfactory fit. Full details are given by Goldman et al. (1984b). An essential feature of explanations of the response is that the rate constants of both the initial detachment and the subsequent reattachment are rapid.

An important conclusion from this experiment is that the ratelimiting step of the cross-bridge cycle is between attached states, and follows a major tension-generating state. This is not totally surprising, given that most bridges are attached in active muscle (Goldman and Simmons, 1977).

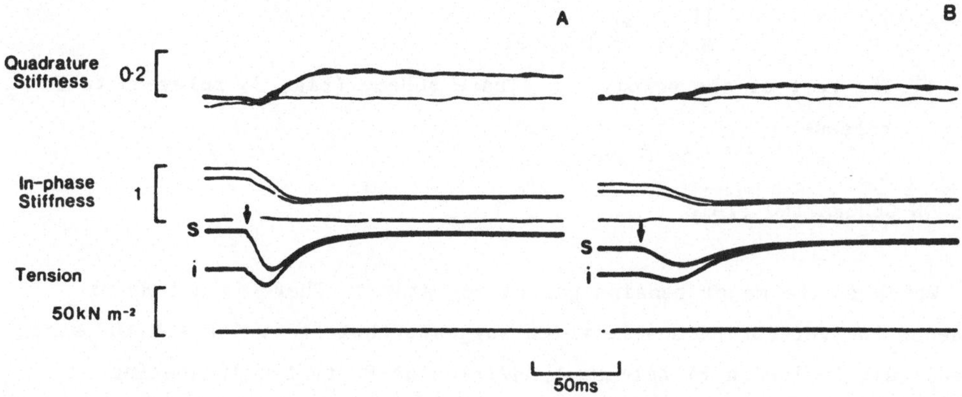

Fig. 9. Response of rabbit fibres to release of ATP from caged-ATP in the presence of Ca^{2+}. A. release of 520 μM ATP. B. release of 170 μM ATP. Traces of tension, in-phase stiffness and quadrature stiffness are shown. The stiffness measurements, made by applying a small high-frequency length oscillation and measuring the in-phase and quadrature tension responses are discussed more fully by Goldman et al. (1984b). They are thought to represent the total fraction of bridges attached, and the fraction of active (cycling through the working stroke) bridges respectively. Two traces are shown in each record. These are obtained with the fibre held isometric (i) and with the fibre pre-stretched (s) as discussed below. From Goldman et al. (1984b).

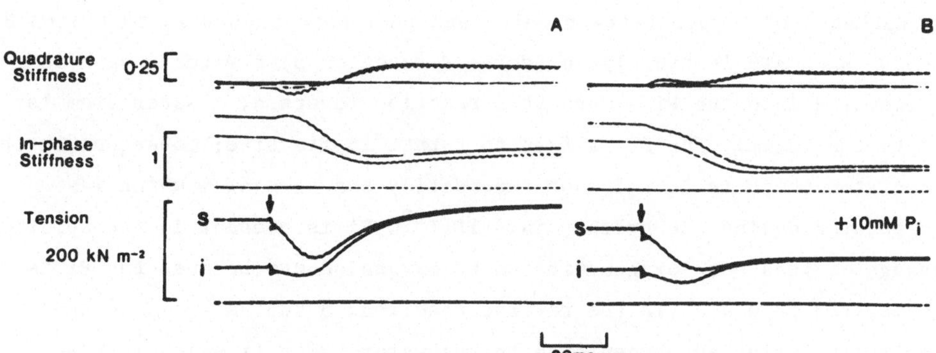

Fig. 10. Effect of repeating the experiment of Fig. 9 in solutions of two [Pi]. A. 0mM Pi. B. 10mM Pi. From Hibberd et al. (1985a)

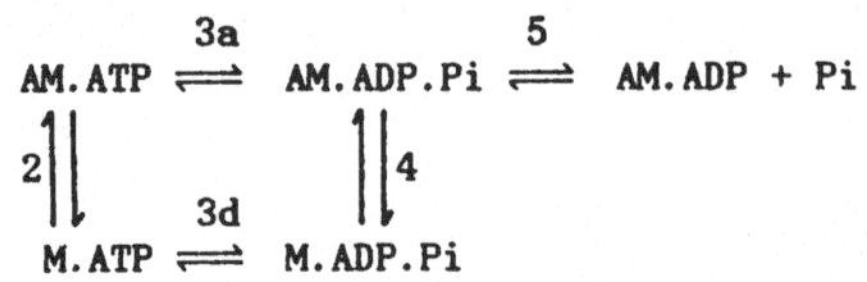

Fig. 11. The part of the actomyosin ATPase scheme (Fig. 3) relevant to Pi release.

TENSION-GENERATING STATE

Which is the major tension generating state? There is a body of evidence for vertebrate muscle which suggests that it is the A.M.ADP state immediately following Pi release. Addition of Pi to the incubating solutions causes a reduction in tension of the active muscle, and following photolysis in caged-ATP experiments the rate constants of the consequent transients are increased (Hibberd, Dantzig, Trentham and Goldman, 1985a, Fig. 10.) The simplest interpretation of these results is that reactions 5 and 4 in Fig. 11 are readily reversible, with the forward rate constant from AM.ADP being slow. Addition of Pi then causes a reduction in the proportion of bridges in the A.M.ADP state, and a lowering of the tension. The apparent rate constant $k'_{-5} = k_{-5}.[Pi]$ is higher with higher [Pi], and thus the rate constant for the transients will be larger.

The reversibility of the steps preceding the A.M.ADP state is shown by oxygen-exchange studies. If Pi, labelled with ^{18}O, is incubated with an active fibre then analysis of the ^{18}O-content of the Pi after incubation reveals a loss of ^{18}O, and its replacement by ^{16}O originating from the water. The explanation for this can be seen by reference to Fig. 11.

Exchange of oxygen between water and phosphate occurs in reactions 3a and 3d. When ATP is hydrolysed (forward reaction 3) a water oxygen is incorporated into the Pi. When this reaction reverses, a water atom is lost to the water. The Pi is free to rotate in its site; consequently the oxygen atom which is lost is not necessarily the same atom which was incorporated during the hydrolysis. If [^{18}O]Pi is present in the solution, exchange of phosphate oxygen with water oxygen occurs because the Pi is incorporated into ATP via the reverse reactions 5 and 3a (or 5, 4 and 3d), necessarily losing an oxygen atom to the water. The Pi which is then released, via the forward pathway, will now incorporate a water oxygen (^{16}O in this case). The ^{18}O-content of Pi in the solution can be measured by

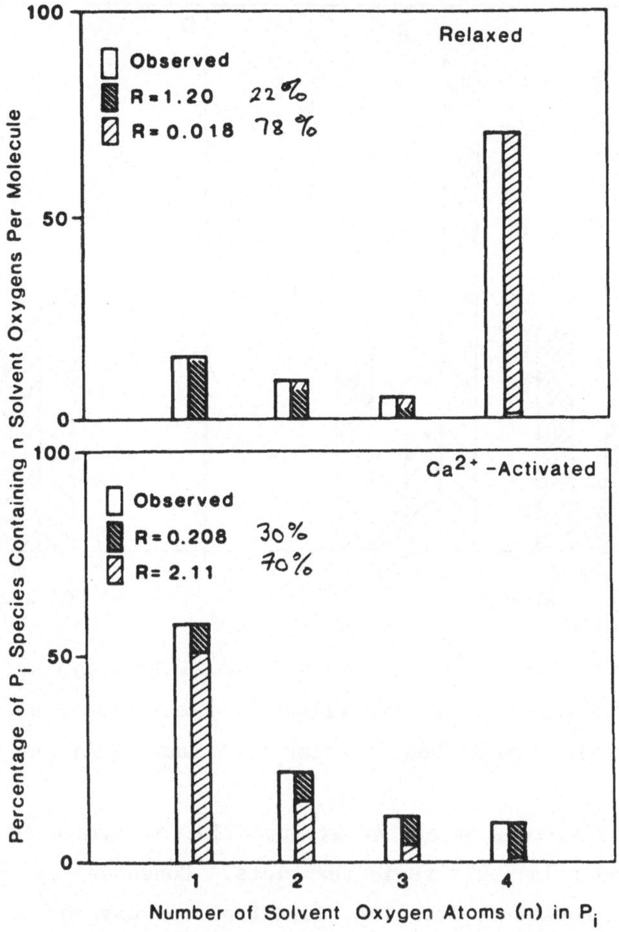

Fig. 12. Distribution of Pi species formed on ATP hydrolysis in relaxed and Ca^{2+}-activated rabbit fibres held isometrically. Incubation in ^{18}O-water. The experimental observations as well as the best theoretical fits are shown. Two independent pathways are required to obtain close fits to the data. From Hibberd, Webb, Goldman and Trentham (1985b).

mass spectrometry or n.m.r. These techniques allow the proportion of Pi containing 0, 1, 2, 3 or 4 ^{18}O to be determined. The rate at which such medium exchange occurs is a measure of the reversibility of these reactions. Studies on both vertebrate striated muscle (Webb, Hibberd, Goldman and Trentham, 1986) and insect flight muscle (Lund, Webb and White, 1985) show that these reactions are readily reversible, with apparent second order rate constants of Pi association of about 120 M^{-1} s^{-1}.

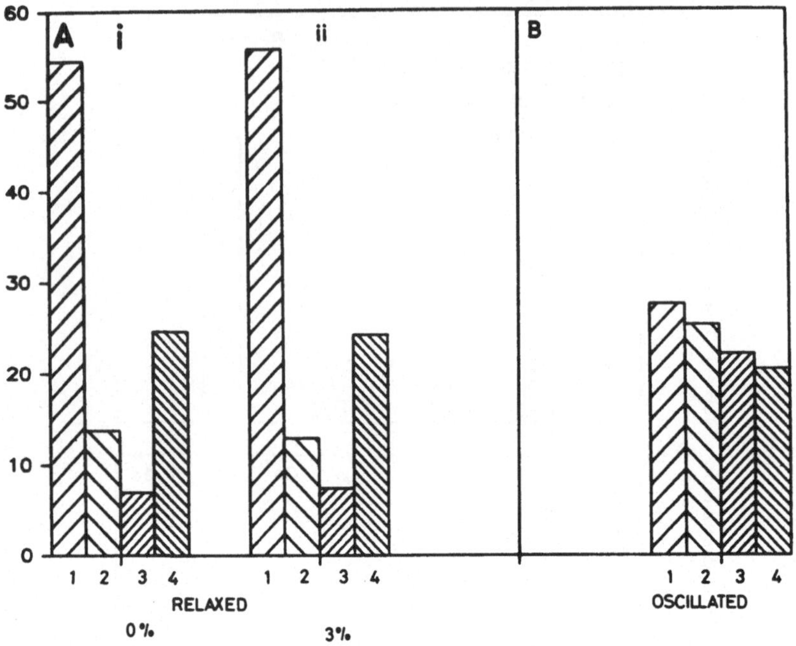

Fig. 13. Distribution of Pi species formed on ATP hydrolysis in maximally activated insect fibrillar flight muscle fibres (oscillation activated). Unpublished experiment of Lund, Webb and White (1986).

The hydrolysis steps 3a and 4d are readily reversible (Rosenfeld and Taylor, 1984) and relatively rapid reactions. Consequently, given the freedom of Pi to rotate in its site, more than one oxygen can be exchanged if the reverse reactions from AM.ADP.Pi are more probable than the forward reaction (reaction +5). Indeed the extent of oxygen exchange gives a measure of the ratio of the probability that AM.ADP.Pi will release Pi to the probability that ATP will be resynthesised. In terms of reaction scheme II, the distribution of solvent oxygens on the released Pi is governed by the ratio (Webb et al., 1986).

$$R = \frac{k_{+5} \cdot K_4}{k_{-3a}K_4 + k_{-3}}$$

In vertebrates the distributions of solvent oxygens on Pi released from ATP during incubations in relaxing and activating solution are shown in Fig. 12. The distributions obtained cannot be derived from a scheme such as scheme II above; it has been conventional to assume that two pathways are involved, with separate values of R, and the best fits to the data are shown in this form in the figure. However, a more appealing explanation has recently been suggested (Webb et al., 1986); this is that the distribution arises because cross bridges whose distortion is different (defining distortion as above) will have different values of rate constant for the tension-generating transition. In particular, bridges whose distortion is positive will have a lower probability of passing into the tension generating state than those whose distortion is negative. If, as we have hypothesized the Pi-release step is the relevant step, this means that there will be a distribution of values of R obtained from bridges with different distortions. This can allow the experimental distributions to be obtained.

Fully activated insect flight muscle gives a distribution (Fig. 13) which is well accounted for in terms of a single pathway (Lund et al., 1986). It is tempting to think that this is related to the distribution of cross-bridge distortions found in this type of muscle, which is not uniform over all values of x (Wray, 1979) as discussed above.

The presence of a single pathway describing the oxygen exchange allows an estimate of the value for the rate constant of Pi release. Given the probable value of k_{-3a} and k_{-3d} of about 10-15s^{-1}, the value of R obtained of about 0.2 for maximally activated insect flight muscle gives a value for k_{+5} of about 2-3s^{-1}. This is approximately equal to the catalytic site activity of the ATPase of these fibres, so for insect flight muscle it is possible that Pi release is the rate limiting step. This is at variance with the above conclusions for vertebrate striated muscle, which suggested that a subsequent step (possible ADP release, possibly an isomerisation of AM.ADP) was rate limiting.

DISTORTION-DEPENDENCE OF THE REACTIONS

As was discussed above, the rate constants of transitions between states whose tension is different will be dependent upon the degree of distortion of the bridges. The transient response of tension to applied length changes (Huxley and Simmons, 1971) has been interpreted in terms of a dependence of this kind. The rate constant of the initial relaxation following a step change of length is much faster for releases than for stretches, and faster the larger the release. Huxley and Simmons interpreted these results in terms of overcoming a contribution to the activation energy from the energy required to strain the compliant element

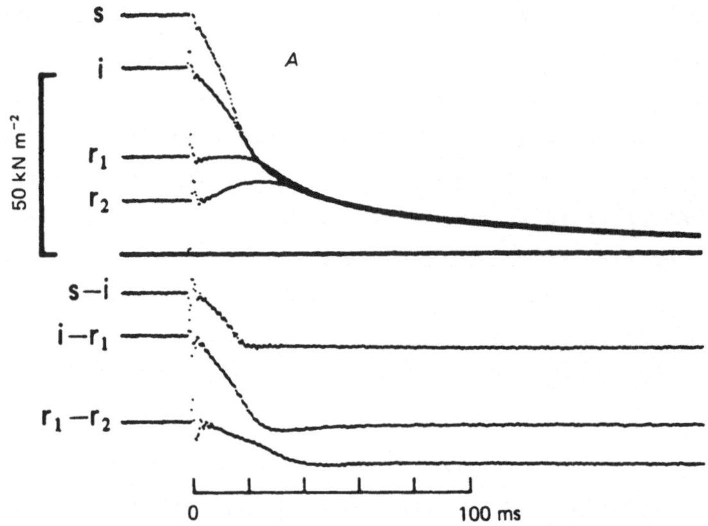

Fig. 14. Tension changes following release of ATP from caged-ATP by photolysis in an experiment similar to that described for Fig. 6. The four traces in the top part of the figure differ in the strain applied to the rigour muscle. For trace (s) the rigour fibre was prestretched by a small amount immediately preceding the trace shown; for traces r_1 and r_2 the fibre was released by small amounts preceding the trace. Trace (i) is for an isometric recording (i.e. no preceding change of length). The three lower traces are difference curves obtained by subtracting the traces indicated. From Goldman et al. (1984a).

of the cross bridge. The more the cross-bridge was strained, the slower the forward rate constant for the tension-generating transition became.

The caged-ATP experiments have allowed preliminary measurements of a distortion-dependent reaction to be made directly. In Fig. 14 the effects of pre-stretching or pre-releasing the fibres in rigour by a small amount preceding the photolysis are shown. The tension transients converge after about 20-40msec, with the final part of the relaxation being common to all the curves. Goldman et al. (1984a) propose that this final relaxation is of bridges which have detached at least once and then reattached again (as discussed above when describing the hump), and which therefore have no memory of the pre-stretch or release. The transients before the convergence then represent traces dominated by the first detachment of the rigour bridges. The three curves in the lower part of the figure are

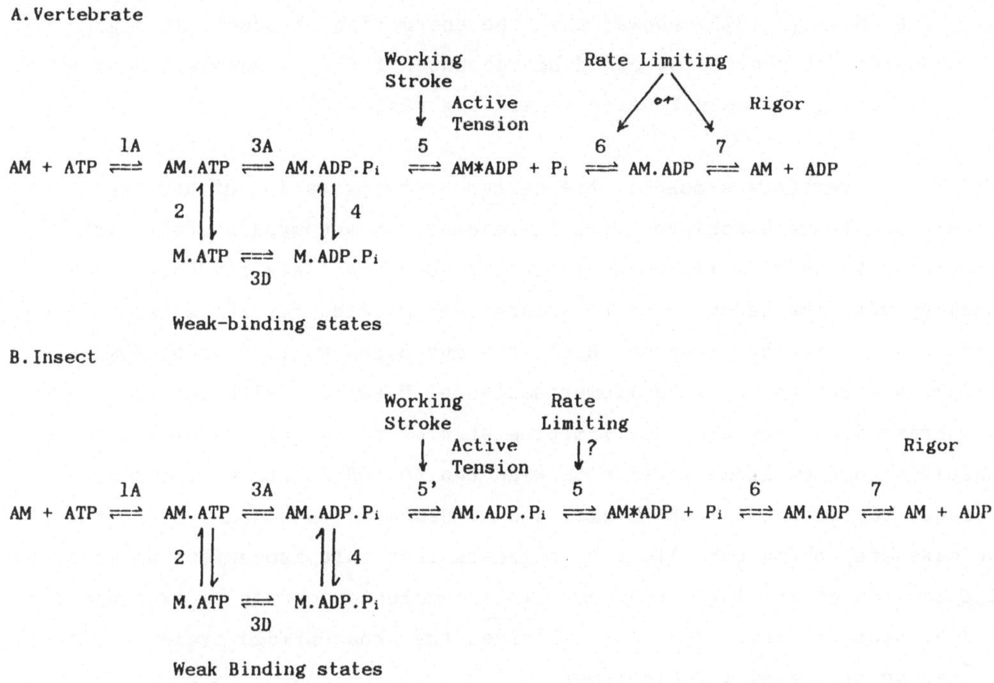

Fig. 15. Summary of the biochemical events thought to occur in the cross-bridge cycle, discussed in this paper.

difference curves, suggesting that detachment is faster for stretched bridges than for released bridges.

Thus, the detachment process would appear to be distortion dependent.

AN ATTACHED STATE PRECEDES THE TENSION-GENERATING STATE

Several lines of evidence show that there is an attached state preceding the tension generating state:

a. An increase of stiffness precedes the increase of tension by about 10msec in frog fibres at $0^{o}C$.

b. X-ray diffraction studies indicate that there is a net movement of cross-bridges towards the I filaments, also about 10msec in advance of the rise of tension.

c. A.F. Huxley (1973) showed that the energetics of muscle at high degrees of shortening could be accounted for by hypothesizing a state preceding the main tension generating state.

Given, for vertebrate muscle, the current interpretation of tension generation being associated with Pi release, an actomyosin state with product ADP and Pi still bound is the obvious candidate for this state. Analogy with the sequence of reactions demonstrated for the myosin ATPase (see e.g. review by Trentham, Eccleston and Bagshaw, 1976 or by Taylor, 1979) in which there is an isomerisation of M.ADP.Pi (with the isomerisation being the rate limiting step of the myosin pathway), it is possible that an isomerisation between two AM.ADP.Pi states occurs. Indeed, for insect flight muscle, the evidence discussed above for the Pi release step being rate limiting suggests that this isomerisation might be the tension-generating mechanism. This conclusion cannot be excluded for the vertebrate data. Fig. 15 summarises the cross-bridge cycle in intact fibres as discussed in this paper.

ACKNOWLEDGEMENT

The work from this laboratory report in this paper was supported by the Medical Research Council, Grant No. G8326496CB.

REFERENCES

Altringham, J.D. and Botinelli, R. (1985) The descending limb of the sarcomere length-force relation in single muscle fibres of the frog. J. Muscle Res. Cell Motil., 6, 585-600.

Cooke, R. (1981) Stress does not alter the conformation of a domain of the myosin cross-bridge in rigour muscle fibres. Nature 294, 570-571.

Eisenberg, E. and Hill, T.L. (1985) Muscle contraction and free energy transduction in biological systems. Science, 227, 999-1006.

Elliott, G.F., Rome, E.M. and Spencer, M. (1970) A type of contraction hypothesis applicable to all muscles. Nature 226, 417 420.

Ford, L.E., Huxley, A.F. and Simmons, R.M. (1977) Tension response to sudden length change in stimulated frog muscles near a slack length. J. Physiol. (Lond.). 269, 441-515.

Ford, L.E., Huxley, A.F. and Simmons, R.M. (1981) The relation between stiffness and filament overlap in stimulated frog muscle fibres. J. Physiol. (Lond.). 311, 219-249.

Ford, L.E., Huxley, A.F. and Simmons, R.M. (1986) Tension transients during steady shortening of frog muscle fibres. J. Physiol. (Lond.). 361, 131-150.

Geeves, M.A., Goody, R.S. and Gutfreund, H. (1984) Kinetics of acto-S1 interaction as a guide to a model for the cross-bridge cycle. J. Muscle Res. Cell Motil., 5, 351-362.

Goldman, Y.E., Hibberd, M.G., McCray, J.A. and Trentham, D.R. (1982) Relaxation of muscle fibres by photolysis of caged ATP. Nature 300, 701-705.

Goldman, Y.E., Hibberd, M.G. and Trentham, D.R. (1984a) Relaxation of rabbit psoas muscle fibres from rigour by photochemical generation of ATP. J. Physiol. (Lond.). 354, 577-604.

Goldman, Y.E., Hibberd, M.G. and Trentham, D.R. (1984b) Initiation of active contraction by photogeneration of ATP in rabbit psoas muscle fibres. J. Physiol. (Lond.) 354, 605-624.

Goldman, Y.E., Simmons, R.M. (1977) Active and rigour muscle stiffness. J. Physiol. (Lond)., 269, 55P-57P.

Gordon, A.L., Huxley, A.F. and Julian, F.J. (1966) The variation in isometric tension with sarcomere length in vertebrate muscle fibres. J. Physiol. (Lond.). 184, 170-192.

Hanson, J. and Huxley, H.E. (1955) The structural basis of contraction in striated muscle. Symp. Soc. exp. Biol., 9, 228-264.

Hibberd, M.G., Dantzig, J.A., Trentham, D.R. and Goldman, Y.E. (1985a) Phosphate release and force generation in skeletal muscle fibers. Science, 228, 1317-1319.

Hibberd, M.G. and Trentham, D.R. (1986) Relationships between chemical and mechanical events during muscular contraction. Ann. Rev. Biophys. biophys. Chem. 15, 119-161.

Hibberd, M.G., Webb, M.R., Goldman, Y.E. and Trentham, D.R. (1985b) Oxygen exchange between phosphate and water accompanies calcium-regulated ATPase activity of skinned fibers from rabbit skeletal muscle. J. Biol. Chem. 260, 3496-3500.

Hill, T.L. (1974) Theoretical formalism for the sliding filament model of contraction of striated muscle: Part I. Progr. Biophys. Molec. Biol., 28, 267-340.

Huxley, A.F. (1957) Muscle structure and theories of contraction. Progr. Biophys. Molec. Biol., 7, 255-318.

Huxley, A.F. (1973) A note suggesting that the cross-bridge attachment during muscle contraction may take place in two stages. Proc. Roy. Soc. B.: 183, 83-86.

Huxley, A.F. (1974) Muscular Contraction. J. Physiol. (Lond.) 243, 1-43.

Huxley, A.F. and Niedergerke, R. (1954) Interference microscopy of living muscle fibres. Nature 173, 971-973.

Huxley, A.F. and Simmons, R.M. (1971) Proposed mechanism of force generation in muscle. Nature 233, 533-538.

Huxley, H.E. (1969) The mechanism of muscle contraction. Science, 164, 1356-1366.

Huxley, H.E. and Hanson, J. (1954) Changes in the cross-striations of muscle during contraction and stretch and their structural interpretation. Nature, 173, 973-976.

Huxley, H.E. and Kress, M. (1985) Cross-bridge behaviour during muscle contraction. J. Muscle Res. Cell Motil., 6, 153-162.

Kimura, M. and Tawada, K. (1984) Is the S2 portion of the cross-bridge in glycerinated rabbit psoas fibers compliant in the rigor state? Biophys. J., 45, 603-610.

Lund, L., Webb, M.R. and White, D.C.S. (1985) Oxygen-exchange studies on insect fibrillar muscle. J. Muscle Res. Cell Motil., 6, 665-666.

Lund, J., Webb, M.R. and White, D.C.S. (1986) Changes in the ATPase mechanism of insect fibrillar flight muscle during Ca^{2+} and strain-activation probed by phosphate-water oxygen exchange. (In preparation).

Reedy, M.K., Holmes, K.C. and Tregear, R.T. (1965) Induced changes in orientation of the cross-bridges of glycerinated insect flight muscle. Nature 207, 1276-1280.

Rosenfeld, S.S. and Taylor, E.W. (1984) The ATPase mechanism of skeletal and smooth muscle acto-subfragment 1. J. Biol. Chem., 259, 11908 11919.

Schoenberg, M., Brenner, B., Chalovich, J.M., Greene, L.E. and Eisenberg, E. (1984) Cross-bridge attachment in relaxed muscle. In: Contractile Mechanisms in Muscle. Pollack, G.H. and Sugi, H., eds., Plenum: New York, pp 269-284

Sleep, J.A. and Smith, S.J. (1981) Actomyosin ATPase and muscle contraction. Current Topics in Bioenergetics, 11, 239-286.

Taylor, E.W. (1979) Mechanism of actomyosin ATPase and the problem of muscular contraction. Crit. Rev. Biochem., 6, 103-164.

ter Keurs, H.E.D., Iwazumi, T. and Pollack, G.H. (1978) The sarcomere length relation in skeletal muscle fibres. J. Gen. Physiol. 72, 565-592.

Trentham, D.R., Eccleston, J.F. and Bagshaw, C.R. (1976) Kinetic analysis of ATPase mechanisms. Quart. Rev. Biophys., 9, 217-281.

Webb, M.R., Hibberd, M.G., Goldman, Y.E. and Trentham, D.R. (1986) Oxygen exchange between Pi in the medium and water during ATP hydrolysis mediated by skinned fibers from rabbit skeletal muscle: evidence for Pi binding to a force generating state. J. Biol. Chem. (in press).

White, D.C.S. and Thorson, J. (1973) The kinetics of muscle contraction. Progr. Biophys. mol. Biol. 27, 173-255.

Woledge, R.C., Curtin, N.A. and Homsher, E. (1985) Energetic aspects of muscle contraction. Academic Press: London

Wray, J. (1979) Filament geometry and the activation of insect flight muscles. Nature 280, 325-326.

Yanagida, T., Arata, T. and Oosawa, F. (1985) Sliding distance of actin filament induced by a myosin cross-bridge during one ATP hydrolysis cycle. Nature 316, 366-369.

SMOOTH MUSCLE ENERGY METABOLISM: CYTOSOLIC COMPARTMENTATION OF METABOLISM AND FUNCTION

Richard J. Paul

Department of Physiology and Biophysics
University of Cincinnati, College of Medicine
Cincinnati, OH 45267-0576 U.S.A.

It was of particular interest to me to see the number of sessions here at the XXX Congress of Physiological Sciences related to the question of how ATP synthesis was coordinated with the energy required for muscle contraction. These dealt almost exclusively with striated muscle, for which it is obvious that muscle performance is dependent on metabolism. On an intuitive level this appears to be less obvious for smooth muscle, however, as for any muscle its metabolic "fitness" is critical to normal function. My research has focussed on the relations between smooth muscle metabolism and function, although there have been several rather different approaches. On the one hand, from the perspective of mechanochemical energy conversion, smooth muscle is unique in its ability to maintain force without fatigue at a remarkably low cost in terms of ATP utilization. One might wonder how this is possible given that the major contractile elements, the actin and myosin filaments, are quite similar to those of striated muscle. On the other hand, smooth muscle energy metabolism *per se* is of interest, particularly in the mechanisms underlying its coordination with contractility. The potential for contractility to be controlled at the level of metabolism is also of interest, particularly as a mechanism for sensing ambient O_2 levels in response to hypoxia. In this brief review based on my plenary lecture, I would like to summarize our current understanding of smooth muscle metabolism. My perspective, however, will be based on recent studies which indicate that energy metabolism is compartmentalized with function in smooth muscle. While not necessarily unique, vascular smooth muscle in particular provides one of the clearest examples of distinct metabolic components organized to support different functions. Smooth muscle thus may serve as a model for a more general biological phenomenon of compartmentation within the cytosol of both the synthetic and energy utilizing enzyme cascades. Given the nature of this work, I do not have the opportunity to credit appropriately my colleagues who worked with me, let alone to cite original works in this area. For more extensive reviews, both in terms of references and the development of ideas, I would suggest Paul (1980, 1986) and from a different perspective, Lundholm, Peterson, Andersson and Mohme-Lundholm (1983). For the most current thought on mechanochemistry or vascular metabolism and oxygen sensing mechanisms, I would recommend the works in the proceedings of satellite symposia to this meeting (Paul, Strauss and Krisanda, 1986; Paul, Ishida and Rubanyi, 1986, respectively). I should also like to add the caveat that I will be making some generalizations which should be taken cautiously. These are based primarily on our studies of hog coronary and carotid arteries and to a lesser extent on rat portal vein and guinea pig taenia coli. Although I will try to be as reasonable as possible, those closely involved in the field will recognize that differences amongst smooth muscles can be as large as those between smooth and skeletal muscle, making any generalization difficult.

One of the clearest ways to assess the "immediacy" of the role of metabolism in supporting contractility is to compare the levels of high energy phosphagen (ATP + PCr) with the rate of ATP utilization associated with contraction. The phosphagen content of smooth muscle, though comparable to many cell types, is up to 10-fold lower than striated muscle. For hog carotid artery, for example, a phosphagen content of about 1 μmol/g could support less than 1 minute of an isometric contraction, which for tonic smooth muscles like this, would not be sufficient to reach the peak of contraction. Thus intermediary metabolism must be closely coordinated with contractile energy requirements for normal function. In fact, under normal aerobic conditions, changes in phosphagen content concomitant with stimulation are difficult to detect in this preparation and there are no differences from the unstimulated control in the steady state. What this implies is that ATP synthesis and ATP utilization are nearly perfectly matched.

This rapid attainment of steady states allows non-destructible, continuous measurements of intermediary metabolism, such as oxygen consumption, to be used to assess tissue energy utilization during contractile responses. Over the past decade or so it has become clear that the level of isometric force is the primary determinant of steady state suprabasal oxygen consumption in stimulated smooth muscle. When stimulated under isometric conditions, the rate of O_2 consumption (J_{O_2}) rapidly increases to about 3 times the basal rate and declines to a steady state increase of about two-fold. Based on concomitant changes in isometric force (F_o) and J_{O_2} with muscle length at constant levels of stimulation, about 80% of this increase in suprabasal J_{O_2} is likely to be associated with the actomyosin ATPase.

Major questions being currently addressed on the energetics level include studies on the pre-steady states. The decline in J_{O_2} while F_o is maintained is translated to an increase in the economy of tension maintenance. The mechanisms underlying this apparent slowing of the cross bridge cycle at constant force and its relation to the overall high economy of tension maintenance, up to 1000-fold higher than skeletal muscle, are of current interest. Along these lines, the efficiency of work production in smooth muscle appears to be up to 5-fold less than skeletal muscle and fitting these observations into a model of mechanochemical energy transduction is the current task of smooth muscle physiologists.

On the metabolic side, a central question revolves about the nature of the signal for the observed increase in J_{O_2} with stimulation. As alluded to above, for several tissues there are no detectable changes in ATP or PCr from basal levels in the steady state when J_{O_2} is doubled. This implies that there is no change in cellular ADP levels, the most obvious candidate for regulation of mitochondrial respiration. This may argue that overall ATP and PCr levels are not representative of local changes, though other metabolites serving as regulators of oxidative metabolism have not been ruled out.

Although oxidative metabolism underlies most of the ATP synthesis in smooth muscle, most smooth muscles are characterized by an unusually large production of lactate under fully oxygenated conditions. Such aerobic glycolysis is limited to relatively few cell types, notably transformed cell lines and those whose lack of oxidative capacity can be explained in terms of their environment, as retinal or some kidney cells. In vascular smooth muscle, aerobic glycolysis has been often suggested to reflect some metabolic defect. Though minor in terms of ATP production, aerobic glycolysis accounts for nearly all of vascular glucose uptake.

TABLE 1

CONDITIONS	ISOMETRIC FORCE	OXYGEN CONSUMPTION	LACTATE PRODUCTION	Na-K TRANSPORT	PHOSPHORYLASE a ACTIVITY	GLUCOSE UPTAKE
(Changes with Respect to Basal, Unstimulated Conditions)						
KCl (80 mM)	↑	↑	↑	↑	↑	↑
KCl + ISOPROTERENOL	↓	↓	↑	↑	↑	
OUABAIN (10^{-5}M)	↑	↑	↓	↓	↑	—
Na^+-FREE (K^+ SUBSTITUTION)	↑	↑	↓	↓		—
K^+-FREE	↑	↑	↓	↓		

Over the past decade, our interest has focussed on the hypothesis that the ATP produced by aerobic glycolysis is specifically coupled to the energetic requirements of membrane ion pumps. The initial evidence was a correlation observed between the activity of the Na-pump and lactate production in porcine coronary arteries (Paul, Bauer and Pease, 1979). Interventions which inhibited the Na pump, such as ouabain or K^+- or Na^+-free solutions, were associated with an inhibition of lactate production, while those which were associated with stimulation of pump activity, high K^+ and isoproterenol, enhanced lactate production. Independent of the nature of the intervention, J_{O_2}, on the other hand, was always correlated with isometric force. A graphic summary of these experiments is presented in Table 1. It is important to note that oxidative metabolism and aerobic glycolysis can be independently varied depending on the conditions. Similar observations have been made for many different smooth muscles and, although the correlation between Na-pump activity and lactate production is not always as absolute as that observed for hog coronary artery, it is quite clear that oxidative metabolism and aerobic glycolysis can be independently varied. Given this independence, the correlation between aerobic glycolysis and membrane ion pump activity and importantly, the absence of trivial explanations such as anoxic core regions or lack of mitochondrial oxidative capacity, we proposed the hypothesis that this

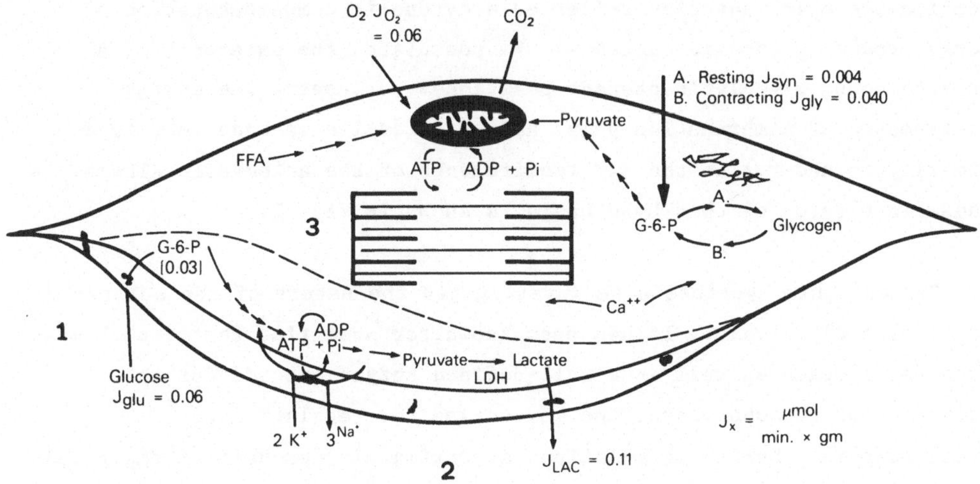

Fig. 1. Diagrammatic representation of carbohydrate metabolism by vascular smooth muscle. 1. Glucose is the sole substrate for aerobic glycolysis in unstimulated mechanically activated porcine carotid arteries. Since glucose transport is rate limiting for its utilization, transport must be regulated under conditions where carbohydrate utilization is increased. 2. Aerobic glycolysis can account for approximately 95% of glucose uptake, and is apparently coupled to Na-K transport at the sarcolemma. The coordination between glucose supply and its utilization at the membrane level, indicates a sensitive and efficient means of regulating carbohydrate metabolism. 3. On the other hand, glycogen utilization is coordinated with the increase in the rate of oxidative metabolism which is associated with the stimulation of mechanical activity. Glycogen breakdown is rapid and substantial during the initial minutes of mechanical activity, which may be particularly important for energy production during the presteady state of tension generation. The functional independence of the oxidative and glycolytic components of intermediary metabolism is expressed as a physical compartmentation of substrate utilization during increases in mechanical activity (represented by the dotted line). Although the rate of glycogenolysis is comparable in magnitude to the rate of glucose uptake under this condition, glucose remains the sole source of substrate for aerobic glycolysis. Rates are approximate values, expressed as μmol/min.g. The stimulated rate of glycogenolysis (J_{gly}) is time averaged over the initial 30 min of treatment. (From Lynch and Paul, 1985).

functional compartmentation reflected a cytosolic compartmentation of energy producing enzymic cascades. We postulated the existence of a membrane bound glycolytic cascade positioned to support the energy requirements of membrane ion pumps and an oxidative cascade localized primarily to underwrite the ATP requirements of the actomyosin ATPase. A schematic version of this hypothesis is shown in Fig. 1.

To test this hypothesis we investigated the nature of the substrate for aerobic glycolysis. It has been known for some time that smooth muscle contains glycogen as well as a well-defined enzyme cascade for the activation of phosphorylase, the key enzyme for regulation of glycogenolysis. During stimulation, approximately one-half of the total vascular content of glycogen is broken down by the time that isometric force achieves a steady state. Glucose uptake, on the other hand, is also comparable to glycogenolysis over this time period, and either substrate would be sufficient to account for the lactate produced. Using ^{14}C labelled glucose, we showed that despite the substantial input of glycosyl moities to the cell from glycogen, all the lactate produced originates from exogenous glucose (Lynch and Paul, 1983). Lactate accounted for over 90% of the glucose uptake with less than 3% of the radiolabel appearing as $^{14}CO_2$. These observations provide strong evidence for a cytosolic compartmentation with at least two distinct enzyme cascades for glycolysis and glycogenolysis. A corollary of this hypothesis is that glycolytic intermediates would also be compartmentalized. Supporting evidence for this was provided by measurements of the specific activity of glucose-6-phosphate (G-6-P). While the specific activity of lactate was shown to be identical to that of exogenous glucose under all conditions, G-6-P did not equilibrate with glucose and in fact, decreased under conditions in which glycogen breakdown was occurring. These observations support the hypothesis that there are multiple pools of glycolytic intermediates in the cytosol (Lynch and Paul, 1986).

Many questions can be raised concerning these observations. How can compartmentation be achieved in these small cells without any obvious cytosolic diffusion barriers? What advantage is such compartmentation to the cell? How is the independent regulation of oxidative metabolism and aerobic glycolysis obtained? What other manifestations of this compartmentation might be anticipated and how general is this phenomenon? We are in the initial stages of formulating approaches to answer some of these questions.

In the absence of glucose, smooth muscle can maintain normal ionic gradients. Because of this and the lack of observable structural barriers in the cytosol, the type of compartmentation we envision is one related to enzyme localization and the concomitant, preferable reaction-diffusion kinetics. Our hypothesis would require a membrane bound glycolytic cascade. We have approached this question in collaboration with Prof. Rik Casteels and his colleagues in Leuven. They have developed a purified plasma vesicle preparation from smooth muscle with the focus on understanding the nature of the membrane Ca^{2+}-pump (Raemaekers, Wuytack and Casteels, 1985). We have identified and quantified the presence of the major glycolytic enzymes in this preparation. Moreover, these enzymes can act in a concerted fashion and catabolize fructose-1,6 diP_i to lactate in the presence of glycolytic cofactors NAD, ADP and P_i. Of more significance perhaps, we have shown in preliminary experiments that Ca^{2+}-uptake itself in this preparation can be fueled by fructose-1,6 diP_i and these cofactors. Though clearly in the early stages, these preliminary findings support the hypothesis that a membrane bound glycolytic cascade supports membrane ion pump activity and promises to help shed light on what advantages the close apposition of energy-providing and energy-utilizing systems may have for the cell.

With respect to regulation, we have shown that glycogen phosphorylase is more strongly correlated with force, rather than the Na-pump (Paul, 1983). For example, ouabain inhibits the Na-pump and lactate production while isometric force, J_{O_2} and the level of phosphorylase _a_ are increased. This presumably reflects the well known dependence of phosphorylase kinase on Ca^{2+}. Regulation of the activity of phosphorylase is however quite complex, with other second messengers, primarily cAMP, playing major roles. To this end we have shown that neither the extent of relaxation nor phosphorylase _a_ activity is uniquely related to the total level of cAMP, suggesting that the cyclic nucleotides or their effects may themselves be compartmentalized (Rubanyi, Galvas, Di Salvo and Paul, 1986).

Glycolysis on the other hand appears to be regulated at the level of its transport across the plasma membrane. We have shown that under basal and KCl-stimulated conditions, the cellular content of glucose is negligible. This implies that the catabolic steps once glucose enters the cells are not rate limiting. We have further reported that glucose uptake is stimulated by high K^+, as would be expected from the increase in lactate observed under these conditions, and the fact that glucose is the sole substrate underlying the production of lactate (Lynch and Paul, 1984). The mechanism by which glucose transport is increased matching the increase in lactate correlated with Na-pump is unknown. However, this coupling raises the possibility that the functional unit includes not only the Na-pump and glycolytic cascade but also a regulated glucose transport system as well.

We have shown that oxidative and glycolytic metabolism are correlated with specific energy requiring processes and that minimally glycolysis and glycogenolysis occur via two distinct enzyme cascades in separate cytosolic compartments in vascular smooth muscle. We also have evidence that glycolytic intermediates, cyclic nucleotides (or their effect) and more recently ATP and PCr (Ishida and Paul, 1986) may be compartmentalized within the cytosol in smooth muscle. The advantages to the cell of such an arrangement are not as yet clear. However there are a number of theoretical arguments that can be made in terms of overall efficiency or reduction of the "transient time" between steady states (Welch, 1977). Compartmentation within the cytosol is not a totally novel idea, with some form of compartmentation proposed for erythrocyte, brain and cardiac muscle metabolism (Lynch and Paul, 1985). What is peculiar to smooth muscle is that both oxidative and glycolytic components are similar in magnitude making this metabolic compartmentation much more obvious. Thus studies of smooth muscle metabolism may be quite useful to our understanding of a more general phenomenon of cytosolic compartmentation.

ACKNOWLEDGEMENTS

I would like to thank Dr. H. McLennan, Chairman of the Programme Committee, for the invitation to be a Plenary Lecturer and for travel support from IUPS. The investigations reported here were supported in part by NIH 23240 and 22619, and an Established Investigatorship of the American Heart Association.

REFERENCES

Ishida, Y. and Paul, R.J. (1986) Lack of equilibrium between PCr and ATP in guinea pig taenia caecum: possible compartmentation of phosphagens. Fed. Proc. 45, 766.

Lundholm, L., Peterson, G., Andersson, R.G.G., and Mohme-Lundholm, E. (1983) Regulations of the carbohydrate metabolism of smooth muscle; some current problems. CRC Biochemistry of Smooth Muscle Vol. II, pp. 85-108.

Lynch, R.M. and Paul, R.J. (1983). Compartmentation of glycolysis and glycogenolysis in vascular smooth muscle. Science 222, 1344-1346

Lynch, R.M. and Paul, R.J. (1984) Glucose uptake by porcine carotid artery: relation to alterations in Na-K transport. Am. J. Physiol. 247, C433-C440.

Lynch, R.M. and Paul, R.J. (1985) Energy metabolism and transduction in smooth muscle. Experientia 41, 970-977.

Lynch, R.M. and Paul, R.J. (1986) Compartmentation of carbohydrate metabolism in vascular smooth muscle: evidence for at least two functionally independent pools of glucose-6-phosphate. Biochim. Biophys. Acta 887, 315-318.

Paul, R.J. (1986) Smooth Muscle: Mechanochemical Energy Conversion, Relations Between Metabolism & Contractility. In: Johnson, L.R. et al. (eds) Physiology of the Gastrointestinal Tract. 2nd Edition, Raven Press: New York, (in press).

Paul, R.J. (1983) The effects of isoproterenol and ouabain on oxygen consumption, lactate production and the activation of phosphorylase in coronary arterial smooth muscle. Circ. Res. 52, 683-690.

Paul, R.J. (1980) The chemical energetics of vascular smooth muscle. Intermediary metabolism and its relation to contractility. In: Handbook of Physiology, Section on Circulation II. Bohr, D.F., Somlyo, A.P. and Sparks, H.V., eds. American Physiological Society. pp 201-235.

Paul, R.J., Strauss, J.D. and Krisanda, J. (1986) The effects of calcium on smooth muscle mechanics and energetics. In: Smooth Muscle Contraction Symposium Siegman, M., Stephens, N.L. and Somlyo, A.P., eds. Liss: New York (in press).

Paul, R.J., Ishida, Y. and Rubanyi, G. (1986) Vascular smooth muscle metabolism and mechanics of oxygen sensing. In: The Pulmonary Circulation in Health and Disease Will, J.A. ed., Academic Press: New York (in press).

Paul, R.J., Bauer, M. and Pease, W. (1979) Vascular smooth muscle: aerobic glycolysis linked to Na-K transport processes. Science 206, 1414-1416.

Raemaekers, L., Wuytack, F. and Casteels, R. (1985) Subcellular fractionation of pig stomach smooth muscle. A study of the distribution of the (Ca^{2+} + Mg^{2+})-ATPase activity in plasmalemma and endoplasmic reticulum. Biochim. Biophys. Acta 815, 441-454.

Rubanyi, G., Galvas, P., Di Salvo, J. and Paul, R.J. (1986) Interactions of vascular prostaglandin biosynthesis and beta-adrenergic mechanism in coronary arterial smooth muscle: Functional compartmentation of cyclic AMP. Am. J. Physiol. 250, C406-C412.

Welch, G.R. (1977) On the role of multienzyme systems in cellular metabolism: A general synthesis. Prog. Biophys. Mol. Biol. 32, 103-191.

THYROID HORMONES, MEMBRANES AND THE EVOLUTION OF ENDOTHERMY

A.J. Hulbert

Department of Biology
University of Wollongong
Wollongong, N.S.W. 2500 Australia

INTRODUCTION

Marking the end of the "age of the reptiles" and the beginning of the "age of the mammals" at approximately 65 million years ago, is a geological layer rich in the unusual element, iridium. It has been proposed that this layer of iridium is the remains of an extraterrestrial body that hit the earth and changed the earth's climate, resulting in both a short term cooling effect (possibly measured in months) as well as long term changes in the climate, such as greater latitudinal variation in temperature and the beginning of greater temporal variation in climate (the ushering in of seasons). Although the first mammals arose 200 million years ago, for the first two-thirds of their evolutionary history they were not the dominant land fauna and it was only after the extinction of the dominant reptile fauna at about 65 million years ago that they began an explosive adaptive radiation that led, among other species, to ourselves, a species through which evolution has begun to become conscious of itself.

One of the fundamental differences between reptiles and mammals is illustrated in Fig. 1. It is that mammals are endotherms and maintain a relatively constant internal temperature by the production of appropriate amounts of internal heat whilst reptiles are ectotherms and do not have this physiological capacity. In ectothermic vertebrates body temperature (T_b) varies directly with, and is slightly greater than, environmental temperature and their metabolic rate varies with body temperature normally with a Q_{10} of 2-3. Mammals, however, show an inverse relationship between

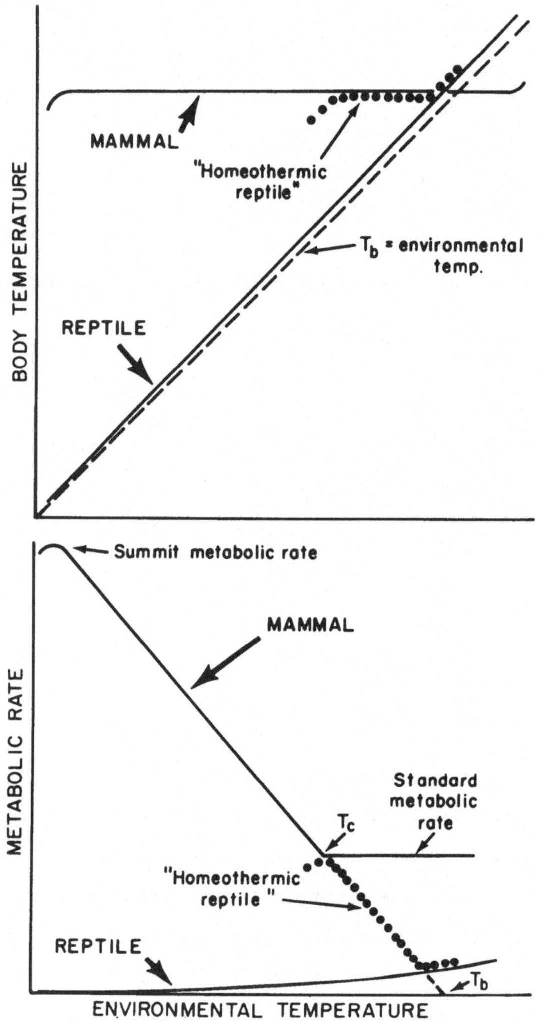

Fig. 1. Relationships of body temperature (T_b) and metabolic rate to environmental temperature in a mammal and a reptile of the same body size. Also shown (circles) are the relationships for a hypothetical "homeothermic reptile". The lower critical temperature for the mammal is shown as T_c.

their metabolic rate and environmental temperature. All groups of mammals are capable of increasing their heat production in the cold by shivering and some are capable of using non-shivering thermogenesis (Hulbert, 1980b). At its lower critical temperature (T_c) the metabolic rate of a mammal reaches a minimal level that is described at its "standard metabolic rate" (SMR). As can be seen from Fig. 1 the SMR of a mammal is about 4-5 times greater than the resting metabolic rate of a similar-sized reptile at the same body temperature. The summit metabolic rate of a mammal appears to be related to this minimal level by a relatively constant ratio. For the sake of comparison I have also drawn in this figure the relationship one might expect from a mammal with a "reptilian" level of metabolism or conversely, a reptile that has the "mammalian" neural machinery capable of maintaining a constant body temperature. I have called this hypothetical animal a "homeothermic reptile".

The evolution of the neural machinery that allows a mammal to sense environmental temperature and create enough internal heat to maintain a constant body temperature is not the subject of this lecture. Rather I will concern myself with the evolution of two other characteristics of endothermic mammals; (1) the increase in the overall level of metabolism in mammals and (2) the inability of most mammals to tolerate low body temperatures. I will postulate that both of these evolutionary changes were due, at least in part, to an evolutionary increase in the activity of the thyroid gland in mammals compared to the gland's activity in reptiles and the consequent effect of the thyroid hormones on cellular membranes, specifically membrane lipids (Hulbert, 1978). However let us first concern ourselves with the evolution of endothermic metabolism in mammals.

THE EVOLUTION OF ENDOTHERMIC METABOLISM IN MAMMALS

The first recorded recognition of the metabolic difference between endotherms and ectotherms (with body size and body temperature taken into account) was by August Krogh, who in the middle of the tragic and futile First World War published results which indicated "that the oxidative energy of the tissues is greater in the warm-blooded than in a cold-blooded organism" (Krogh, 1916). Since that time many people have measured and compared the metabolic rates of endotherms and ectotherms. Standard metabolism varies with both body size and body temperature and the general conclusion from all these studies is that the level of standard metabolism of an endothermic mammal is approximately 4-5 times that of a similar sized ectothermic reptile at the same body temperature (Hemmingsen, 1960; Hulbert, 1980a). This difference also applies when the maximal aerobic activity metabolism of the two groups is compared (Bennett, 1978).

In order to understand the cellular basis of this large difference in the level of energy metabolism, a few years ago we undertook a detailed comparison of a mammal and a reptile of the same body size and the same body temperature. Our initial comparison was between the mouse (<u>Mus</u>

musculus) and the central netted dragon (Amphibolurus nuchalis), a small lizard with a preferred body temperature of about 37°C and the same size as the mouse. Later the comparison was changed to the larger mammal, the rat Rattus norvegicus and the bearded dragon (Amphibolurus vitticeps) which also had a preferred body temperature of about 37°C and was the same size as the rat (Else and Hulbert, 1968 and in preparation).

In both of these comparisons the internal organs (liver, kidney, heart and brain) were significantly larger (almost twice as big) in the mammal compared to the reptile (Else and Hulbert, 1981 and in preparation). In humans, these tissues are responsible for approximately 72% of the resting heat production although they only account for about 8% of the total body weight. We took a large number of random electron micrographs from these tissues and measured by stereological techniques both the amount of cell volume occupied by mitochondria and the amount of mitochondrial membrane surface area. The results showed that one cc of each of these internal organs from the mammal contained about twice the amount of mitochondrial membrane as the equivalent reptilian tissue. When the effect of larger organs in the mammal is taken into account we can conclude that the mammalian tissues have a total mitochondrial membrane surface area that is about 4-5 times that found in the reptile. As well as comparing cellular structure we also measured the activity of cytochrome oxidase, the mitochondrial enzyme that is responsible for the cellular consumption of oxygen. When given excess reduced cytochrome C we found that, at 37°C, the total cytochrome oxidase activity of the mammalian tissues is also about 4-5 times that of the reptilian tissues. From both a structural and a functional comparison we concluded that mammalian tissues had evolved an increased capacity to produce ATP, approximately of the order of 4-5 fold greater than that found in the reptiles.

However these were single species comparisons and to be sure that it was a valid generalization we extended our measurements to a diverse range of both mammals and reptiles. The basic logic has been that any parameter common to lizards, crocodiles and tortoises is a fundamental "reptilian" characteristic that was also present in the original reptilian ancestor. Similarly, any parameter common to monotremes, marsupials and eutherian mammals is a "fundamental" mammalian characteristic also present in the earliest mammals. By this method and the more detailed single species comparison we can hopefully deduce the changes that took place during the reptile-mammal evolutionary transition. When this diverse range of reptiles and mammals was compared we found that the generalizations from the single species comparison also hold true for the broader comparison. That is, the internal organs of mammals are generally larger than those of reptiles and the mitochondrial membrane surface area is on average about four times that found for equivalent reptilian tissues. This generalization applied to skeletal muscle as well as the internal organs and thus may help explain the difference in maximal aerobic metabolism as well as the difference in standard metabolism. Another interesting finding is that the allometric slopes relating the mitochondrial parameters to body weight are not dissimilar to those relating whole organismal metabolic rates of these species to body size (Else and Hulbert, 1985).

Examination of mitochondrial parameters looks at only one side of the energy cycle. In other words, although mammals evolved an increased capacity to produce metabolic energy, that does not necessarily mean that they use this extra capacity. The fact that they have a higher standard metabolism than reptiles certainly implies that they use considerably more energy even when at rest and at the same body temperature. But for what cellular activities do they use this 4-5 fold increase in metabolic energy? This is not an easy area to examine. Cellular energy is thought to be required for three main classes of activity; biosynthetic work, transport work and mechanical work. We have attempted to examine two of these areas. For example, comparison of the initial species of mammal and reptile revealed that the growth rate of the young mammal is an order of magnitude greater than that measured in the similar sized reptile kept at the same body temperature (Hulbert and Else, 1981). This large difference in growth rates of endotherms and ectotherms has been reported by other workers (Case, 1978) and is probably a reflection of several-fold greater biosynthetic rates in endotherms compared to ectotherms.

The reptile-mammal difference in level of metabolism is also manifest when the oxygen consumption of tissue slices is measured by manometry. Mammalian tissues use considerably more energy in vitro at 37°C than do the equivalent reptilian tissues at the same temperature (Hulbert and Else, 1981). The use of ouabain, a specific inhibitor of Na^+/K^+ ATPase (the "sodium pump"), has enabled us to estimate and compare the energy requirements for sodium-potassium pumping in reptilian and mammalian tissues. Although not dissimilar proportions of cellular metabolism are used for transmembrane sodium-potassium pumping in mammalian and reptilian tissues, the much greater total rates of oxygen consumption of mammalian tissues mean that they use several times more energy for this process than does the same amount of equivalent reptilian tissue at the same temperature (Hulbert and Else, 1981; Else and Hulbert, 1986).

There are a number of possible explanations for this large difference: (i) the mammalian pumps may be more inefficient than the reptilian pumps (i.e. they may consume more ATP in pumping the same amount of Na^+ and K^+), (ii) the gradients of Na^+ and K^+ may be greater across the mammalian cell membrane than across the reptilian cell membrane, (iii) the mammalian tissues may have more membrane surface area across which to maintain these ion gradients, and (iv) mammalian cell membranes may be "leakier" than reptilian cell membranes thus requiring greater pumping activity to maintain the same transmembrane gradients. Neither of the first two explanations seems likely in that the scientific literature suggests that "sodium pumps" from ectotherms and endotherms are basically very similar and that transmembrane Na^+/K^+ gradients are similar in a wide range of vertebrates. For the third alternative to be the explanation the mammalian cell would have to be an order of magnitude smaller than its reptilian counterpart or its surface would have to be highly convoluted. From our extensive electron microscopic survey of mammalian and reptilian tissues this does not seem to be the case. Very little is known about the final possibility so we decided to compare the passive permeability of the reptilian and mammalian cell membrane to both Na^+ and K^+.

Liver and kidney slices from both the reptile and mammal were loaded with $^{42}K^+$, transferred to nonradioactive medium and the efflux of the $^{42}K^+$ was followed at 37°C and in the presence of ouabain (to block sodium pump activity). When this was done it was observed that the mammalian cell membranes were considerably "leakier" to potassium than were the corresponding reptilian cell membranes (Else and Hulbert, 1986). For an increased "leakiness" partly to explain the increased energy requirements, the mammalian cell would also need to be leakier to Na^+. Since it is difficult to measure the passive inward leak of Na^+ in tissue slices it was necessary to prepare isolated cells. Hepatocytes were prepared from the reptile and mammal and when uptake of $^{22}Na^+$ was measured at 37°C in the presence of ouabain it was observed that the mammalian cell was eightfold more permeable to sodium ions than was the reptilian cell (Else and Hulbert, 1986). That the mammalian cell membrane was several-fold "leakier" to both of these ions than was the reptilian cell membrane meant that the membrane pumps in the mammalian cell must use more energy to maintain the same transmembrane ion gradients. In a way, this can be regarded as a type of "futile cycle" and is a prime candidate for explaining, at least in part, the large difference in the standard metabolism between ectotherms and endotherms. Presently investigations are under way to see if this same difference holds for other tissues, other species, and possibly other molecules.

The question now becomes why should mammalian cell membranes be more passively permeable to ions than reptilian cell membranes? One intriguing possibility is that this difference is due to different membrane fatty acids in mammalian cell membranes. Increasing unsaturation of membrane lipids has been correlated with increased permeability to ions and nonelectrolytes in some mammalian systems (De Gier, Mandersloot and van Deenen, 1968; Hendriks, Klompmakers, Daemen and Bonting, 1976), whilst the addition of lipids containing polyunsaturated fatty acids to frog skin epithelia preparations has been shown to result in an increased passive sodium permeability (Yorio, Torres and Tarapoom, 1983). We have isolated the phospholipids from the tissues used in the $^{42}K^{+}$ experiments mentioned above and analyzed their fatty acid composition. For both of the tissues examined, the mammalian phospholipids contained significantly more polyunsaturated fatty acids than the reptilian phospholipids (Else and Hulbert, in preparation). I would like to propose that the increased permeability of the mammalian cell membrane is possibly due to their changed fatty acid composition compared to reptilian cell membranes.

To summarize so far, our findings suggest that during the reptile-to-mammal evolutionary transition there was an increase in the size of the metabolically active organs. These organs accrued a greater amount of ATP generating machinery, that is, an increase in mitochondrial membrane surface area. The mammalian tissues grew faster than the reptilian tissues and spent more energy on sodium/potassium transport because their cell membranes became more "leaky" to these ions than were the reptilian cell membranes.

In mammals, the thyroid hormones are known to affect all of these parameters. The thyroid is a gland which takes up iodide from the circulation, makes thyroid hormones and then either stores them or secretes the thyroid hormones into the circulation. In the plasma, the hormones exist in both a free and a bound form and travel to the tissues where they are metabolized and exert their action. We have examined the activity of the thyroid in a variety of reptiles. In the crocodile, tortoise and lizard there is an area of ^{125}I uptake that corresponds to the thyroid gland and when we follow this area over time we observe that although there is uptake there is no release of ^{125}I at 20-22°C. At their preferred body temperature of 30-32°C there was both uptake and release of ^{125}I from this area in all of these reptiles, which indicates some secretory activity of the thyroid gland at this temperature (Hulbert, 1985; Hulbert, Williams and Grigg, 1986). However when these radioiodine release rates (and others from the literature) are compared to those for a range of mammals (including monotremes, marsupials and eutherians), they indicate that the thyroid is considerably less active in reptiles than in mammals (Hulbert, 1985; Hulbert et al., 1986). This is not due to a difference in body temperature since the mammals have body temperatures that range from 32-39°C and, as well, when thyroid activity was measured in our initial reptile-mammal single species comparison at 37°C the same difference was apparent (Hulbert and Else, 1981). Similarly, the plasma levels of both triiodothyronine and thyroxine are much less in reptiles than they are in mammals (Hulbert et al., 1986).

In mammals, an increase in thyroid activity has been shown to result in an increase in tissue mitochondrial membrane surface area (Gustaffson, Tata, Lindberg and Ernster, 1965; Jakovcic, Swift, Gross and Rabinowitz, 1978; Reith, Brdiscka, Nolte and Staudte, 1973), an increase in ion permeability and consequently ion transport (Asano, 1977; Haber and Loeb, 1984; Ismail Beigi and Edelman, 1971), an alteration in membrane fatty acid composition (Hoch, Subramanian, Dhopeshwarkar and Mead, 1981; Hulbert, 1978; Patton and Platner, 1970; Shaw and Hoch, 1977) as well as increased growth rate and metabolic rate in general (Evans, Taurog, Koneff, Potter, Chaikoff and Simpson, 1960). In fact the similarity in growth rate of a thyroidectomized rat and a reptile is uncanny (Evans et al., 1960).

The fact that all these parameters previously shown to be different in mammals and reptiles are influenced by thyroid status suggests that it was the evolutionary increase in thyroid activity that was at least in part responsible for the evolution of endothermic metabolic level in mammals.

As mentioned previously, another change took place during the reptile-mammal transition. Mammals, in general, lost the ability to tolerate low body temperatures. Although relatively little work has been done comparing this difference between ectothermic and endothermic vertebrates, it is likely that the inability of mammals to tolerate low body temperatures is related to their inability to maintain normal membrane function at low temperatures, specifically cellular ion gradients (Hochachka, 1986; Willis, 1979). Hochachka (1986) has suggested that this and an inability of mammalian tissues to tolerate hypoxia may both be related to their "leakier" membranes. Time does not permit a full discussion of this area but I want to concentrate on one membrane-associated activity, that is, mitochondrial function. When isolated mitochondria from mammals are cooled they generally show an abrupt decrease in their enzymic activity at approximately 20-22°C, whilst mitochondria from ectothermic vertebrates, such as toads, generally show no such abrupt change (McMurchie, Raison and Cairncross, 1973). This is normally of no importance to the mammal because its body temperature is usually considerably higher than 22°C. This property of mammalian mitochondria is related to changes in the physical state of the mitochondrial membrane lipids (McMurchie et al., 1973) and the temperature at which this change takes place (T^*) is in turn related to the fatty acid composition of these membrane lipids. Although no work has been done directly on reptiles and mammals examining this difference it probably also applies to the reptile-mammal transition.

There is a group of mammals which during part of the year adopt a poikilothermic strategy and tolerate very low body temperatures. These are the hibernating mammals and some of them, notably the ground squirrels, allow their body temperatures to drop to levels near 0°C. I want to concentrate on one particularly good and interesting recent study by Augee, Pehowich, Raison and Wang (1984). These authors examined the enzymic activity, at temperatures ranging from 2°C to 35°C, of liver mitochondria isolated from the ground squirrel Spermophilus richardsonii. They found that the ground squirrels fell into three groups; some had a T^* of about 22°C, whilst others had a T^* of about 13°C and the remaining group had T^* less than 4°C. The value of T^* depended on the animals' history. Animals killed in summer had a T^* of 22°C which is typical of mammals in general. Animals killed during the hibernating season but which had been kept at an ambient temperature of 19°C and had remained homeothermic all showed T^* of

13°C, whilst animals killed during the hibernating season but kept at 4°C showed T^* of either 13°C or <4°C. Animals killed in torpor all showed a T^* of <4°C. They concluded that lowering of T^* was a two stage process and that the lowering of T^* to 13°C was a seasonal event that was a prerequisite for hibernation. The second lowering was either a response to the low body temperature experienced during hibernation or low ambient temperatures.

What is interesting is that the first membrane change is probably due to a seasonal decrease in thyroid activity. We have shown that following thyroidectomy the T^* of rat liver mitochondria drops from the normal value of about 23°C to about 10°C and that this is correlated with changes in mitochondrial membrane fatty acid composition (Hulbert, Augee and Raison, 1976). This is of interest because in a related species of ground squirrel the thyroid gland becomes very inactive prior to the hibernating season (Hulbert and Hudson, 1976). Is it too much to suggest that just prior to the return to the "reptilian-like" condition of a low body temperature that the thyroid returns to "reptilian-like" level of activity and this results in changes in the membrane fatty acids and consequent changes in membrane function? These changes in mitochondria may not be related to survival at low temperature as such but may be related to the need to maintain mitochondrial function for arousal from the hibernating state. However, changes in membrane composition and function with hibernation are not just restricted to mitochondria. We have found seasonal changes in the T^* of red blood cell membranes from the hedgehog that are correlated with changes in thyroid activity (Augee, Raison and Hulbert, 1979) and similar changes in T^* of Na^+/K^+ ATPase in the brains of hamsters have been related to changes in the composition of the membrane fatty acids in the hamster's brain prior to hibernation (Goldman, 1975).

CONCLUSIONS

In conclusion, although more work needs to be done, a picture is emerging of some of the physiological changes that took place during the evolution of endothermic mammals from their ectothermic reptilian ancestors. This is summarized in Fig. 2. I would postulate that one of the primary changes that took place during the reptile-mammal transition was that there was an increase in thyroid activity. This resulted in an increase in metabolic activity to the endothermic level, which is characterized by larger internal organs, increased capacity to produce ATP

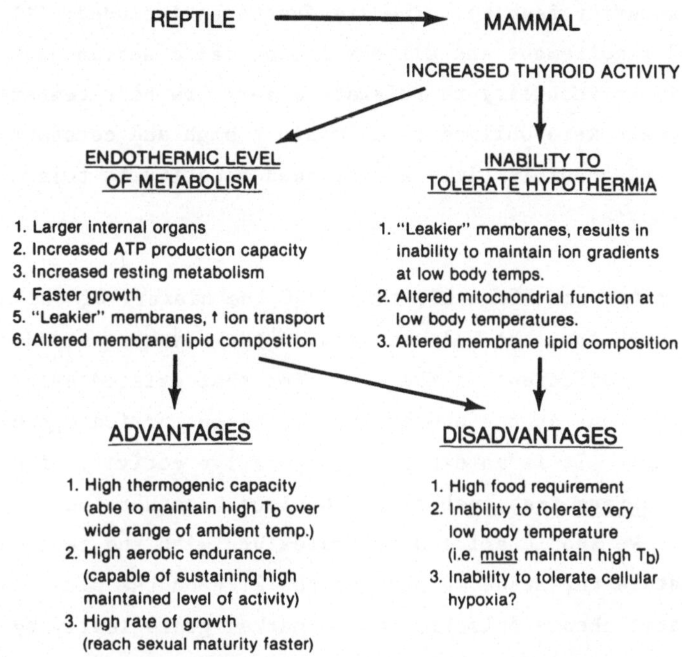

Fig. 2. Summary of proposed metabolism and body temperature changes that took place during the reptile-mammal transition and some of their evolutionary advantages and disadvantages.

(greater mitochondrial membrane surface area), increased resting metabolism and faster growth. The increase in metabolic rate is due in part to "leakier" cell membranes which resulted in an increased energy requirement for ion transport. Many of these changes may be related to a change in the fatty acid composition of membrane lipids.

The other consequences of this evolutionary increase in thyroid activity was the loss of the ability to tolerate low body temperatures. This was probably because the "leakier" membranes were unable to maintain ion gradients at low body temperatures and mitochondrial function at these temperatures was severely depressed. Both of these problems may also have been due to a thyroid-induced change in membrane fatty acid composition.

Like much evolutionary change there were both disadvantages and advantages for the organism involved. The advantages included; (i) a higher thermogenic capacity (which enabled the mammal to maintain a high body temperature over a wide range of ambient temperature), (ii) a high aerobic capacity (which enabled the mammal to sustain a high level of maintained activity), and (iii) a faster growth rate (which enabled it to reach sexual maturity faster). The disadvantages included; (i) a much increased food requirement and all the ecological disadvantages that go with that, (ii) an inability to tolerate a very low body temperature (which meant that mammals were obliged to maintain a high and constant body temperature), and possibly (iii) a decreased capacity to tolerate periods of cellular hypoxia.

Presumably for the first two thirds of the history of the mammals, the biosphere was such that the balance between the advantages and the disadvantages was different to the situation that existed after the reptilian extinctions at the end of the Cretaceous period approximately 65 million years ago. It is an exciting cooperative activity of a very recent species, Homo sapiens that enables us to piece together the history of life on our planet. We should all concern ourselves with the fact that our competitive nature threatens so much of the life on the planet with another possible climatic change this time to be marked geologically by a layer rich in the man-made element, plutonium.

ACKNOWLEDGEMENTS

I would like to express my appreciation to my coworkers, specifically Drs. P.L. Else, M.L. Augee, J.K. Raison and G.U. Inness. Much of this work was supported by grants from the University of Wollongong and the Australian Research Grants Scheme.

REFERENCES

Asano, Y. (1977) Increased cell membrane permeability to Na^+ and K^+ induced by thyroid hormone in rat skeletal muscle. Experientia Suppl. 32, 199-203

Augee, M.L., Raison, J.K. and Hulbert, A.J. (1979) Seasonal changes in membrane lipid transitions and thyroid function in the hedgehog. Am. J. Physiol. 236, 589-593

Augee, M.L., Pehowich, D.J., Raison, J.K. and Wang, L.C.H. (1984) Seasonal and temperature-related changes in mitochondrial membranes associated with torpor in the mammalian hibernator Spermophilus richardsonii. Biochim. Biophys. Acta 776, 27-36

Bennett, A.F. (1978) Activity metabolism of the lower vertebrates. Ann. Rev. Physiol. 40, 447-469

Case, T.J. (1978) On the evolution and adaptive significance of postnatal growth rates in the terrestrial vertebrates. Quart. Rev. Biol. 53, 243-282

De Gier, J., Mandersloot, J.G. and van Deenen, L.L.M. (1968) Lipid composition and permeability of liposomes. Biochim. Biophys. Acta 150, 666-675

Else, P.L. and Hulbert, A.J. (1981) A comparison of the "mammal machine" and the "reptile machine": energy production. Am. J. Physiol. 240, R3-R9

Else, P.L. and Hulbert, A.J. (1985) An allometric comparison of the mitocchondria of mammalian and reptilian tissues: the implications for the evolution of endothermy. J. Comp. Physiol. B. 156, 3-11

Else, P.L. and Hulbert, A.J. (1986) The evolution of mammalian endothermic metabolism: "leaky" membranes as a source of heat. Am. J. Physiol. (submitted)

Else, P.L. and Hulbert, A.J. A biochemical comparison of metabolism in mammalian and reptilian tissues (in preparation).

Evans, E.S., Taurog, A., Koneff, A.A., Potter, G.D., Chaikoff, I.L. and Simpson, M.E. (1960) Growth response of thyroidectomized rats to high levels of iodide. Endocrinol. 67, 619-634

Goldman, S.S. (1975) Cold resistance of the brain during hibernation. III. Evidence of a lipid adaptation. Am. J. Physiol. 228, 834-838

Gustaffson, R., Tata, J.R., Lindberg, O. and Ernster, L. (1965) Relationship between structure and activity of rat skeletal muscle mitochondria after thyroidectomy and thyroid hormone treatment. J. Cell. Biol. 26, 555-578

Haber, R. and Loeb, J.N. (1984) Early enhancement of passive potassium efflux from rat liver by thyroid hormone: relation to induction of Na,K-ATPase. Endocrinol. 115, 291-297

Hemmingsen, A.M. (1960) Energy metabolism as related to body size and respiratory surfaces and its evolution. Reps. Steno Mem. Hosp. 9, 1-110

Hendriks, Th., Klompmakers, A.A., Daemen, F.J.M. and Bonting, S.L. (1976) Biochemical aspects of the visual process. Biochim. Biophys. Acta 433, 271-281

Hoch, F.L., Subramanian, C., Dhopeshwarkar, G.A. and Mead, J.F. (1981) Thyroid control over biomembranes: VI Lipids in liver mitochondria and microsomes of hypothyroid rats. Lipids 16, 328-335

Hochachka, P.W. (1986) Defence strategies against hypoxia and hypothermia. Science 231, 234-241

Hulbert, A.J. (1978) The thyroid hormones: a thesis concerning their action. J. Theor. Biol. 73, 81-100

Hulbert, A.J. (1980a) The evolution of energy metabolism in mammals. In: Comparative Physiology: Primitive Mammals Schmidt-Nielsen, K., Bolis, L., Taylor, C.R., Bentley, P.J. and Stevens, G.E., eds. University Press: Cambridge, pp 129-139

Hulbert, A.J. (1980b) Evolution from ectothermia towards endothermia. In: Advances in Physiol. Sci. Szelenyi, Z. and Szekely, M., eds. vol. 32, Pergamon: London, pp 237-247

Hulbert, A.J. (1985) A comparative study of thyroid function in reptiles and mammals. In: The Endocrine System and the Environment Follett, B.K., Ishii, S. and Chandola, A., eds. Springer: Berlin, pp 105-115

Hulbert, A.J., Augee, M.L. and Raison, J.K. (1976) The influence of thyroid hormones on the structure and function of mitochondrial membranes. Biochim. Biophys. Acta 455, 597-601

Hulbert, A.J. and Else, P.L. (1981) A comparison of the "mammal machine" and the "reptile machine": energy use the thyroid activity. Am. J. Physiol. 241, R350-R356

Hulbert, A.J. and Hudson, J.W. (1976) Thyroid function in a hiberator, Spermophilus tridecemlineatus. Am. J. Physiol. 230, 1211-1216

Hulbert, A.J., Williams, C.A. and Grigg, G.C. (1986) Temperature and thyroid function in an Australian lizard, tortoise and crocodile and a comparison with mammals. Gen. Comp. Endocrinol. (submitted)

Ismail Beigi, F. and Edelman, I.S. (1971) The mechanism of the calorigenic action of thyroid hormone: stimulation of Na^+ and K^+ activated adenosinetriphosphate activity. J. Gen. Physiol. 57, 710-722

Jakovcic, S., Swift, H.H., Gross, N.J. and Rabinowitz, M. (1978) Biochemical and stereological analysis of rat liver mitochondria in different thyroid states. J. Cell. Biol. 77, 887-901

Krogh, A. (1916) The Respiratory Exchange of Animals and Man. Longman Green: London

McMurchie, E.J., Raison, J.K. and Cairncross, K.D. (1973) Temperature-induced phase changes in membranes of heart: a contrast between the thermal response of poikilotherms and homeotherms. Comp. Biochem. Physiol. 44B, 1017-1026

Patton, J.F. and Platner, W.S. (1970) Cold acclimation and thyroxine effects on liver and liver mitochondrial fatty acids. Am. J. Physiol. 218, 1417-1422

Reith, A.D., Brdicska, D., Nolte, J. and Staudte, H.W. (1973) The inner membrane of mitochondria under the influence of triiodothyronine and riboflavin deficiency in rat heart muscle and liver. Exp. Cell Res. 77, 1-14

Shaw, M.J. and Hoch, F.L. (1977) Thyroid control over biomembranes: IV Rat heart muscle mitochondria. J. Mol. Cell. Cardiol. 9, 749-761

Willis, J.S. (1979) Hibernation: cellular aspects. Ann. Rev. Physiol. 41, 275-286

Yorio, T., Torres, S. and Tarapoom, N. (1983) Alteration in membrane permeability by diacylglycerol and phosphatidylcholine containing arachidonic acid. Lipids 18, 96-99

MODERN CONCEPTS OF REGULATION OF CEREBRAL MACRO- AND MICROCIRCULATION

George Mchedlishvili

I. Beritashvili Institute of Physiology
Georgian Academy of Sciences
14 Gotua Street
380060 Tbilisi, U.S.S.R.

The present lecture is an attempt to show modern concepts related to the physiological mechanisms of the regulation of cerebral circulation. Progress in this important physiological field proceeded rather specifically. Owing to their extreme significance for medical practice, the cerebral blood flow problems became subjects of numerous applied studies in animals and humans, while comparatively few fundamental investigations related to the physiological mechanisms of cerebral blood flow regulation were being carried out. The lack of deep physiological insight into the structural and functional aspects of the regulation could only have hampered the actual progress in this important field.

HISTORICAL SKETCH

It is well known that the "Monro-Kellie doctrine" dominated physiological thinking in the course of the 19th Century and maintained some of its influence even up to the early 1930s. The doctrine implied that due to the presence of three incompressible elements inside the rigid skull - blood, cerebrospinal fluid, and brain substance - the cerebral blood flow is always constant, and hence defies any physiological regulation.

However, as a result of amassing new experimental data during the first half of the 20th Century this view changed gradually. Currently there is a large body of evidence indicating that the cerebral blood flow is perfectly regulated, and we are even able now to trace back the reasons accounting for the development of such a refined regulating system in animals and humans. Firstly, the brain is highly important for the performance of vital activities of the whole body, and therefore in extreme situations the operation of the regulatory mechanisms of the whole circulatory system is directed so as to provide the brain with blood even at the expense of other parts of the body. Secondly, it is quite clear at present that this perfect regulating system has developed as a result of a very long evolutionary period, from primordial vertebrates to primates. One important conclusion which can be drawn from the above is that such a refined regulatory system operative in the brain, is an ideal object for the physiological investigation of both the cerebral macro- and microcirculation.

Proceeding from my own 30 year-long experience, I can add that the anatomical arrangement of the cerebral vascular bed provides perfect opportunities for analyzing its functional behaviour, which is of crucial importance for understanding the physiological mechanisms regulating the cerebral macro- and microcirculation. However, unfortunately, not many physiologists engaged in the study of these exciting problems seem to be making the best use of this advantageous opportunity.

The conceptual shift in the physiological views on cerebral blood flow regulation is schematically shown in Fig. 1.

NOTION OF CEREBRAL BLOOD FLOW REGULATION

The notion of regulation of physiological functions has undergone noticeable transformation throughout the recent three decades. The notion originated long ago, when it became evident to physiologists that the bodily vital functions are perfectly coordinated with, and adjusted to, each other. However, at that period the traditional physiological approach was confined to investigating various effects on some physiological

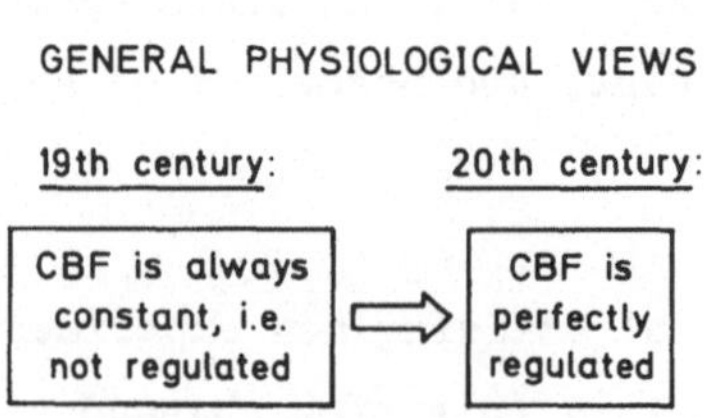
GENERAL PHYSIOLOGICAL VIEWS
19th century:
20th century:
CBF is always constant, i.e. not regulated
CBF is perfectly regulated

Fig. 1.

functions, e.g. the effect of stimulation of cardiac nerves on heart function, or the action of various endogenous vasoactive substances on vessel diameters, etc. Correspondingly the notion of regulation came to imply nothing but just such physiological effects on various functions of the body. Even now one may occasionally come across the use of the word "regulation" with such meaning.

In the meanwhile, with the appearance of cybernetics in the middle of this century, it became evident that the principles of control and regulation are common to all the diverse phenomena in nature, living things, social life and technology. Careful insight into various types of the physiological control of cerebral blood flow has provided evidence that the latter fully conforms to the general principles of regulation. Thus, currently it has become evident that owing to the natural physiological mechanisms of regulation some of the circulatory parameters of the brain are actively maintained constant, whereas others are adjusted to certain functional changes. Further, it has become evident that any physiological regulating system operates via feedback or feedforward mechanisms. This certainly applies to the macro- and microcirculation of the brain, as to any other physiological function.

Transformation of the notion of physiological regulation since the 1960s is schematically presented in Fig. 2.

CEREBRAL CIRCULATORY PARAMETERS BEING REGULATED

These parameters can be divided into two basic groups. The first group embraces those which are actively maintained at a constant level, with the second group involving those circulatory parameters which adjust to, i.e. change in concert with, other physiological processes.

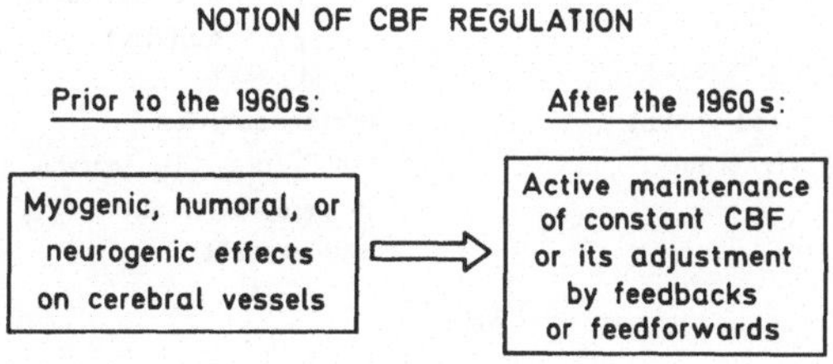
NOTION OF CBF REGULATION
Prior to the 1960s:
After the 1960s:
Myogenic, humoral, or neurogenic effects on cerebral vessels
Active maintenance of constant CBF or its adjustment by feedbacks or feedforwards

Fig. 2.

Proceeding from our present-day knowledge, it can be stated that those parameters of cerebral circulation which are actively maintained constant, belong to the circulatory phenomena of the whole brain (i.e. macrocirculation) and are kept unaltered owing to the operation of certain physiological mechanisms of regulation. Thus the cerebral blood pressure and cerebral blood flow are actively maintained constant despite changes in the systemic arterial pressure within a certain range. The same applies to the cerebral blood volume, which is also maintained constant despite occasional impediments of venous blood outflow from the skull. Therefore the venous blood stagnation does not easily occur in the brain.

On the other hand, those circulatory parameters which are invariably adjusted to some others pertain to cerebral microcirculation. Thus the blood flow rate in minor brain regions is regularly changed to conform to the functional, and hence the metabolic activity of the tissue elements surrounding the cerebral microvessels. This in turn is provided by the operation of a specific physiological mechanism of regulation. Variations in the cerebral blood flow rate are also adjusted to the tension of blood gases, i.e. of oxygen and carbon dioxide, in the cerebral blood and tissue. This regulation serves to maintain the tension of blood gases in the brain tissue at a constant level.

All the principal circulatory parameters regulated in the brain are schematically presented in Fig. 3.

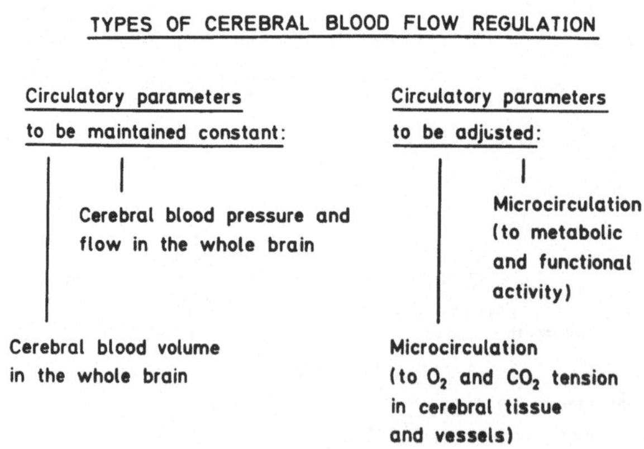
TYPES OF CEREBRAL BLOOD FLOW REGULATION
Circulatory parameters to be maintained constant:
Cerebral blood pressure and flow in the whole brain
Cerebral blood volume in the whole brain
Circulatory parameters to be adjusted:
Microcirculation (to metabolic and functional activity)
Microcirculation (to O_2 and CO_2 tension in cerebral tissue and vessels)

Fig. 3.

OPERATION PRINCIPLES OF PHYSIOLOGICAL MECHANISMS OF CEREBRAL BLOOD FLOW REGULATION

As I have already mentioned, physiological mechanisms regulating any function in the living body, in particular the cerebral circulation, operate either by feedback or feedforward, as is the case in any other sphere involving controlling mechanisms. To find out how such mechanisms regulate the circulatory parameters within the brain, one will have to answer the following questions: a) what are the triggers of the mechanism (e.g. stretch of vessel walls, metabolic changes producing some vasoactive substance, or stimulation of nervous receptor structures) and how do they operate; b) how and where is the information spread about specific changes (i.e., diffusion of vasoactive substances towards the blood vessel walls or afferent nervous impulses) in order to trigger the controlling mechanism; c) where and how is the information processed (e.g. in nervous centres) to produce controlling signals and to coordinate them with the adjoining functions; d) what kind of physiological controlling signals (i.e. humoural and/or neurogenic) reach the specific effectors of the regulation; and e) what are the effectors (specific blood vessels in the present case) of this type of cerebral blood flow regulation? The feedback mechanism regulating the cerebral blood flow is schematically shown in Fig. 4.

To analyze particular physiological mechanisms regulating specific parameters of macro- and microcirculation of the brain the most rational way is to start with the identification of effectors of the regulation, and to trace back all the links of the regulating mechanism to specific triggers of the regulation.

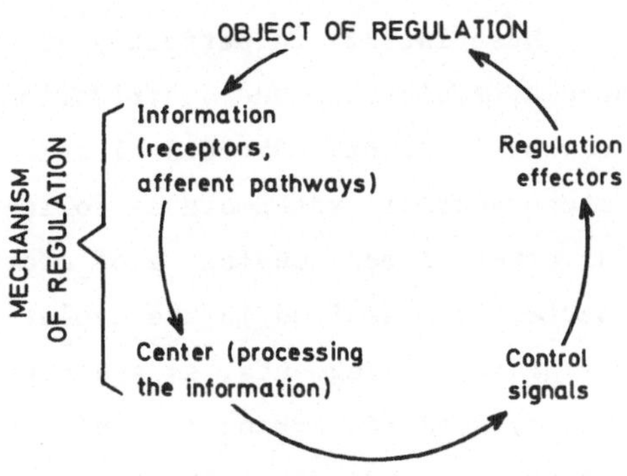

Fig. 4.

VASCULAR EFFECTORS OF CEREBRAL BLOOD FLOW REGULATION

It is common knowledge that any regulation of regional circulation in any organ of the body is fulfilled by changes in the peripheral resistance. Under physiological conditions this latter is primarily controlled by active diameter changes of arterial portions of the peripheral vascular bed. Formerly it was believed that these active portions were arterioles, i.e. the smallest precapillary vessels with a single sheet of smooth muscle cells in their walls. Later on, with gradual accumulation of new physiological data, it became evident that larger arterial branches of organs might also be involved as vascular effectors in the regulation of their blood supply.

As to specific types of cerebral blood flow regulation, it has been established that under natural conditions the constancy of circulatory parameters belonging to cerebral macrocirculation (i.e. the averaged cerebral blood pressure and flow as well as the blood volume in the whole brain) is provided for by active operation of the system of major brain arteries, i.e. by the internal carotids and vertebrals (as well as partly by larger cerebral arteries originating from the circle of Willis). These vascular effectors are abundantly supplied with adrenergic and cholinergic nerves. In addition, they are perfectly arranged from the anatomical point of view: located outside the cranial space and hence independent of intracranial pressure changes; interconnected, to provide the collateral blood supply when necessary; possessing curvature along their course, which effectively changes vascular resistance even at minor vasomotor responses.

Vascular effectors of regulation of cerebral microcirculation have been identified for the cerebral cortex and turned out to be primarily the smallest pial arteries. These latter are perfectly arranged to provide adequate control of vascular resistance and distribution of blood among the smallest areas of the cerebral cortex. Structural and functional peculiarities of this microvascular system are as follows: a) modular organization of the pial arterial bed consisting of microcircles, which became gradually more perfectly organized in the evolutionary process of vertebrates; b) presence of active segments, as sphincters at offshoots and precortical arteries; and c) specific arrangement of vascular bifurcations, which provide that the most effective regulation of vascular resistance changes is fulfilled by the smallest pial arteries with the diameter approximately under 100 μm.

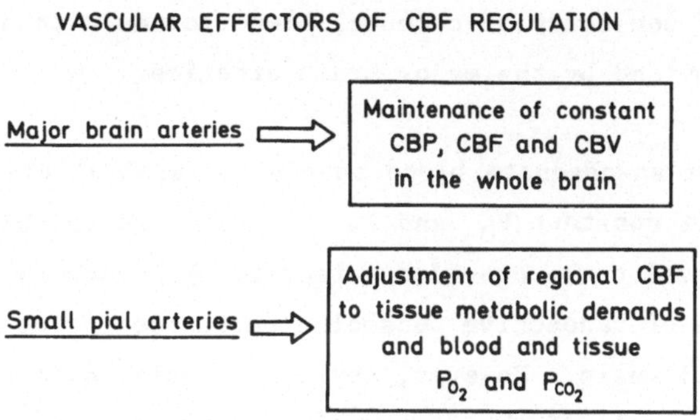

Fig. 5.

As to further ramifications of the pial arteries inside the cerebral cortex, in accordance with the anatomical and physiological evidence they are not as active as the regulators of microcirculation within the cerebral cortex.

Function of the major brain arteries and the pial arteries as vascular effectors of various types of cerebral blood flow regulation is schematically shown in Fig. 5.

PHYSIOLOGICAL MECHANISMS OF VASOMOTOR RESPONSES DURING REGULATION OF CEREBRAL CIRCULATION

It had been held traditionally up to the 1960s that the cerebral blood flow regulation is deprived of a neurogenic vasomotor control. At that time physiologists believed that there existed only myogenic regulation (in case of primary changes of intravascular pressure) and humoural or metabolic control (in case of adequate blood supply to the tissue) of cerebral blood vessels.

However, identification of specific vascular effectors of every type of the cerebral blood flow regulation (see above) provided new opportunities for detailed experimental analysis of the physiological mechanisms of this regulation. Thus, starting from the early 1960s convincing experimental evidence has been gradually accumulated proving that the maintenance of constant cerebral blood pressure and flow by the major brain arteries is primarily accomplished via the neurogenic mechanism. As to the myogenic responses of these arteries, their role is insignificant and might be of importance only at acute intravascular pressure changes. The neurogenic mechanism is presumably responsible for the regulation of constancy of cerebral blood volume within the skull, which is also provided by the major brain arteries.

Regulation of an adequate blood supply to cerebral tissue, as well as the maintenance of constant P_{O_2} and P_{CO_2} in the cerebral blood and tissue, is commonly believed to be provided primarily by a humoural mechanism, i.e. by direct effects of vasoactive metabolic substances or of the blood gases on cerebral vessel walls. However, new experimental data give all grounds to assume that the role of the neurogenic vasomotor mechanism is equally important in these types of regulation of the cerebral blood flow.

COORDINATING SYSTEMS OF CEREBRAL BLOOD FLOW REGULATION AND THEIR TRIGGERING MECHANISMS

It seems to cause no doubt at present that the operation of physiological mechanisms of the cerebral blood flow regulation (as it is actually in a healthy organism) requires special "centres". Their function consists in processing all the information about changes which need either appropriate regulation or stabilization of the cerebral blood flow, and producing the corresponding controlling signals reaching specific vascular effectors of the cerebral blood flow regulation. Another function of these centres is to coordinate the circulatory changes occurring with the related processes in order to exclude various unfavourable effects, e.g. the "steal phenomenon", etc. However, the existence of the "centres" with such a complex function is completely excluded in cases of operation of the myogenic and humoural mechanisms of the cerebral blood flow regulation. Undoubtedly the centres might exist only when the neurogenic mechanism is involved in any type of cerebral blood flow regulation.

So far there are few available physiological data related to this problem. What seems to be out of the question today is that there exist more than a single nervous centre in the brain, which process the information and send controlling signals to specific vascular effectors of the cerebral blood flow regulation. It can be conjectured that the centres regulating the cerebral blood pressure, flow, and volume inside the whole brain are located somewhere in the hypothalamus or in the brain stem. As to the triggers of these types of the cerebral blood flow regulation, they are in all probability mechanoreceptors, located either in the cerebral arterial bed (i.e. the baroreceptors, in case of regulation of a constant

CENTERS OF CBF REGULATION

1. General center(s):	2. Local centers:
Maintaining the constancy of CBP, CBF, and CBV in the whole brain	Adjusting the regional CBF to tissue metabolic demands and blood P_{O_2} and P_{CO_2}

Fig. 6.

cerebral blood pressure and flow), or in the walls of cerebral veins (for the regulation of a constant cerebral blood volume within the skull).

On the other hand, the coordinating systems processing the information (e.g. about an inadequate blood supply, etc.) during the regulation of adequate blood supply to particular areas of cerebral tissue seems to be sited locally, i.e. in just those cerebral tissue regions where the microcirculation is regulated. As to the specific information triggering these centres, this problem remains almost uninvestigated and is in need of further careful and detailed studies.

The two types of centres responsible for cerebral blood flow regulation are schematically presented in Fig. 6.

CONCLUDING REMARKS

The above-mentioned consideration indicates a significant conceptual shift in the physiological knowledge concerning the mechanisms of regulation of the cerebral circulation in the course of the recent 25 years. The concepts of the cerebral blood flow regulation changed from almost a complete negation of the active regulation to the recognition of its superior perfection, as compared with any other part of the body. Besides, many details of the regulating system have been elucidated at present. However, many findings related to these mechanisms still do not seem to have been given due attention so far. Physiological data substantiating the concepts presented in this report are given in greater detail in the present author's book entitled "Arterial Behavior and Blood Circulation of the Brain" published by Plenum Publishing Corporation in New York in 1986.

ROLE OF CHEMICAL SUBSTANCES IN THE CONTROL OF FOOD INTAKE

Yutaka Oomura

Department of Physiology
Faculty of Medicine
Kyushu University
Fukuoka 812, Japan

The dependence of body functions on the process of bioassay has been implied if not evident since the earliest study of biology. We discuss here the contributions of several endogenous chemicals to the control of feeding, and relations of neuronal and endocrine processes to this control. I will describe the feeding-related effects of several chemicals, some recently discovered or synthesized, and some well known but only recently evaluated. I then hope to show that bioassay is a viable and effectual function of the nervous system, and the system's hierarchy of control.

Blood borne materials such as glucose, insulin, free fatty acids (FFA) (Oomura, 1976), glucagon (Inokuchi, Oomura, Shimizu and Yamamoto, 1986a), calcitonin (Shimizu and Oomura, 1986), and estrogen (Nabekura, Oomura, Minami, Mizuno and Fukuda, 1986; Oomura, Minami and Nabekura, 1986a), all of which are related to metabolism, are known to be involved in the induction of hunger or satiety. Behavioural and electrophysiological experiments have shown that these and other humoural factors might act as hunger and satiety signals. Infusion of glucose, glycerol, 3-hydroxybutyrate (Davies, Wirthshafter, Asir and Brief, 1981) or insulin (Woods and Porte, 1983; Plata-Salaman, Oomura and Shimizu, 1986) into the third cerebral ventricle has depressed feeding with consequent loss of body weight which persisted for at least the duration of the infusion.

Blood from rats in various stages of deprivation was analyzed to identify blood factors that might affect feeding and body weight (Shimizu, Oomura and Sakata, 1984). Two short chain sugar acids, 3,4-dihydroxybutanoic acid (2-deoxytetronic acid, 2-DTA) and 2,4,5-trihydroxypentanoic acid (3-deoxypentonic acid, 3-DPA), were isolated. The blood serum levels of 3-DPA increased and reached a peak about 12 h after the start of food deprivation, and 2-DTA decreased slightly 24 h after the start of deprivation but then increased to a peak at about 48 h. Continued deprivation further increased both 2-DTA and 3-DPA. The variations in the blood levels of 2-DTA and 3-DPA suggest that they may affect food intake by mediating hunger and satiety. Since the blood level of ketone bodies rises significantly during starvation, 3-hydroxybutyric acid (3-HBA) might also affect food intake.

EFFECT OF INTRA-THIRD VENTRICULAR INJECTION OF 2 DTA AND 3-DPA ON FEEDING BEHAVIOUR OF RATS

At 1030, just before a period when little or no feeding is normally evident, 2-DTA or 3-DPA was injected into the third ventricle and ingestive behaviour was measured for the next 2 h. Injections of 1.25 or 2.5 μmole 2-DTA did not induce feeding or drinking in any of 17 rats tested. Injection of 2.5 μmole 3-DPA or 3-HBA induced eating and drinking for 6 to 8 min, after latencies of 8 to 25 min, in more than 60% of the rats tested. Control injections of NaCl or mannitol did not induce feeding. Feeding was induced in two of nine rats by 1.25 μmole doses of 3-DPA. Non-specific stereotyped jumping was induced by 5.0 μmole of either 2-DTA and 3-DPA. Thus, physiological quantities of 3-DPA and 3-HBA might enhance, and 2-DTA might suppress food intake.

Feeding Facilitation by 3-DPA

Depending on the time of injection, 3-DPA induced increases of food intake. After 2.5 μmole 3-DPA was injected at 1030, the consumption of pellets increased transiently, but the increase over 24 h was not significant. Injection of 3-DPA at 1900 increased food intake and meal duration between 2000 and 2300 to 1.5 times the preinjection values. That increase was significant, but food intake in the 9 h period, from 2300 to 0800 was not significantly greater.

Feeding Inhibition by 2-DTA

Food intake of pelleted chow was measured for 24 h after 2.5 μmole injections of 2-DTA into the third ventricle at 1630. In 9 rats, the mean number of pellets consumed in 24 h decreased significantly to one half. Consumption recovered in the period from 24 to 48 h. Similar injections of saline had no effect. Since motor activity was the same during the test period as before, the suppression of food intake was apparently not due to depression of general behaviour.

Injection of 2-DTA into the third ventricle at 1900 significantly reduced food intake from 2000 to 0100 in 72 h deprived rats. Deprived rats treated with 2-DTA ate the same amount as non-deprived rats, and deprived rats injected with artificial cerebrospinal fluid (ACSF) ate more than twice as much. Food intake from 0 to 12 h, and from 0 to 24 h after 2-DTA injection was significantly less than that of the ACSF controls in the corresponding periods. Food intake from 24 to 48 h after injection was almost the same for both groups.

Dual Action of 3-HBA

Injection of 2.5 μmole of 3-HBA into the third ventricle at 1030 induced one transient episode of feeding, and then depressed further feeding for 24 h. Normal dark time feeding was almost completely suppressed.

NEURONAL ACTIVITY MODULATION IN THE LHA BY 2-DTA, 3-DPA AND 3-HBA IN CHRONIC RATS

Third ventricle injection of 2-DTA, 3-DPA or 3-HBA affected both single LHA neurone activity and feeding behaviour. Availability of food at 1800, after 24 h of food deprivation, induced immediate simultaneous increases in the activity of a single neurone being recorded in the LHA, in both food and water intake. Injection of 2-DTA at 1830 abolished consumption of food and prandial water, and depressed neuronal activity within about 5 min. At 1850 neuronal activity was significantly depressed. There was usually no further feeding, and neuronal activity was depressed for the next 4 h. Neurones identified as feeding-related (Katafuchi, Oomura and Yoshimatsu, 1985) by correlation between their activity changes and food intake, also responded to 2-DTA injection, but not to motor activity. Injection of 3-DPA into a non-deprived, ad lib fed rat at 1410 induced both consumption of food and water, and a 40 min increase in neuronal activity.

Injection of 3-HBA at 1130 induced an increase in activity of a single LHA neurone that began within 5 min, and continued for more than 50 min before returning to the original level. Neuronal activity did not decrease during feeding suppression at night.

Effects of 2-DTA, 3-DPA and 3-HBA on single LHA neurone activity and feeding behaviour are summarized for 9 rats. The activity of 64% of LHA neurones was decreased by 2-DTA with 2 to 12 min latency, and no feeding behaviour was observed. Activity was increased by 3-DPA (67%), and feeding behaviour accompanied 4 of the increases. In two cases, 3-DPA affected neither neuronal activity nor feeding behaviour. Activity of 60% of the LHA neurones tested was increased by 3-HBA, and feeding behaviour was induced in 70% of the tests. None of the neurones tested decreased in activity during the depression of feeding by 3-HBA. Neuronal responses to these factors are discussed in more detail later.

DEPENDENCE OF BLOOD GLUCOSE, INSULIN AND FREE FATTY ACIDS ON 2-DTA, 3-DPA AND 3-HBA

Intra-third ventricular injection of 2-DTA increased blood glucose about 15%, first transiently, and again persistently for 8 h. Insulin levels did not change significantly. FFA initially increased 25%, but after 8 h its normally expected pre-feeding increase did not occur. Glucose was depressed 20% by 3-DPA, but recovered later; insulin level increased to 2.6 times control for several hours and FFA increased initially and was then suppressed, but not as much as by 2-DTA. Glucose and insulin were transiently increased by 3-HBA (30% and 1.8 times, respectively), and hunger-induced FFA increase was only slightly affected. The 3-DPA and 3-HBA increases in insulin level agree with the increase in vagal efferent activity and decrease in splanchnic efferent activity to the pancreas when 3-DPA or 3-HBA increases LHA activity (Oomura, 1983; Oomura and Kita, 1981).

PERIPHERAL APPLICATION OF 3-DPA AND 2-DTA

Injection of 100, 250 or 500 μmole of 3-DPA into the jugular vein at 1900, the time of its greatest effect, enhanced cumulative food consumption dose-dependently from 2000 to 0100. Cumulative consumption in 5 h was increased 17% by 100, 30% by 250, and 39% by 500 μmole 3-DPA. The 5 h and 24 h food consumption were both significantly increased by the 500 μmole dose of 3-DPA.

Similar doses of intravenous (i.v.) 2 DTA injected at 1900 did not significantly affect food consumption by non-deprived animals. This might be explained by failure of i.v.-injected 2-DTA to reach the brain through the blood-brain barrier, or 2-DTA adhesion to some organ tissue. Endogenous 2-DTA in the blood may be conjugated with some lipid-soluble fraction that facilitates passage through the blood-brain barrier, and deproteinization of the serum for the analysis might have removed some lipid-soluble fraction from the total sample of 2-DTA in the ethanol solvent. To test this last possibility, 2-DTA was encapsulated in liposome vesicles and injected intraperitoneally (i.p.) (Terada, Sakata, Oomura, Fukimoto, Arase, Osanai and Nagai, 1986). Injection of 1.25 and 2.5 mmole encapsulated 2-DTA significantly suppressed feeding dose-dependently. It was estimated that 8% of the liposome treated 2-DTA was actually encapsulated and that 1% of the encapsulated 2-DTA, i.e. 1 and 2 μmole, respectively, reached the brain. This dose was comparable to that diffusing from the 2.5 μmole injected into the third ventricle. Relatively massive doses of 2-DTA (2.5 mmole) injected into the carotid artery have been reported to suppress feeding. This suggests that entrapment of systemically applied 2-DTA within the blood circulatory system also might diminish its effectiveness.

In the production of synthetic 2-DTA, the immediate precursor of the final product is 2-butene-4-olide (2B40). It is not yet known if this derivative of 2-DTA is also a precursor in the metabolic cycle in which 2-DTA is produced, but its similarity to 2-DTA suggested that it be tested for its effectiveness in suppression of food intake. When injected into the third ventricle at 1.25, 2.5 and 5 μmole concentrations, 2B40 dose-dependently suppressed food intake with a potency similar to that of 2-DTA. It did not affect water intake. Furthermore, unlike 2-DTA, 2B40 was effective when injected i.p. or intragastrically (i.g.). When administered at 1800 (2 h before dark) by either of these routes, it also dose-dependently (i.p., 30, 50 and 100 mg/kg; i.g., 50, 100 and 300 mg/kg) suppressed food intake. Thus, the suppressive effects of 2-DTA can be realized in either of two ways that permit peripheral administration: encapsulation of 2-DTA in a liposome; or use of its derivative, 2B40. Use of the derivative seems to be by far the more convenient and economical procedure.

The lactone ring arrangments of 2-DTA (it exists in blood in this form), its derivative and its analogues are shown below:

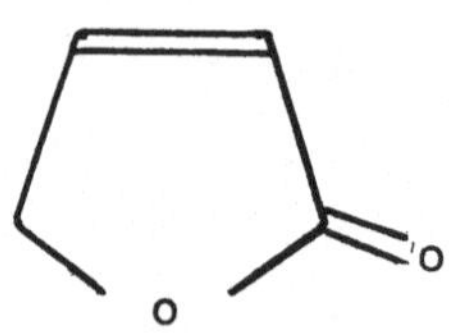

2-butone-4-olid

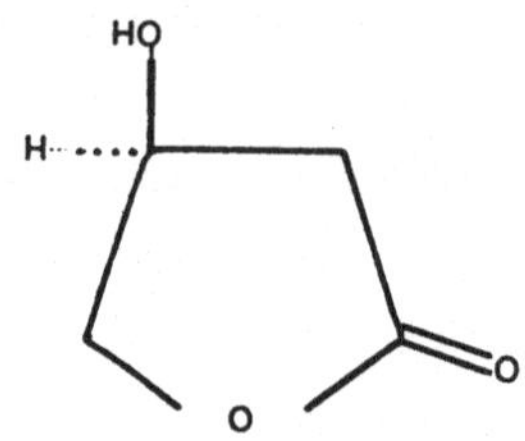

3,4-dihydroxy-
butanoic acid

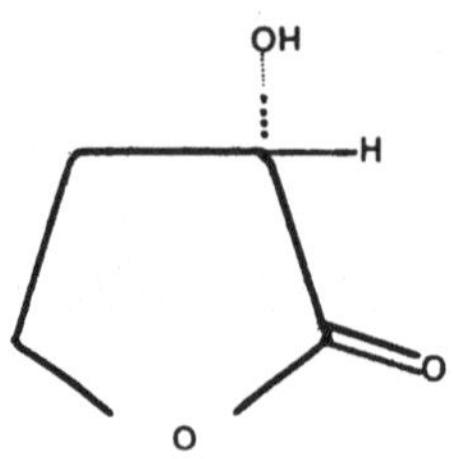

2,4-dihydroxy-
butanoic acid

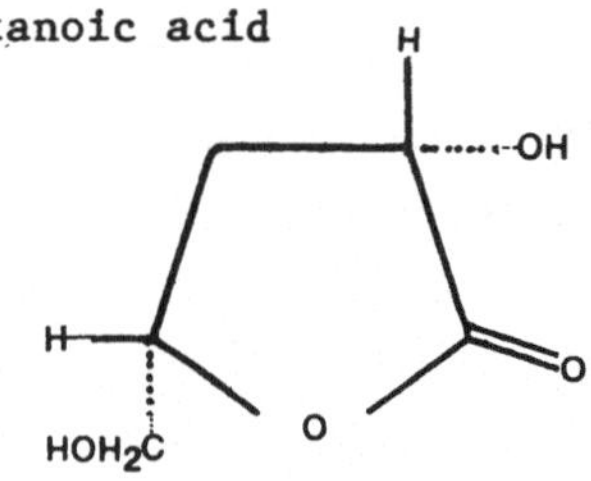

2,4,5-trihydroxy-
pentanoic acid

The upper two behave similarly, and the two below behave similarly; the effects of the upper two are essentially opposite to those of the other two.

The metabolites and hormones mentioned earlier (glucose, FFA, insulin, glucagon, estrogen, calcitonin) are all agents with well known effects on feeding behaviour. However, their neuronal effects are probably less generally appreciated, so we will discuss their neuronal relations here in more detail and assume acceptance (within the limits of possible controversy) of their effects on feeding behaviour. Before proceeding with this, we should describe two groups of neurones that are diametrically related to the same behaviour, feeding.

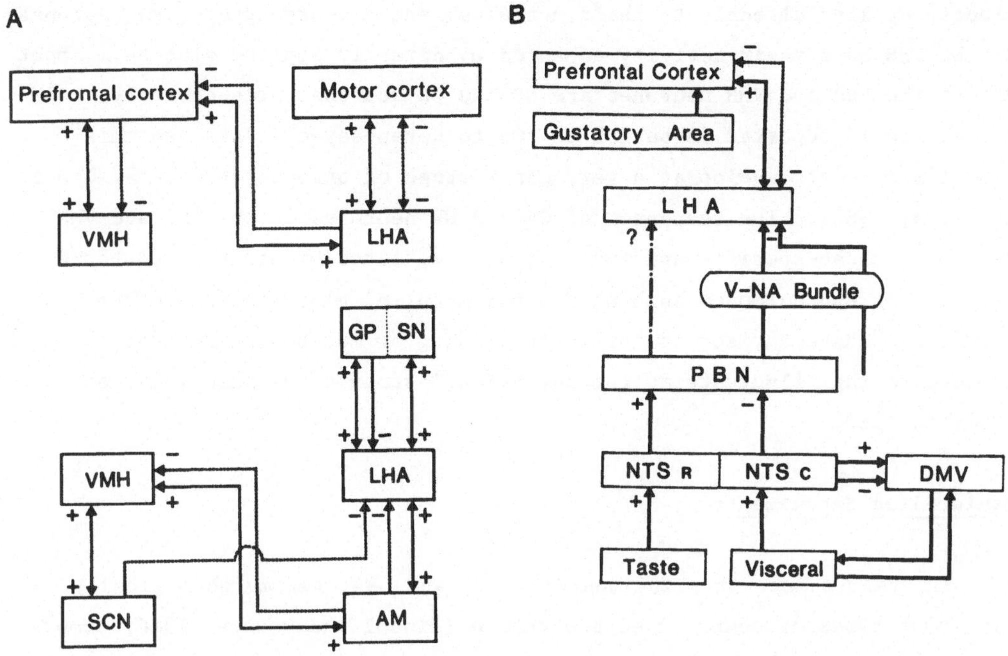

Fig. 1. A: Summary of functionally verified limbic, prefrontal cortex, pyramidal (motor cortex), and extrapyramidal (GP, SN) connections related to feeding control. Arrows show orthodromic directions. B: Hierarchical organization of sensory and chemical information processing. Broken line is known to exist, but is not yet characterized. Information from gustatory area goes to LHA through prefrontal cortex; taste also relayed through NTS and DMV as are visceral afferents and efferents. In A and B: Signs, excitatory (+) and inhibitory (-) connections. Solid lines, monosynaptic connections; dotted line, polysynaptic. AM, amygdala; GP, globus pallidus; SCN, suprachiasmatic nucleus; SN, substantia nigra. DMH, dorsomedial hypothalamic nucleus; DMV, dorsal motor nucleus of the vagus; NTS_R and NTS_C, rostral and caudal parts of nucleus tractus solitarius; PBN, parabrachial nucleus; V-NA bundle, ventral noradrenergic bundle (Oomura, 1985).

GLUCOSE-SENSITIVE (GS) AND GLUCORECEPTOR (GR) NEURONES

Glucose effects on neurones in the central nervous system were first observed in the LHA and VMH (Fig. 1). Precise placement of glucose applications as small as 0.1 pmole within the hypothalamus by electrophoresis verified that neurones in the hypothalamus respond directly to glucose. The activity of GS neurones in the LHA is suppressed by glucose applied directly to their surfaces, and glucoreceptor (GR) neurones in the VMH have their activity enhanced by directly applied glucose. About 25% of the LHA and VMH neurones are GS and GR neurones, respectively. The percentage is reported to be 16% in two-to-seven-day-old rats, so this sensitivity must develop at a very early stage of ontogeny (Shibata, Oomura and Kita, 1982). The responses of GS and GR neurones in the LHA and VMH are not glucose-specific and most are also affected by other blood-borne metabolites and hormones such as FFA and insulin, glucagon, and other substances characterized as intrinsic feeding or metabolism-related components that fluctuate as the nutritional state of an animal varies (Oomura, 1983).

Brain Slice Experiments

In brain slices that included the VMH and LHA, two morphologically different types of neurones were observed (Minami and Oomura, 1986; Oomura, 1983). One of these was bipolar with few dendrites, and the other multipolar with extensive dendritic arborization and many spines. It was electrophoretically determined that the multipolar neurones were GR and GS neurones, while the bipolar neurones were not. Stimulation with depolarizing electrical pulses induced action potentials in both GR and non-GR types, but with different wave shapes. Hyperpolarization following the action potential induced in bipolar neurones was slight or absent, but definite hyperpolarization appeared after the action potential in multipolar neurones. The after-hyperpolarization reached about -90mV, the K^+ equilibrium potential E_K. The multipolar neurones had I_A current after application of an hyperpolarizing pulse. Both I_A current and after-hyperpolarization tend to reduce the firing rate. Non GR neurones fire at a higher rate, and exhibit the properties of interneurones (Minami et al., 1986).

In the presence of the Na^+ channel blocker tetrodotoxin (TTX), a depolarizing pulse produced a graded response only, and no spike potential in the bipolar neurone. A slow action potential with a higher threshold, a Ca^{2+} spike, was evoked in the multipolar neurone. In the presence of TTX plus tetraethylammonium, a K^+ channel blocker that facilitates Ca^{2+} spikes, slow action potentials were induced in both neurone types. This indicates that the Ca^{2+} spike is easily produced in the dendrites, i.e. in the multipolar neurones. These effects, plus the report by Llinás and Sugimori

(1980) that Ca^{2+}-induced K^+ permeability is much greater in dendrites than in other neurone membrane areas, indicate that the after-hyperpolarization is due to a Ca^{2+}-induced K^+ current.

Glucose application to a GR neurone depolarizes the membrane and decreases the membrane conductivity. This increases the frequency of the spikes. The depolarization and decrease in conductivity are due to decreased K^+ permeability since the reversal potential for the glucose depolarization is -90mV, close to E_K. On the other hand, GS neurones are hyperpolarized by glucose without membrane conductance change. This is due to enhanced activity of the ATP-dependent Na^+-K^+ pump caused by energy available from the additional glucose. These effects can not be observed during ouabain application.

Distal Responders

There are glucose responsive neuronal elements in hepatic portal vagal afferents (Niijima, 1982), duodenal vagal afferents (Mei, 1978), jejuno-ileal splanchnic afferents (Perrin, Crousillat and Mei, 1981), and the nucleus tractus solitarius (NTS) and the dorsal motor nucleus of the vagus (DMV) (Mizuno and Oomura, 1984) as well as in the LHA and VMH (Fig. 1). The caudal region of the NTS contains GS neurones in proportions similar to those in the LHA (this similarity might be significant), and a few GR neurones. Fewer glucose-responsive (GR and GS) neurones are found in the rostral region. The information about peripheral blood glucose, which influences the activity of autonomic efferent nerves, might reflect sensing functions in both the hypothlamus and these lower structures.

Although the NTS is a relay nucleus in the network through which glucose, metabolites, and peptides are analyzed there is no obvious reason why it should contain chemosensitive neurones unless it acts as a more distal counterpart of the hypothalamus. Its proximity to the fourth ventricle can be compared to the LHA proximity to the third ventricle, so it could integrate peripheral information with intraventricular and systemic blood chemical information, and transmit the results to the LHA and VMH, or it could transmit back to the DMV to elicit reflex secretion, motility, and hepatic enzyme activity in the approprite visceral organs. All peripheral information except that from the gustatory area passes through the NTS where it can be integrated and processed.

Behavioural and electrophysiological studies suggest that synaptic release of NA might mediate central mechanisms of feeding (Leibowitz, Hammer and Chang, 1981). Most GS neurones are found to be inhibited by ventral NA bundle stimulation which induces monosynaptic IPSPs in the LHA. These IPSPs are blocked by electrophoretic application of phenoxybenzamine, an α-adrenoceptor blocking agent, but not by a β-adrenoceptor blocking agent (Miyahara and Oomura, 1982). This indicates that the inhibition is mediated through α-adrenoceptors on the GS neurones. Electrophoretic NA inhibits GS neurones in the LHA, and phenoxybenzamine almost blocks the effect of the NA in that region. At the same time, suppression caused by portal injection of glucose also disappears although neuronal sensitivity to local glucose application remains intact.

Myers and McCaleb (1980) reported that glucose infused directly into the rat duodenum increased NA in the LHA and decreased it in the VMH. They suggested that duodenal receptor signals are conveyed to the rat brain by way of NA inhibitory neurones in the hypothalamus to terminate feeding. It has been suggested that the NA pathway transmits signals related to the blood glucose level from the portal system to GS neurones in the LHA (Shimizu, Oomura, Novin, Grijalva and Cooper, 1983) through GS neurones in the NTS (Adachi, Shimizu, Oomura and Kobashi, 1984). Selective lesions of the ventral NA bundle produce hyperphagia and obesity, and NA injection into the LHA suppresses food intake. The participation of α-adrenoceptors in the regulation of food intake has been demonstrated by behavioural studies (see Oomura, 1983).

Discharge rates of GS units in the hepatic branch of the vagal afferent nerve decrease when glucose is injected into the hepatic portal vein (Niijima, 1982) and hepatic GS units influence regulation of food intake. Effects of glucose, 2-DG, and insulin on the firing rate of vagal hepatic afferents and vagal pancreatic afferents in rabbits, guinea pigs, and rats have been reported (Niijima, 1981). The activity of hepatic and pancreatic afferents is decreased by i.v. glucose, and increased by 2-DG and FFA. Lesions of the VMH, the dorsomedial hypothalamic nucleus (DMH), or the paraventricular nucleus do not generally affect vagal pancreatic or splanchnic pancreatic efferent responses to i.v. glucose infusion, and there is no correlation between response to glucose infusion and the site or size of a lesion. This means that the glucose induces a reflex response within the visceral-NTS-DMV loop (Fig. 1), possibly through glucose responding elements in one or more of the viscera or the NTS or all of these. Insulin and 3-DPA increase and glucagon and 2-DTA decrease the activity of hepatic GS units and pancreatic afferents, (Niijima, personal communication), so some characteristics of these afferent elements are similar to those of GS neurones in the LHA.

ENDOGENOUS CHEMICALS THAT AFFECT FEEDING BEHAVIOUR

Having identified GS and GR neurones in central and peripheral sites, it is of interest to see how these neurones respond to various chemicals other than glucose and how their responses can be related to feeding behaviour. We also compare some of these responses to responses of neurones that are not significantly affected by glucose.

Insulin

Activity of GR neurones in the VMH is slightly inhibited by insulin alone, but is facilitated by simultaneous application of insulin and glucose more than by glucose alone. Activity of GS neurones in the LHA is facilitated dose-dependently by insulin (Oomura, 1976). Insulin-induced feeding is inhibited by injection of glucose directly into the rat LHA and is also suppressed by LHA lesions. These results are consistent with the characteristics of GS neurones in the LHA. Insulin binding sites in the brain (Oomura and Kita, 1981) are abundant in the hypothalamus and unaffected by peripheral insulin concentration. This stability in the hypothalamus suggests that constant brain insulin might provide a control point against which other materials such as glucose are compared in order to regulate body weight.

Glucagon

Glucagon-like immunoreactivity (GLI) has been observed in the brain although immunoreactive pancreatic glucagon (IRG) has not been found in any brain region (Inokuchi et al., 1986a,b; Tager, Hohenboken, Markese and Kinerstein, 1980). Nerve fibres react with anti-GLI antibody in the periventricular region, PVN, supraoptic nucleus, anterior hypothalamus, DMH, and VMH. During deprivation, the GLI concentration in the VMH rises to 1.5 times the control level, but does not change in the LHA. Intracerebroventricular infusion of glucagon induces hyperglycemia and suppresses insulin secretion. GLI in the brain may be related to sugar metabolism as well as to feeding behaviour.

Electrophoretic application of glucagon inhibits GS neurones by hyperpolarization without membrane conductance change. This effect is blocked by ouabain and enhanced by glucose at low concentration, so glucagon may regulate cyclic-AMP and, consequently, Na^+-K^+ pump activity. Glucagon injected into the carotid artery suppresses GS activity (Inokuchi et al., 1986a), and when infused into the third ventricle it suppresses food intake (Inokuchi, Oomura and Nishimura, 1984). Pancreatic glucagon might hematogenically suppress GS neurones in the LHA to terminate feeding, and GLI peptide in the brain might act as an inhibitory neurotransmitter or neuromodulator in the hypothalamus.

Calcitonin

Calcitonin suppresses activity of GS neurones in the rat LHA, an effect that is independent of Na^+-K^+ pump activity, and of noradrenergic or serotonergic mechanisms. Calcitonin hyperpolarizes GS neurones and reduces membrane conductivity indicating a decrease in Na^+ and/or Ca^{2+} permeability. This suppression of GS neurones is mediated by cyclic-AMP within the cell, because phosphodiesterase inhibitors augment and adenylate cyclase inhibitors reduce the suppression. Calcitonin injected into the third ventricle reduces food intake (Shimizu and Oomura, 1986).

Estrogen

Estradiol (ES) is concentrated in the medial amygdala (Med-AMG), medial preoptic area, ventromedial hypothalamus (VMH) and arcuate nucleus. When implanted in the Med-AMG and the VMH, it affected feeding behaviour, gonadotropin release, and ovulation. To investigate the anorexic effect of ES inhibition of Med-AMG neurones and its excitation of VMH neurones, intracellular recordings were made from rat brain slices. Preparations were made after ovariectomy and a single priming by ES. The excitability of Med-AMG neurones was rapidly inhibited, dose-responsively, with hyperpolarization and increased membrane conductance by 17β-ES (10^{-9} to 10^{-7} M) but not by 17α-ES. The reversal potential for ES hyperpolarization shifted 53 mV when the external K^+ concentration was increased ten-fold. The Hill coefficient of the relation between membrane conductance and ES-dose was about 1. The suppression of excitability persisted after elimination of synaptic input and after suppression of protein synthesis by actinomycin D or cyclohexamide. Thus, 17β-ES directly changes the K^+ conductance of the postsynaptic membrane. The numbers of responses by male

(8%) and female (27%) rats were significantly different. Intracellular horseradish peroxidase staining showed the ES responding neurones to be mainly bipolar (14 μm diameter) with few dendrites, and non-responding neurones to be pyramidal (17 μm) with more than 4 dendrites extending in various directions (Nabekura et al., 1986). This morphology of the ES non-responding neurones in the Med-AMG appears to be similar to that of GR and GS neurones in the VMH and LHA.

GR neurones in the VMH were depolarized by 17β-ES with a decrease in membrane conductance. The reversal potential of the depolarization was approximately -90mV, close to the E_K. This depolarization was greatly augmented and prolonged by application of a phosphodiesterase inhibitor, or the adenylate cyclase activator, forskolin. The evidence suggests that ES depolarization might be mediated through an increase in the intracellular cyclic-AMP level. Other neurones (not GR neurones) in the VMH were hyperpolarized along with an increase in membrane conductance. The reversal potential of the hyperpolarization was approximately -90 mV, E_K. Neither phosphodiesterase inhibitor nor forskolin affected these neurones (Oomura et al., 1986a). The Med-AMG and VMH are ES-sensitive target tissues, and are involved in the integration of chemosensory information.

Other Substances

A substance (FS-T) isolated from the feces of deprived rats, but otherwise still unidentified except for molecular weight, has been found potently to depress appetite and has produced compatible results on GS neurones in the LHA (Tsuda, Katsunuma, Shiraishi, Fujimoto and Sakata, 1985).

Neuropeptide Y, which coexists with NA, is reported to enhance feeding, especially carbohydrate intake, when administered into the VMH and PVN (Morley; Stanley and Leibowitz personal communication). General activity also increases. Morley also mentioned that central injection of corticotropin-releasing factor and VIP decrease food intake. The neuronal mechanism by which these peptides work is not yet clear.

2-DTA, 3-DPA AND 3-HBA EFFECTS ON LHA GS NEURONES

The spontaneous activity of single GS neurones was dose-dependently suppressed by 2-DTA and excited by 3-DPA when electrophoretically applied.

Absence of response to Na^+ or Cl^- verified that none of the results were due to non-specific osmosensitivity or current effects. Of 148 LHA neurones tested, 34% were GS. Activity of 49% of the GS neurones tested was decreased by 2-DTA, and its effects on glucose-insensitive (GIS) neurones were negligible (6%). Suppression of GS neurone activity by 2-DTA and by glucose was blocked by preapplication of ouabain. These suppressive effects recovered reversibly after termination of ouabain application. GS neurones were excited by 3-DPA (56%), and most GIS neurones did not respond (90%).

Electrophoretic applications of 3-HBA to 79 LHA neurones excited a significant number (37%) of the GS neurones while negligible numbers of GIS neurones (4%) were affected.

2-DTA, 3-DPA AND 3-HBA ON GR NEURONES

GR neurones in the VMH were dose-dependently facilitated by 2-DTA (63%) and inhibited by 3-DPA (56%). The effect of 3-HBA was almost the same as that of 2-DTA on the GR neurones. The effects of these three compounds were specific and were seen on a significant number (16%) of GR neurones. Of 46 non-GR neurones, only 6% were affected by one of the three compounds. None of the cortical neurones tested responded to electrophoretic application of 2-DTA, 3-DPA or 3-HBA.

MEMBRANE MECHANISM OF 2-DTA, 3-DPA AND 3-HBA

The effects of 2-DTA, 3-DPA and 3-HBA were studied by measuring changes in the membrane potential and conductance of neurones in guinea pig and rat hypothalamic slice preparations. Measurements were made on a total of 103 VMH and 160 LHA neurones. The data are based on 58 VMH and 60 LHA cells that could be held long enough to determine all of their membrane properties. It was required that stable recording be maintained for at least 20 min with a resting potential more negative than -50 mV, the membrane input resistance exceeding 80ΩM , and a membrane time constant of at least 4 msec. Neurones meeting these criteria could often be studied for up to 5 h. LHA neurones were hyperpolarized by 2-DTA with no change in membrane conductance. Attenuation of the 2-DTA suppression by ouabain indicates that its effect, like that of glucose, is due to Na^+-K^+ pump activation (Oomura, Ooyama, Sugimori, Nakamura and Yamada, 1974). VMH neurones were depolarized by 2-DTA with a decrease in membrane conductance.

This effect was also the same as that of glucose, and glucose depolarization of GR neurones is caused by a K^+ permeability decrease (Minami et al., 1986). The reversal potential of 2-DTA depolarization is also close to E_k. This again indicates that 2-DTA will be metabolized in the brain in the same way as ketone bodies if there is a shortage of glucose (Shimizu et al., 1984). Membrane depolarization with membrane conductance increase of GS neurones was induced by 3-DPA. The reversal potential for this depolarization was close to +10 mV, indicating Na^+ and/or Ca^{2+} permeability increase. The effect of 3-HBA on GS neurones has not yet been analyzed.

The inhibitory effect of 3-DPA on GR neurones was due to membrane hyperpolarization with membrane conductance increase. The reversal potential for the hyperpolarization was approximately -90 mV which was close to E_K. Thus 3-DPA causes a membrane K^+ permeability increase. The effects of 3-HBA were almost the same as those of 2-DTA on GR neurones. This facilitation of 3-HBA on GR neurones may explain its dual action on food intake.

SOME CHEMICAL DEPENDENT FUNCTIONS AND RELATIONS

Important information relevant to the regulation of feeding goes to the hypothalamus from internal and external sources (Fig. 1). Visual (Ono, Oomura, Nishino, Sasaki, Fukuda and Muramoto, 1981; Yamamoto, Oomura, Nishino, Auo, Nakano and Nemoto, 1984), olfactory (Takagi, 1984), and gustatory (Kita and Oomura, 1981; Norgren, 1978) signals, plus internal visceral information such as that from hepatic and intestinal glucose sensors (Mei, 1978; Shimizu et al., 1984) and stomach mechanoreceptor afferents converge in the VMH and the LHA, to integrate sensory and effector information. The LHA, VMH and other parts of the limbic system then mediate feeding processes.

INVOLVEMENT OF GS NEURONES IN FOOD INTAKE

Monkeys restrained in a chronic stereotaxic frame facing a panel equipped with a cue lamp, bar and pellet box were trained to perform a bar press (fixed ratios, 30) feeding task. Neuronal activity was recorded and drugs were applied through 10-barreled micropipettes. Single neurone activity was recorded through the central barrel, and drugs were applied electrophoretically from the surrounding pipettes.

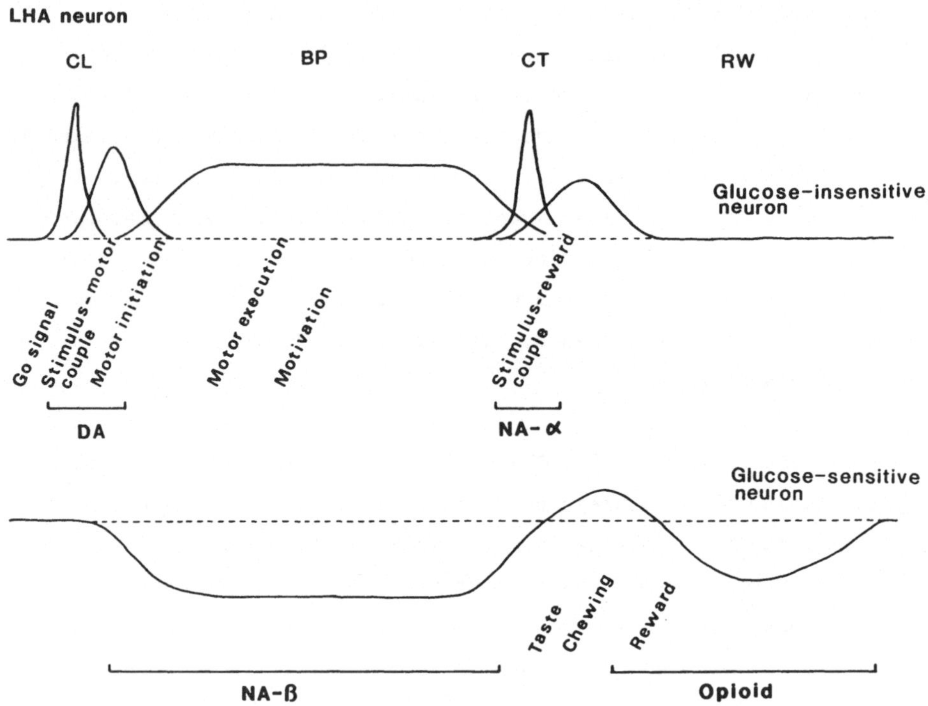

Fig. 2. Schema of neurone responses in monkey LHA in each phase of bar press feeding task, and sequence of possible related events. Upper: sections of response patterns of glucose-insensitive neurones. Complete response pattern for one neurone in any one trial might consist of one, any, or all parts. Lower: response pattern of glucose-sensitive neurones. Underlined sections might be controlled by agent indicated at each underline. CL, cue light; BP, bar press; CT, cue tone; RW, ingestion reward; DA, dopamine (blocked by spiperone); NA-α, noradrenaline (blocked by phenoxybenzamine, an α blocker); NA-β, noradrenaline (blocked by sotalol, a β blocker); Opioid (blocked by naloxone). Catecholamine and opiate modulated responses in different phases of feeding task (Oomura, 1986).

In the feeding task out of 312 neurones tested, 61% responded in one or more phases (Fig. 2). The responses varied according to the reward situation. Activity change evident in a task for the standard food, bread, was enhanced in trials for more palatable food. Responses diminished and finally disappeared during extinction, upon satiation, or at high blood glucose levels induced by intravenous glucose injection (Aou, Oomura, Lénárd, Nishino, Inokuchi, Minami and Misaki, 1984; Nishino, Oomura, Aou and Lénárd, 1986).

Of 198 neurones tested with noradrenaline (NA), 3% were excited, 45% were inhibited (Fig. 2), and 52% were not affected. Of 150 neurones tested with dopamine (DA), almost equal numbers were excited (16%) or inhibited (23%), and 61% were not affected. Morphine had almost the same effect as NA: excitation (1%), inhibition (38%), and no effect (61%). Most neurones that were suppressed by NA were also suppressed by morphine (Fig. 2). Of 213 LHA neurones tested, 26% were GS.

Feeding responsiveness was correlated with chemical sensitivity. NA and DA sensitive neurones responded more often in the feeding task than NA insensitive neurones (χ^2 test, $p<0.05$). NA sensitive neurones responded with decrease during bar press and reward periods. Neurones facilitated by DA responded more often to the cue light (facilitation) than DA nonresponding-neurones (χ^2 test, $p<0.01$). Neurones that were inhibited by DA responded more often during the bar press (inhibition) (χ^2 test, $p<0.05$). Morphine sensitive neurones responded (suppression) more often during bar press and reward than morphine insensitive neurones (χ^2 test, $p<0.05$). GS neurones, which were not affected by cue light or cue tone but were affected more often by NA and morphine than glucose insensitive (GIS) neurones (_t_ test, $p<0.05$), were significantly suppressed during bar press and reward (χ^2 test, $p<0.05$) (Oomura, Nishino, Aou and Lénárd, 1986b; Nishino et al., 1986). GIS neurones responded to the cue light and cue tone. This was specific to reward relations since the same neurones did not respond to non-relevant light or tone stimuli such as a red light in other than the cue light position, or a green light in the cue light position, or tones that mimicked the cue tone but occurred at random times.

Spiperone, a DA antagonist, blocked cue light related activity increase and/or activity increase seen in the early part of the bar press period, i.e. motor initiation, but did not affect responses that continued through the bar press and reward phases (Fig. 2). The attenuating effects of spiperone were reversible. Sotalol, a β-adrenoceptive antagonist, blocked responses of 5 of 19 neurones whose firing rate decreased during bar press. Neurone responses related to the cue tone with excitation were attenuated by PBZ, an α-adrenoceptive antagonist, while those with inhibition were attenuated by sotalol. Responses of morphine sensitive neurones were modulated by naloxone. Naloxone did not affect bar press responses, but blocked inhibitory responses in the reward period. Naloxone did not affect the excitatory responses of neurones that increased in activity during the reward period. GIS neurones have DA inputs from the substantia nigra and the ventral tegmental area, and in some cases they have α-adrenoceptors. The NA system may originate from the ventral NA bundle. GS neurones have β-adrenoceptors and opioid receptors. The opioid system may come from the intrahypothalamus (paraventricular nucleus, median eminence) and/or centromedian nucleus of the amygdala.

ODOURS

Responses were elicited in 100% of GS neurones by at least one odour (mainly excitation by orange and one fecal odour, skatole; and inhibition by borneol, a camphor odour). Less than 50% of the GIS neurones responded to odours, and no tendency was apparent.

Orbitofrontal area (area 11) stimulation elicited responses in 58% of the GS neurones. This proportion was much higher than GS neurone responses to dorsolateral prefrontal cortex stimulation, and GIS neurones responded only to dorsolateral prefrontal cortex but not to orbitofrontal area stimulation.

CONCLUSIONS

Chemosensitive neurones in the VMH and LHA that respond to glucose, sugar acids, free fatty acids, catecholamine, opioid, insulin, glucagon, calcitonin and other endogenous compounds regulate feeding behaviour through integration of endogenous humoural and exogenous sensory stimuli. Most endogenous chemicals that modify feeding behaviour do so through their effects on GR neurones in the VMH and GS neurones in the LHA.

LHA neurones that are concerned with actual feeding behaviour are probably the most active single population of cells in the hypothalamus. The known sequence of events in a bar press task for food is: perception of a light stimulus, stimulus-motor coupling or motor initiation, motor execution to procure food, perception of tone, attention, stimulus-reward association, and movement to pick up and ingest food. Since DA sensitive neurones responded only to reward-coupled cue light and cue tone signals, and not to other visual or auditory stimuli, and spiperone blocked responses to the cue light and in the early stage of motor initiation, and PBZ blocked the cue tone responses, it seems reasonable that DA sensitive cells contribute to external cue recognition and motor execution, and α-adrenoceptive receptors are involved in stimulus-reward association. Neurones that do this are mostly GIS neurones. On the other hand, GS neurones have β-adrenoceptive and opioid receptors, and are suppressed during bar press and reward. Since naloxone blocked the inhibitory reward response, it is suggested that the endogenous opioid system is concerned with this type of response and is activated only in the reward period. Thus, NA, DA and opioid inputs to the LHA are all important to monkey operant feeding, but in different ways. GIS neurones act to integrate external information, while GS neurones are involved in the integration of internal information and in reward perception. Afferents from mechanical and chemical sensors in the viscera, and chemical sensors in the LHA are basic to the control of feeding behaviour. GS neurones in the LHA communicate with the orbitofrontal and dorsolateral prefrontal cortices and have olfactory, NA-β, and opioid inputs to evaluate actual and potential ingestive reward situations. GIS neurones communicate with the DL (but not the orbitorontal cortex) and receive exogenous (except olfactory) inputs through DA and NA-α receptors eventually to control specific behaviour related to feeding. Integration of these functions in the LHA might be important in accomplishing motivated feeding behaviour.

REFERENCES

Adachi, A., Shimizu, N., Oomura, Y., and Kobashi, M. (1984). Convergence of hepatoportal glucose-sensitive afferent signals to glucose-sensitive units within the nucleus of the solitary tract. Neurosci. Lett. 46, 215-218.

Aou, S., Oomura, Y., Lénárd, L., Nishino, H., Inokuchi, A., Minami, T., and Misaki, H. (1984). Behavioral significance of monkey hypothalamic glucose-sensitive neurons. Brain Res., 302, 69-74.

Davies, J.D., Wirthshafter, D., Asir, K.E. and Brief, D. (1981). Sustained intracerebroventricular infusion of brain fuels reduced body weight and food intake in rats. Science, 212, 81-83

Inokuchi, A., Oomura, Y., and Nishimura, H. (1984). Effect of intracerebroventricular infused glucagon on feeding behavior. Physiol. Behav., 33, 397-400.

Inokuchi, A., Oomura, Y., Shimizu, N. and Yamamoto, Y. (1986a). Central action of glucagon in rat hypothalamus. Am. J. Physiol., 250, R120-126.

Inokuchi, A., Tomita, Y., Yanaihara, C., Yui, R., Oomura, Y., Kimura, H., Hase, T., Matsumoto, Y., and Yanaihara, N. (1986b). Identification, localization and characterization of glucagon-related peptides in the rate hypothalamus. Cell Tiss. Res., in press.

Katafuchi, T., Oomura, Y., and Yoshimatsu, H. (1985). Single neuron activity change in the rat lateral hypothalamus during 2-deoxy-D-glucose induced and natural feeding behaviour. Brain Res., 359, 1-9.

Leibowitz, S.F., Hammer, N.J., and Chang, K. (1981). Hypothalamic paraventricular nucleus lesions produced overeating and obesity in the rat. Physiol. Behav., 27, 1031-1040.

Llinás, R. and Sugimori, M. (1980). Electrophysiological properties of in vitro Purkinje cell somata in mammalian cerebellar slices. J. Physiol. (Lond.), 305, 171-195.

Mei, N. (1978). Vagal glucoreceptors in the small intestine of the cat. J. Physiol. (Lond.), 282, 485-506.

Minami, T. and Oomura, Y. (1986). Electrophysiological properties and glucose responsiveness of guinea-pig ventromedial hypothalamic neurons in vitro. J. Physiol. (Lond.), in press.

Miyahara, S. and Oomura, Y. (1982). Inhibitory action of the ventral noradrenergic bundle on the lateral hypothalamic neurons through alpha-noradrenergic mechanisms in the rat. Brain Res., 234, 459-463.

Mizuno, Y., Nabekura, J. and Oomura Y. (1985). Electrophysiological study of input-output organization in the dorsal motor nucleus of the vagus in vitro. J. Physiol. (Lond.), in press.

Myers, R.D. and McCaleb, M.L. (1980). Feeding: satiety signal from intestine triggers brain's noradrenergic mechanism. Science, 209, 1035-1037.

Nabekura, J. Oomura, Y. Minami, T. Mizuno, Y. and Fukuda, A. (1986). Mechanism of the rapid effect of 17 estradiol on medial amygdala neurons. Science, in press.

Niijima, A. (1982). Glucose-sensitive afferent nerve fibres in the hepatic branch of the vagus nerve inthe guinea-pig. J. Physiol. (Lond)., 332, 315-323.

Nishino, H., Oomura, Y., Aou, S., and Lénárd, L. (1986). Catecholaminergic mechanisms on feeding related lateral hypothalamic activity in the monkey. Brain Res., in press.

Ono, T., Oomura, Y., Nishino, H., Sasaki, K. Fukuda, M., and Muramoto, K. (1981). Neural mechanisms of feeding behavior. In: Brain Mechanisms of Sensation, Katsuki, Y., Norgren, R. and Sato, M., eds. Wiley: New York, pp 271-286.

Oomura, Y. (1976). Significance of glucose, insulin, and free fatty acid on the hypothalamic feeding satiety neurons. In: Hunger: Basic Mechanisms and Clinical Implications Novin, D., Wyrwicka, W. and Bray, G.A., eds. Raven Press: New York, pp 145-157.

Oomura, Y. (1983). Glucose as a regulator of neuronal activity. In: Advances in Metabolic Disorders Szabo, A.J., ed., Academic Press: New York, Vol. 10, pp 31-63.

Oomura, Y. (1985). Feeding control through bioassay of body chemistry. Jap. J. Physiol. 35, 1-19.

Oomura, Y. (1986). Modulation of the prefrontal and hypothalamic activity by chemical sense in chronic monkey. In Umami: The Physiology of its Taste, Kawamura, Y. and Kare, M.R., eds., Dekker: New York, in press.

Oomura, Y. and Kita, H. (1981). Insulin acting as a modulator of feeding through the hypothalamus. Diabetologia (suppl)., 20, 290 298.

Oomura, Y., Ooyama, H. Sugimori, M., Nakamura, T., and Yamada, Y. (1974). Glucose inhibition of the glucose-sensitive neuron in the rat lateral hypothalamus. Nature, 247, 284-286.

Oomura, Y., Minami, T., Nabekura, J. (1986a). Effect of estradiol on the amygdala and ventrolmedial hypothalamic neuron. Soc. Neurosci. Abstracts 12, in press.

Oomura, Y., Nishino, H., Aou, S., and Lénárd, L. (1986b). Opiate mechanism in reward related neuronal responses during operant feeding behavior of the monkey. Brain Res., 365, 335-339.

Perrin, J., Crousillat, J. and Mei, N. (1981). Assessment of true splanchnic glucoreceptors in the jejuno-ileum of the cat. Brain Res. Bull., 7, 625-628.

Plata-Salaman, C.R., Oomura, Y. and Shimizu, N. (1986). Dependence of food intake on acute and chronic ventricular administration of insulin. Physiol. Behav., in press.

Shibata, S., Oomura, Y., and Kita, H. (1982). Ontogenesis of glucose sensitivity in the rat lateral hypothalamus: a brain slice study. Develop. Brain Res., 15, 114-117.

Shimizu, N. and Oomura, Y. (1986). Calcitonin induced anorexia in rats: evidence for its inhibitory action on lateral hypothalamic chemosensitive neurons. Brain Res., 367, 128-140.

Shimizu, N., Oomura, Y., Novin, D., Grijalva, C., and Cooper, P. (1983). Functional correlations between lateral hypothalamic glucose-sensitive neurons an hepatic portal glucose-sensitive units in rat. Brain Res., 265, 49-54.

Shimizu, N., Oomura, Y., and Sakata, T. (1984). Modulation of feeding by endogenous sugar acids acting as hunger or satiety actors. Am. J. Physiol., 246, R542-R550.

Tager, H., Hohenboken, M., Markese, J. and Kinerstein, R.J. (1980). Identification and localization of glucagon-related peptides in rat brain. Proc. Nat. Acad. Sci., USA, 77, 6229-6233.

Takagi, S.F. (1984). The olfactory nervous system of the old world monkey. Jap. J. Physiol., 34, 561-573.

Terada, K., Sakata, T., Oomura, Y., Fukimoto, K., Arase, K., Osanai, T. and Nagai, Y. (1986). Evidence that hypophagia is induced by endogenous or liposome encapsulated 3,4-dihydroxybutanoic acid through central action. Proc. Soc. Exp. Biol. Med., in press.

Tsuda, T.T., Katsunuma, T., Shiraishi, T., Fujimoto, K., and Sakata, T. (1985). Feeding suppression induced by a fecal anorexigenic substance (FS-T). Physiol. Behav., 34, 791-798.

Woods, S.C. and Porte, D. Jr. (1983). The role of insulin as a satiety factor in the central nervous system. In: Advances in Metabolic Disorders Szabo, A.J., ed., Academic Press: New York, Vol. 10, pp 457-468.

Yamamoto, T., Oomura, Y., Nishino, H., Auo, S., Nakano, Y., and Nemoto, S. (1984). Monkey orbitofrontal activity during emotional and feeding behavior. Brain Res. Bull., 12, 441-443.

PULMONARY OEDEMA

Francis P. Chinard

Departments of Medicine and Physiology
University of Medicine and Dentistry
New Jersey Medical School
Newark, NJ, U.S.A.

INTRODUCTION

Pulmonary oedema, a common and major clinical event, can occur as a result of cardiac failure and as a result of injury to the tissues of the lungs. It represents a problem that is vast and complex and that has been extensively reviewed in recent years (see, for example, Kazemi, Hyman and Kadowitz, 1986; Parker, Guyton and Taylor, 1979; Prichard, 1982; Said, 1985; Staub, 1984; Zapol, 1985). Rather than attempt what could only be a cursory review of the whole field, I have instead presented here a brief summary of the classical dogma, focused on some structural aspects and examining the factors regulating or possibly determining the passage from blood to the interstitium of the two major constituents of oedema fluid: water and sodium. What I report is then in part review, in part work in progress and in part speculation.

HISTORICAL NOTE

It is to Laennec (1819) that we owe the first clear definition of pulmonary oedema:

> L'oedème pulmonaire est une infiltration de serosité dans le tissu pulmonaire portée à un degré tel qu'elle diminue notablement sa perméabilité à l'air.

FACTORS INVOLVED IN THE CAUSATION OF DROPSY

I. Factors causing increased transudation:—

A. Increased intra-capillary pressure:
- *a.* Venous obstruction.
- *b.* Vasodilatation.
- *c.* Plethora.

B. Increased permeability of vessel wall:
- *a.* Local injury by mechanical irritants.
  - ,, ,, ,, thermal ,,
  - ,, ,, ,, chemical ,,
- *b.* Malnutrition.
- *c.* General injury by circulating poisons (?).

C. Watery condition of blood (hydræmia).

D. Increased molecular concentration of tissues.

II. Factors causing diminished absorption:—

A. By lymphatics:
- *a.* Paralysis of limbs.
- *b.* Obstruction of lymphatic trunks.

B. By veins:
- *a.* Venous obstruction.
- *b.* Watery condition of blood.
- *c.* Concentrated transudations (*i.e.* in protein).

Fig. 1. "Factors involved in the causation of dropsy". Table, slightly modified, from Starling's Herter Lectures (Starling, 1909).

This can be freely translated as:

> Pulmonary oedema is an infiltration of serous fluid into the pulmonary tissue to such an extent that the permeability to air is significantly decreased.

We would modify this a bit today in recognizing that hypoxia or haemoglobin unsaturation appears at an advanced rather than at the now detectable early stages of pulmonary oedema.

A major milestone was reached through the studies of Ernest Starling. Fig. 1 shows tabular material from his Herter Lectures (Starling, 1909). The list of factors includes not only what we now call the four Starling factors, i.e. hydrostatic and oncotic pressures in blood and in tissues, but also lymphatic drainage and vascular permeability.

THE CLASSICAL DOGMA

The relations among the Starling factors are indicated schematically in Fig. 2. At equilibrium, we would have

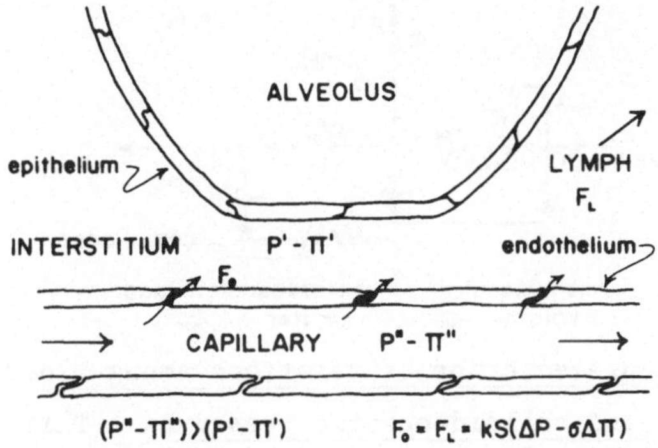

Fig. 2. Schematic representation of relations between the capillary, the interstitium and the alveolar gas phase. See text for details.

$$P'' - \Pi'' = P' - \Pi' \qquad (1)$$

where the hydrostatic pressures are denoted by P, the colloid osmotic pressure by Π and the blood and interstitial compartments are indicated by double and single primes respectively. Ordinarily, equilibrium does not in fact obtain, as is evident from the formation of lymph: there is imbalance of the Starling factors such that the difference of the hydrostatic and "oncotic" pressures in the microvasculature exceeds that difference in the tissues. There ensues net passage of water and solutes from blood to tissues. The current dogma is that fluid passes through pores or slits at endothelial cell junctions, the fluid having the composition of plasma except that it is nearly free of proteins and that it follows the constraints of the Gibbs-Donnan equilibrium. The presence of proteins in the fluid means that the full effect of the proteins in plasma in restraining the passage of water is not exhibited. This is indicated by the reflection coefficient, denoted σ in the expression which follows, describing the rate of passage of fluid F_o from blood to tissues:

$$F_o = k.S(\Delta P - \sigma\Delta\Pi) = F_L \qquad (2)$$

where k is a proportionality coefficient incorporating barrier characteristics, S is surface area and F_L is the rate of lymph formation. The reflection coefficient takes on values between 0 and 1, being highest for macromolecules such as albumin and smallest for small solutes such as sodium and chloride ions. If the rate of outward filtration, F_o, is greater than the rate of outflow of lymph, F_L, there will necessarily be accumulation of fluid in the interstitium as described by Starling.

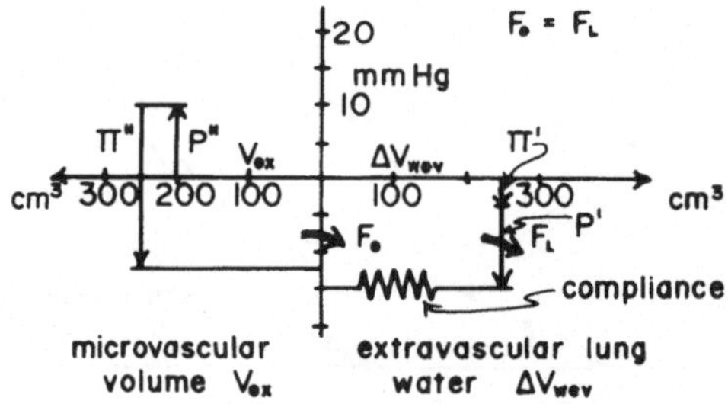

Fig. 3. Schematic representation of relations among blood and tissue hydrostatic and colloid osmotic pressures. This is the normal steady state situation with hydrostatic and colloid osmotic pressures of about 10 and 22 mmHg respectively acting in opposite directions, as indicated by the arrows, with respect to the direction of movement of water. Colloid osmotic pressure in the tissues is small and hydrostatic pressure is is negative, i.e. less than atmospheric. Compliance is suggested by a spring: volume increases of the interstitium are followed by pressure increases (less negativity) in that compartment. The heavy curved arrows indicate the direction of net movement of water and solutes. Outward filtration rate from blood, F_o, is equal to lymphatic outflow, F_L. The volume of the microvascular blood volume, here denoted V_{ex} and referred to later as the vascular exchange volume, is given by the intersection of the downward line for Π" with the abscissa. The volume of the extravascular compartment, here denoted ΔV_{wev} and referred to later as the extravascular lung water, is given by the intersection with the abscissa of the line summing P' and Π'.

Another type of representation (Figs. 3, 4 and 5) may be found useful as it provides quantification of some of the variables, includes compartment volumes and, in addition, introduces tissue compliance as a possible regulatory factor in the rate of outward filtration. Note the upward direction of the vascular hydrostatic pressure arrow and the downward direction of the corresponding tissue pressure arrow because the latter is less than atmospheric (see, for example, Guyton, Granger and Taylor, 1971). What Fig. 3 represents is a steady state, not an equilibrium. We are dealing with an open system which in this instance, the normal situation, is invariant in time.

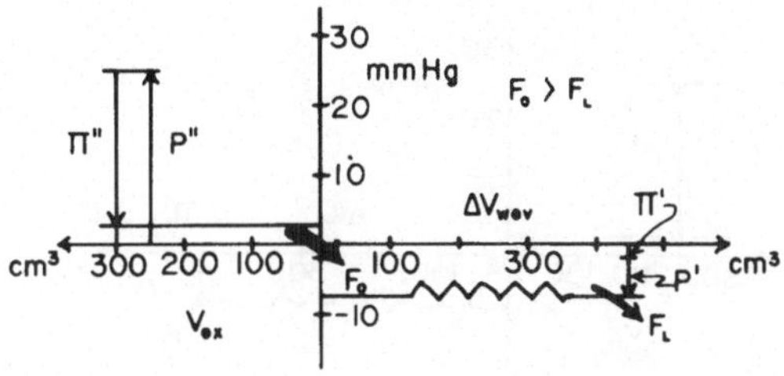

Fig. 4. Schematic representation of the development of "cardiac oedema". Microvascular pressure, P", has increased without change of colloid osmotic pressure, Π". There is an increase of outward filtration, F_o, which now exceeds the lymphatic outflow, F_L. Both microvascular volume, V_{ex}, and the extravascular lung water, ΔV_{wev}, have increased, the latter much more than the former. In the interstitial compartment, the colloid osmotic pressure, Π', has decreased because of protein washout from the interstitium. The interstitial pressure, P', has become less negative, i.e. it has increased. The compliance has decreased so that further increases of P' can be expected to be associated with relatively larger increases of ΔV_{wev}.

We can impose a change in this steady state system by changing one variable, the microvascular pressure, P". The results of an increase of P" (the equivalent of "cardiac" oedema) are shown in Fig. 4. Following this increase the net resultant between the microvascular and colloid osmotic pressures is increased (eqn. 2) and there is an increase of the rate of outward movement of water and solutes from blood to tissues. There is expansion of the volume of the microvasculature (congestion) and, because the increased rate of filtration is not matched by a corresponding increase of lymph flow, expansion of the interstitium also occurs. The latter increase of volume results in an increase of tissue pressure (it becomes less negative) because of its normally low compliance (or high resistance to stretching). In addition, because of the increased flow of relatively protein-poor fluid from blood to interstitium, there is a decrease or wash-down of protein concentration in the interstitium and an attendant decrease of colloid osmotic pressure. A new steady state can ensue but at the cost of a net accumulation of fluid in the interstitium. If the increased rate of filtration cannot be accommodated by the lymphatic system, there can be system breakdown and alveolar flooding with interference with gas exchanges as adumbrated by Laennec. Protein wash-down and increased tissue pressure moderate the outward filtration.

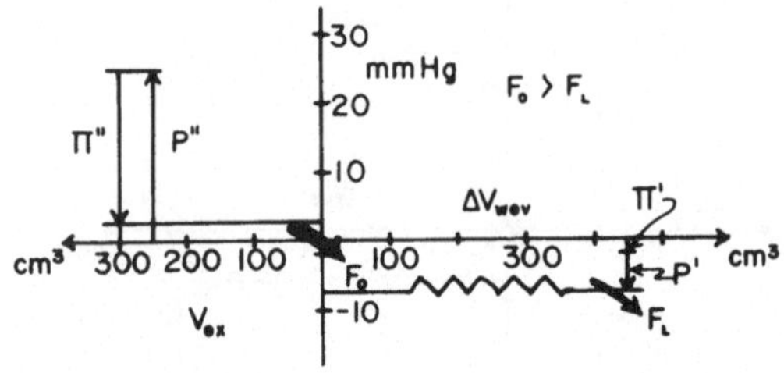

Fig. 5. Schematic representation of the development of a permeability type of pulmonary oedema. There is little change in the parameters of the microcirculation. However, in the interstitium there has been a marked increase in the colloid osmotic pressure, Π', because of the leakage of protein from the plasma. The tissue pressure, P', has become less negative but because tissue compliance may have increased, as indicated by the question mark, the increase of P' may not be as great as it would have otherwise been. As in "cardiac" oedema fluid accumulation in the tissues continues as long as F_O is greater than F_L.

Permeability oedema, as considered to occur in the adult respiratory distress syndrome is illustrated in Fig. 5. It is thought to result from an increase of the permeability of the endothelium to proteins so that what has been called the "osmotic gradient", Π" - Π', is markedly decreased if not abolished with the decrease of the reflection coefficient to proteins at the endothelial barrier. The tissues may become far more compliant through the loosening of the submicroscopic attachments (tethering) of elastin and collagen attendant on the leakage of the enzymes elastase and collagenase into the tissues from the plasma or from leucocytes and macrophages. The volume of the interstitium may increase tremendously without increase of the microvascular compartment. The lungs may become very wet indeed.

The above are essentially theoretical constructs that are more or less classical and generally accepted. The following features are to be noted:

a. the outward filtration is assumed to occur at the level of the alveolar capillaries;
b. the pathway for the passage of water and small solutes (and large solutes in permeability oedema) is at cell junctions and is not transcellular;
c. small solutes such as sodium and chloride ions are ignored as

regulating or moderating factors in the passage of fluid from blood to the interstitium.

SEQUENCE IN THE DEVELOPMENT OF PULMONARY OEDEMA

The classical sequence is that proposed by Staub, Nagano and Pearce (1967), on the basis of experiments involving circulatory overload or administration of alloxan, of perivascular cuffing or sleeving, alveolar septal oedema, presumably in the thick portions of the alveolar septa, and finally alveolar flooding. We have carried out several series of experiments in which data relevant to this sequence have been obtained. In one series, in isolated dog lungs perfused with plasma/dextran mixtures, we did find, following induction of oedema either by increasing microvascular pressure or by decreasing colloid osmotic pressure, cuffing followed by septal oedema and finally alveolar flooding (DeFouw and Berendsen, 1978, 1979). As shown in Table 1, the oedema of the alveolar walls was interstitial and not cellular. However, in another series carried out _in vivo_ with circulatory overload, we have found somewhat different results. At the light microscopic level (Chinard and DeFouw, unpublished) the sequence is clearly (1) vascular engorgement and distension and increase of overall alveolar septal thickness, (2) cuffing or sleeving of the extra-alveolar arterioles and venules with peribronchiolar cuffing less frequent and less pronounced than the perivascular cuffing, (3) alveolar flooding, in the early stages occurring usually in close proximity to cuffed extra-alveolar arterioles. As noted, increase of the alveolar septal thickness was found but at the resolution available with light microscopy we could not clearly distinguish between engorgement and distension on the one hand and septal oedema on the other. With electron microscopy however, resolution was adequate to answer the question. As shown in Table 2, there was no interstitial (or cellular) oedema in the alveolar septa in these experiments in which the overload was produced _in vivo_. In these _in vivo_ studies, the increased alveolar wall thickness is thus clearly not the result of oedema but of engorgement/distension of the microvasculature. We have never seen oedema or splitting of the basal lamina in the thin portion of the alveolar septa in overload or "cardiac oedema. We can now propose the revised sequence illustrated in Fig. 6 for "cardiac" or overload oedema. Our ancillary thesis is that septal oedema occurs in situations where the endothelial integrity has been damaged, following deliberate introduction of an injurious agent or the effects of the trauma involved in preparing an isolated perfused lung preparation. We have found splitting

Table 1

Structural parameters of isolated perfused dog lung preparations

thicknesses in μm

	interstitium	epithelium	endothelium
controls	0.64	0.32	0.34
"oncotic" oedema	0.83	0.35	0.33
hydrostatic oedema	0.89	0.35	0.37

of the basal lamina following the administration of injurious agents such as ethchlorvynol and oleic and other unsaturated fatty acids.

Does the increased outward filtration in "cardiac" oedema actually occur at the level of the alveolar capillaries, sites for which there is no direct supporting experimental but only inferential evidence, or does it occur at the level of the extra-alveolar arterioles and venules, sites at which the first evidence is seen? We have no definitive answer to that question at this time.

ENDOTHELIAL AND EPITHELIAL VESICLES

Since the early studies of vesicles by Palade and his associates (see Simionescu and Simionescu, 1984 for a review) these structures have attracted considerable attention as possible pathways for macromolecules and perhaps smaller solutes and water across the endothelium. Experiments

Table 2

Structural parameters of lungs from <u>in vivo</u> experiments

thicknesses in μm

	interstitium	epithelium	endothelium
controls	0.63	0.26	0.27
plasmapheresis	0.64	0.28	0.25
overload	0.59	0.25	0.23

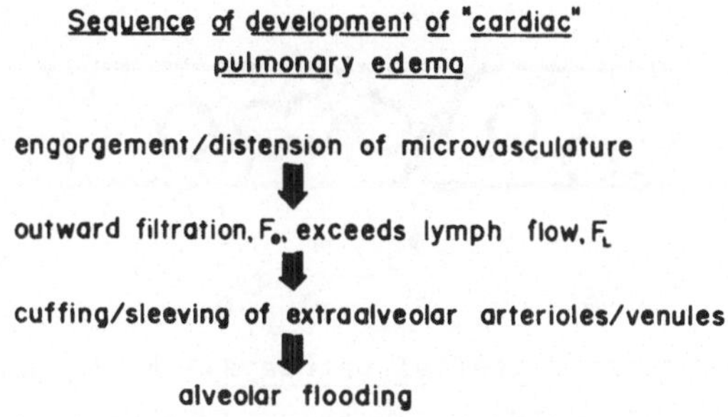

Sequence of development of "cardiac"
pulmonary edema
engorgement/distension of microvasculature
outward filtration, F_o, exceeds lymph flow, F_L
cuffing/sleeving of extraalveolar arterioles/venules
alveolar flooding

Fig. 6. Proposed sequence for the development of "cardiac" pulmonary oedema.

with peroxidase-reacting macromolecules showed that following their injection into the blood stream, initially the vesicles close to the luminal border contain the macromolecules before the interstitium itself shows evidence of their presence. Minutes later, all the vesicles in the endothelium are stained whatever their position between the lumen and the interstitium and the latter now contains the macromolecules (Simionescu, Simionescu and Palade, 1975).

This type of data could be interpreted in a number of ways. The hypothesis of a vesicular shuttle had been proposed earlier (Shea, Karnovsky and Bossert, 1969) and suggested to some that vesicles ping-ponged or barged across the endothelial cytoplasm, hindered to some degree by the endoplasmic reticulum and the cytoskeleton but happily picking up on one side and unloading on the other. For some investigators, the vesicular balloon was punctured by Bundgaard, Frøkjer-Jensen and Crone (1979) in their demonstration, by serial sections viewed by electron microscopy, of the continuity of transcellular channels formed by fusion of the vesicles of which the origin may have been the racemose clusters occasionally seen in the pulmonary endothelium. Whether the vesicles are always in contact, thus forming continuous transendothelial channels, or are caught, flagrante delicto, in transient osculation by the techniques of fixation has not yet been determined. It does seem, however, that there are few if any free or cytoplasmic vesicles. Fig. 7 indicates some of the possible vesicular and other transcellular and transendothelial pathways.

The existence of the vesicles and the role proposed for them in macromolecular transport did suggest that the following question should be considered: do the vesicles play a significant role in the development of pulmonary oedema? DeFouw and Berendsen (1978, 1979) working in my

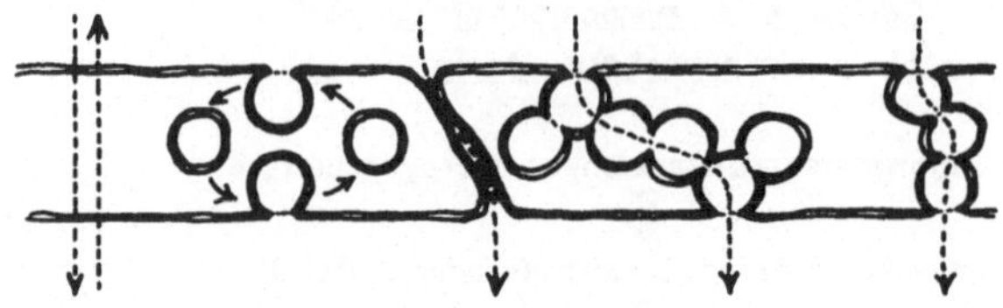

Fig. 7. Possible transendothelial pathways. A, transcellular diffusion; B, vesicular shuttle; C, junctional slits or pores; D, transient transcellular channel from osculating vesicles; E, fixed transcellular channels. All pathways except A are extracellular, i.e. do not involve contact with endothelial cell contents as does A, the transcellular diffusional pathway.

laboratory did find that in association with the development of pulmonary oedema in isolated perfused dog lung preparations there was a very substantial increase in the number of vesicles seen in both the endothelium and the epithelium (Table 3). A possible mechanism (differences of intra- and extracellular pressures) was proposed for the generation of new vesicles from redundant plasmalemma surface available in endothelial and epithelial folds (Chinard, 1980). Our initial enthusiasm for the hypothesis that oedema and vesicle formation were perhaps even causally associated was soon shattered.

We argued that neoformation of vesicles would require that the plasmalemma, in major part a lipid bilayer, be fluid. Decreasing the temperature of the isolated lung preparations by perfusion with cold fluids could be expected to decrease that fluidity significantly. In line with that reasoning we carried out experiments in preparations perfused at 15°C

Table 3

Structural parameters of isolated dog lung preparations

perfused at 36°C

vesicles, volume density

	epithelium	endothelium
control	7.0	26.0
"oncotic" oedema	15.0	42.0
hydrostatic oedema	17.0	46.0

and produced oedema in these. In contrast to what we had found in the preparations perfused at 36°C there was no increase in vesiculation in the cold perfused preparations despite extensive oedema which included marked septal fluid accumulation (Chinard and DeFouw, 1981). Oedema had occurred without an increase in the number or density of endothelial (or epithelial) vesicles (Table 4).

Finally, in the lungs of animals in which we produced pulmonary oedema in vivo through overloading of the circulation (or plasmapheresis) we found, as shown in Table 5, no increase in vesiculation in the endothelium of the alveolar capillaries despite extensive cuffing or sleeving of the extra-alveolar vessels and the production of some alveolar flooding (DeFouw and Chinard, 1983). However, in the walls of the extra-alveolar vessels, both arterioles and venules, with cuffing or sleeving we have found a very marked increase in vesiculation (DeFouw and Chinard, unpublished).

I cannot dismiss vesicles as vehicles or pathways between the vascular lumen and the interstitium. If the vesicles are connected from one plasmalemma to the other, as Crone and his colleagues have suggested, then they offer a transcellular pathway, possibly size selective, that is just as extracellular as the pores or slits at the cell junctions. In instances where they increase in number or volume density, I believe that they are secondary to the development of the oedema rather than primary determinants. Thus, the intravenous administration of ethchlorvynol is followed not only by extremely rapid development of pulmonary oedema but also by an equally rapid and marked increase in vesicle density. Cause and effect cannot yet be separated. At the present, one can just as well postulate that a membrane destabilizing agent can lead to the neoformation

Table 4

Structural parameters of isolated dog lung preparations perfused at 15°C

vesicles, volume density

	epithelium	endothelium
control	10.5	20.7
moderate oedema	11.0	21.2
severe oedema	9.0	22.1

Table 5

Structural parameters of lungs from dogs subjected to severe vascular overloading or plasmapheresis in vivo

vesicles, volume density

	epithelium	endothelium
control	8	26
plasmapheresis	9	17
overload	8	17

of vesicles as postulate that the interstitial oedema itself can induce the formation of vesicles.

FUNCTIONAL ASPECTS OF THE ENDOTHELIUM

The particular aspects examined in this section are the permeabilities of the endothelial barrier to water and to sodium ion since these two substances are the major constituents of oedema fluid. Answers are sought to two questions:

a. is the passage of water modulated or restricted by the endothelium?

b. is the passage of sodium ion modulated or restricted by the endothelium?

In seeking answers to these questions we have made use of the bolus injection multiple indicator dilution technique as described elsewhere previously (e.g. Chinard, 1981; Chinard and DeFouw, 1984). Extractions are calculated from the ratios of the areas under the concentration-time curves of the extracted substances to the corresponding area under the curves of the appropriate vascular reference up to the time of the maximum of the reference substance assumed not to be extracted, i.e., to be completely barrier-restricted. The extent of the extractions of the permeating substances may be limited in the simplest model by a permeability restriction at the endothelium (that barrier being in fact defined functionally by the vascular reference) or by the volume of distribution available to the tracer or indicator beyond the barrier. A volume (or, as more conventionally referred to, flow) limited distribution does not mean that the permeability coefficient has a value of infinity. It does mean that, in the transit time required for the blood to traverse the vascular exchange volume (volume of blood in the microvasculature from which exchanges with the extravascular volume take place), the tracer or indicator distributes itself between blood and interstitium and in the interstitium so rapidly that no barrier limitation is detectable. Theoretically, if one could increase the flow indefinitely one would eventually reach a flow (decrease in time of exposure to permeable portion of the microvasculature) at which a barrier or permeability limitation became apparent. In effect, we are saying that Crone's elegant formulation of the permeability-surface area, PS, and flow, F, and extraction, E, relation

$$PS = F \ln 1/(1 - E) \qquad (3)$$

will not provide actual values of PS unless we are demonstrably dealing with a barrier limitation.

The distribution of tracer water

What tests can be used to determine whether we are in fact dealing with a barrier limitation? We have used three:

1. we have examined the effects of lowering the temperature in isolated perfused lung preparations since this manoeuvre will decrease fluidity of the membrane and its permeability if permeability is limiting. If the distribution of a substance is volume limited or determined lowering of the temperature should have little or no effect. A caveat is necessary here: we are considering a barrier limitation within the time of transit of blood through the microcirculation.
2. we can examine the effects of increasing the volume of distribution as by alveolar flooding (by adding isotonic solutions to the gas phase side of the alveolar capillary barrier). If the substance is barrier-limited, the extraction will remain unchanged. If the distribution is volume related and not barrier determined, the extraction will increase with increase of volume.
3. we can also examine the effects of flow changes on extraction. If we are dealing with a barrier-limited substance, extraction will vary in an inverse relation to

flow and PS will remain constant. If we are dealing with a flow-limited substance, then PS may increase as flow increases since the extraction should not be flow dependent, i.e., not determined by the time of exposure in the exchange vessels. Another caveat is required here: either S, the surface area, must remain constant in the face of changes of flow or the changes must be known and accounted for.

Effects of decrease in perfusion temperature

In a series of experiments in which isolated dog lungs were perfused at 36°C and then at 15°C we have found the extractions (and calculated PS values) of tracer water, THO, to be essentially unchanged (Chinard and Cua, 1986). In another series in isolated rat lungs perfused over a temperature range from 36°C to 8°C we found similar results. With antipyrine and iodoantipyrine we found temperature related decreases of the PS products below 24°C (Chinard, Basset, Saumon, Garrick and Cua, 1985; Cua, Basset, Saumon, Garrick and Chinard, 1985). Antipyrine and iodoantipyrine are much more lipophilic than THO. We consider the decrease of the extractions and PS products of the antipyrines as indications of decreases of permeability secondary to a decrease of membrane fluidity. The interpretation that the distribution of tracer water is flow or volume limited is consistent with these results.

Effects of alveolar flooding

Alveolar flooding was produced in vivo by the introduction of isotonic, protein-containing fluid at a dose of approximately 20 $cm^3.kg^{-1}$ into the airway of anaesthetized, artificially ventilated dogs. As shown in Fig. 8, there is clearly an increase of the extraction of tracer water as the extravascular lung water, ΔV_{wev}, is increased (Chinard, Cua, Tice and Bower, 1986; Chinard and Cua, unpublished). The latter quantity was obtained as described previously (Chinard, 1975). Similar results were obtained with isolated perfused lung preparations in which much larger amounts of intratracheal fluid were introduced. In neither the in vivo nor in the isolated perfused preparations did the PS product for sodium ion (as ^{22}Na) increase as ΔV_{wev} was increased. Again, the interpretation that the distribution of tracer water is flow or volume limited is consistent with these results obtained with alveolar flooding. The distribution of sodium ion, in contrast, would appear to be barrier limited since it was unchanged

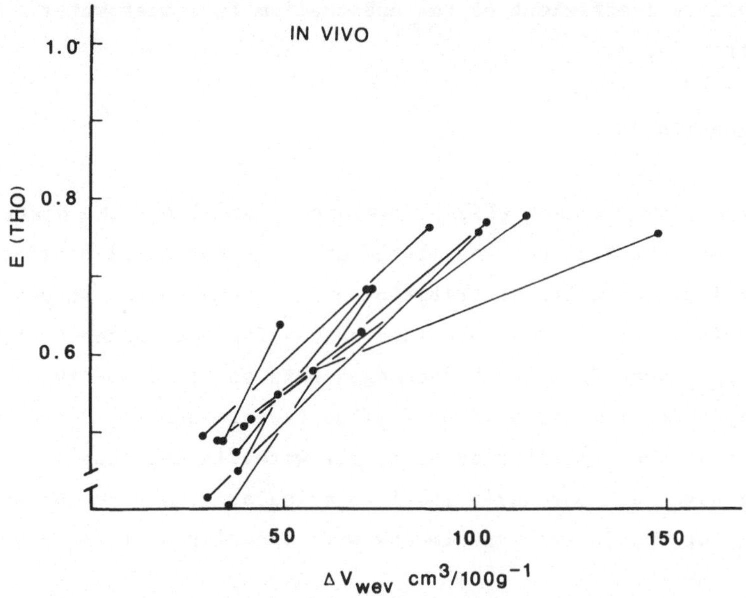

Fig. 8. Relation of tracer water extraction, E(THO) and extravascular lung water, ΔV_{wev}, in vivo. The extraction increases as ΔV_{wev} increases.

despite the marked increase of volume that was potentially accessible to it.

We have obtained other evidence to support the interpretation that the distribution of tracer water is flow or volume determined. Thus, the extractions and PS products of THO are not affected by mass exchanges of deuterium oxide and ordinary water substance (Chinard, Tice, Bower and Cua, 1984). Also, extractions and PS products of THO are not affected by net movement of ordinary water brought about by step inputs of hypertonic or hypotonic solutions of small solutes (Chinard and Cua, unpublished).

On the basis of the evidence presented or cited above, we feel that we can state with some confidence that the distribution of tracer water is flow or volume limited in the lungs. In other words, the distribution of tracer water between blood and interstitium (presumably endothelial and epithelial cell contents) is quasi-instantaneous relative to the velocity of blood in the microcirculation.

An important consequence of this conclusion is that PS values for tracer water can lead at best only to lower bounds of the actual values of the permeability coefficient of the endothelium to tracer water. We return to this matter later.

VASCULAR RECRUITMENT

Some years ago Goresky (1963) developed a model for the distribution of a tracer or indicator not restricted at the endothelial barrier but effectively limited in its distribution only by the volume accessible to it. With this model, Goresky obtained values for the volume of the vascular compartment from which exchanges between blood and the extravascular compartments of flow-limited tracers occur. Based on our conclusion that the distribution of tracer water in the lung is volume limited, we have used Goresky's model to calculate the vascular exchange volume, V_{ex} for THO in the experiments with vascular overload described above.

We had found that the calculated values for ΔV_{wev}, the extravascular lung water, for $(PS)_{THO}$ and for $(PS)_{Na}$ increased as the flow increased (Chinard and Cua, unpublished; Chinard and Defouw, 1984; Cua, Tice, Bower and Chinard, 1986). The increases of ΔV_{wev} could represent oedema or

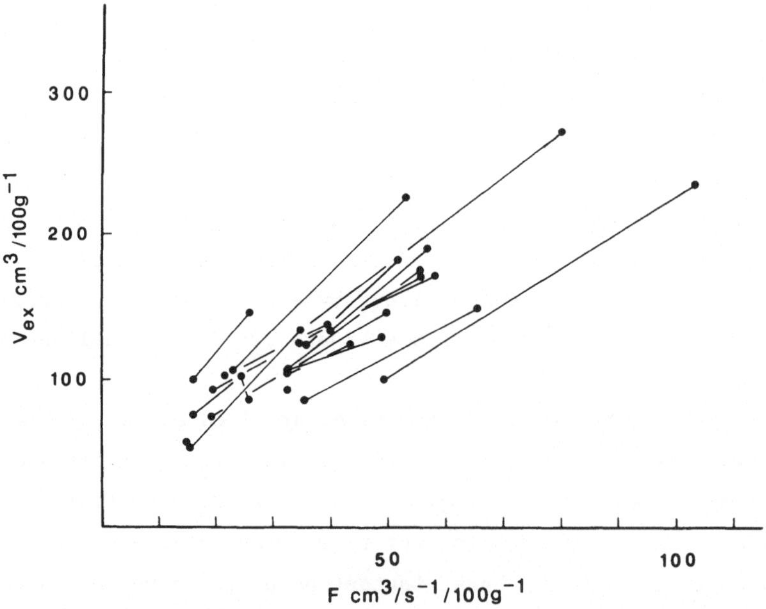

Fig. 9. Relation of vascular exchange volume, V_{ex}, to flow, F, in vivo in anaesthetized dogs. Flow increases were produced by loading the circulation with isotonic fluids.

recruitment of not previously perfused exchange vessels. Similar considerations applied to the increases of $(PS)_{THO}$ and $(PS)_{Na}$. When we examined the relation of the vascular exchange volume, V_{ex}, to flow we found a striking increase of V_{ex} as flow increased (Fig. 9). We considered that this increase of V_{ex} represented exchange vessel recruitment and re-examined the increases of ΔV_{wev}, $(PS)_{THO}$ and $(PS)_{Na}$ on the basis of the assumption that increases of S, the surface area, were proportional to the increases of V_{ex}. Specifically, we divided the ΔV_{wev} and PS values by the appropriate V_{ex}. The increases of $\Delta V_{wev}/V_{ex}$ and of $(PS)_{THO}/V_{ex}$ were somewhat less pronounced than the uncorrected values but not obliterated. In marked contrast, the values of $(PS)_{Na}/V_{ex}$ were essentially flat over the range of flows that was achieved.

Our conclusions that the distribution of THO was flow or volume limited while that of Na was barrier limited were thus supported by these results.

PERMEABILITY COEFFICIENTS TO THO OF ISOLATED ENDOTHELIAL AND OTHER CELLS; COMPARISONS WITH RESULTS OBTAINED BY THE MULTIPLE INDICATOR DILUTION TECHNIQUE

Extensive studies of segments of the pulmonary microvasculature let alone of isolated segments of alveolar capillaries have not yet been possible. We have turned instead to studies of isolated cell systems by means of the linear diffusion method developed by Redwood, Rall and Perl (1974). A summary of some of the results obtained to date for THO and for antipyrine in studies carried out under the direction of Dr. Rita A. Garrick in this laboratory are given in Table 6 (from Garrick, Ryan and Chinard, in press and as indicated in the footnote). The higher values for the permeability to THO (and to antipyrine) of mixed lung cells than of cultured endothelial cells may be the result of the use of enzymes in the isolation of the mixed lung cells with possible alteration of the glycocalyx and of the cell plasmalemma. We will therefore use the value of 304×10^{-5} $cm.s^{-1}$ as the more significant value for the permeability of the cultured endothelial cells to THO.

In earlier multiple indicator dilution studies from this laboratory (Perl, Silverman, Delea and Chinard, 1976) we had obtained values for the permeability coefficient of the endothelium to THO of approximately 300×10^{-5} $cm.s^{-1}$ from a modified Crone equation after making major corrections

Table 6[1]

Isolated cell permeability coefficients, P_d

P_d, x 10^{-5} $cm.s^{-1}$

	THO	^{14}C-antipyrine
endothelial cells (cultured bovine pulmonary artery)	304	181
alveolar macrophages (dog)	110	232
mixed lung cells (rabbit)	755	444
erythrocytes (dog)	908	318

1 Data for cultured endothelial cells are from Garrick, Ryan and Chinard, in press; for alveolar macrophages from Garrick, Polefka, Cua and Chinard, in press; for fixed lung cells from Garrick and Chinard, 1982; and for erythrocytes from Garrick, Patel and Chinard, 1982.

for back diffusion. The corrections and calculations must now be considered of doubtful validity in view of the lack of evidence of a barrier restriction to the distribution of THO. However, we have obtained lower bound values for P_{THO} very close to these values, namely about 308 x 10^{-5} $cm.s^{-1}$, in the alveolar flooding experiments when we use for surface area S = 500 $cm^2.g^{-1}$ lung. If we use the currently more favoured value S = 3.5 x 10^3 $cm^2.g^{-1}$, the value for P_{THO} drops to about 44 x 10^{-5} $cm.s^{-1}$. Given the probability for derecruitment in the flooded isolated perfused lungs (V_{ex} decreases on flooding), the lower value for S seems appropriate. In brief, the indicator dilution lower bound values for P_{THO} are close to those found by the linear diffusion method for cultured endothelial cells. Accordingly, we use the value P_{THO} = 300 x 10^{-5} $cm.s^{-1}$ in the considerations which follow.

ENDOTHELIAL PERMEABILITY COEFFICIENTS TO SODIUM AND COMPARISONS WITH PERMEABILITY COEFFICIENTS TO WATER

In earlier indicator-dilution studies we had found values for P_{Na} of approximately 3.0 x 10^{-5} $cm.s^{-1}$ based on surface area S = 500 $cm^2.g^{-1}$ (Perl et al., 1976). If we use S = 3.5 x 10^3 $cm^2.g^{-1}$, P_{Na} is about 0.43 x 10^{-5} $cm.s^{-1}$, in close accord with the value of 0.38 x 10^{-5} $cm.s^{-1}$ found in isolated dog lung lobes by Tancredi and Yipintsoi (1980). Values for P_{Na} in the <u>in vivo</u> alveolar flooding experiments average about 3.2 x 10^{-5} $cm.s^{-1}$ for S = 500 $cm^2.g^{-1}$ and 0.40 x 10^{-5} $cm.s^{-1}$ for S = 3.5 x 10^3 $cm^2.g^{-1}$.

With P_{THO} = 300 x 10^{-5} cm.s^{-1}, values for the ratio P_{Na}/P_{THO} are therefore approximately 1.3 x 10^{-3}, indicating a far higher diffusional permeability of the endothelium to water than to sodium. Extractions and calculated endothelial permeabilities of other small solutes such as chloride ion, urea and glucose are quite close to the corresponding values for sodium ion. This being granted, we can generalize and conclude that P_{THO} is far larger than small solute permeability coefficients.

TRANSPORT OF WATER AND SOLUTES ACROSS THE PULMONARY ENDOTHELIUM; REGULATING FACTORS IN PULMONARY OEDEMA

As stated above, the conventional or classical dogma is that net transfer of fluid from blood to interstitium occurs in bulk, i.e., without separation of water and solutes, only through pores or slits at the endothelial cell junctions. Calculations of filtration coefficients of the pulmonary endothelium based on this dogma have been made mainly by measuring the weight increment per unit time of isolated perfused lung preparations following stepped increments of microvascular pressure. Values of approximately 0.7 $cm^3.min^{-1}.mmHg^{-1}.100\ g^{-1}$ wet tissue are reported (e.g. Parker, Guyton and Taylor, 1979) for the filtration coefficients and correspond to values of 2.5 x 10^{-5} cm.s^{-1} for the filtration permeabilities. In injured lungs, values as high as 7 x 10^{-1} $cm^3.min^{-1}.mmHg^{-1}.100\ g^{-1}$ and 3.6 x 10^{-2} cm.s^{-1} can be calculated. We are concerned mainly with the lower normal values.

A major objection can be raised to these values. Filtration across the pulmonary endothelium, or ultrafiltration, does not differ fundamentally from reverse osmosis. Reverse osmosis is the process by which pure water is obtained from salt solutions, e.g. from Dead Sea water, by imposition of a pressure difference across a membrane permeable to the solvent water but impermeable to the solutes. The higher the solute concentration, the higher the pressure difference that must be imposed to obtain a given rate of pure water production (filtration). Conversely, in a system in which the concentration of the solute is not kept constant but is allowed to increase, the filtration rate will decrease as the solute concentration increases. Clearly the properties of the membrane (i.e. the filtration coefficient for water) cannot be characterized in the presence of non- or poorly permeating solutes. An engineer reporting membrane filtration coefficients obtained while filtering a solution would probably be sacked on the spot. Valid coefficients can be obtained only with pure water. This condition has not and probably can not be met in most biological systems without their disruption. The filtration permeability values which are indicated above represent functional parameters and not endothelial filtration permeabilities. The latter are in fact underestimated by these values because the lesser permeabilities of the barrier to small solutes than to water are not properly taken into account although the permeabilities of the macromolecules are.

With these considerations we can now describe what would happen in the event of an increment of microvascular pressure. An increment of pressure would lead to net movement of water across the endothelium because of the gradient of the chemical potentials of water. This movement would be more or less countered by an increment of small solute concentration on the vascular side of the endothelium. An increment of small solute concentration of one milliosmole would offset an increase of pressure of about 19 mm Hg. Thus, the rate of net movement of water is limited by non- or poorly permeating solutes as in other reverse osmosis membranes. This is a short term regulation occurring in a period of the order of the mean transit time of blood through the vascular exchange vessels, i.e. 0.8 to 4.0 seconds. Transfer of small solutes will occur as a result of the concentration gradient and in the time frame of small solute equilibration between blood and interstitium which is of the order of 1 to 3 minutes (Chinard, 1962). In the long term, of the order of minutes to hours, small solutes have equilibrated and the regulation is at the macromolecular level ("oncotic" pressure). Changes of the permeability to small solutes may be the major determinants in the development of the adult respiratory distress syndrome.

In substance, in this view we consider the pulmonary endothelium to be freely permeable to water with the net transfer of water from blood to interstitium being in the short term modulated or regulated by the low permeability to small solutes and in the long term by the much lower permeability to macromolecules.

This is admittedly a speculative proposal. It does provide a reconciliation of data obtained in two time domains - milliseconds and seconds for water and small solute interactions and minutes and hours for "filtration" and macromolecular interactions. At the least, the proposal has the virtue of being testable.

ACKNOWLEDGEMENTS

Supported in part by grants HL 12879 and HL 12974 from the National Heart, Lung and Blood Institute and in part by a Biomedical Research Support Grant to the New Jersey Medical School. I am particularly grateful to Dr. William O. Cua, Ms. Vivien Bower and Ms. Cheryl Tice for their collaboration in the more recent studies from this laboratory. Without them, the studies could not have been carried out.

REFERENCES

Bundgaard, M., Frøkjer-Jensen, J. and Crone, C. (1979) Endothelial plasmalemmal vesicles as elements in a system of branching invaginations from the cell surface. Proc. Natl. Acad. Sci. USA 76, 6439-6442

Chinard, F.P. (1975) Estimation of extravascular lung water by indicator dilution techniques. Circulation Res. 37, 137-145

Chinard, F.P. (1980) Pulmonary endothelial and epithelial vesiculation as a response to increased blood to tissue filtration. Physiologist 23, 62-66

Chinard, F.P. (1981) Capillary exchanges: small solutes. In: Microcirculation. Effros, R.M., Schmid-Schonbein, H., Ditzel, J., eds. Academic Press: New York, pp 33-50.

Chinard, F.P., Basset, G., Saumon, G., Garrick, R.A. and Cua, W.O. (1985) Transition from flow to barrier limited distribution of antipyrine and iodoantipyrine with decreased temperature in isolated perfused rat lungs. Microvascular Res. 29, 212, Abstract.

Chinard, F.P. and Cua, W.O. (1986) Endothelial extraction of tracer water is independent of temperature in dog lungs. Am. J. Physiol. 250, H1017-H1021

Chinard, F.P., Cua, W.O., Tice, C. and Bower, V. (1986) Alveolar flooding and pulmonary endothelial extractions of tracer water, sodium ion and antipyrine in vivo in dogs. Fed. Proc. 45, 1150

Chinard, F.P. and DeFouw, D.O. (1981) Ultrastructural changes of the air-blood barrier after spontaneous development of oedema in isolated dog lungs perfused at 15°C. Microvascular Res. 21, 48-56

Chinard, F.P. and DeFouw, D.O. (1984) Microcirculation of the lungs. In: Physiology and Pharmacology of the Microcirculation. Mortillaro, N., ed. Academic Press: New York. vol. 2. pp 1-42.

Chinard, F.P., Enns, T. and Nolan, M.F. (1962) The permeability characteristics of the alveolar capillary barrier. Trans. Assoc. Am. Phys. 75, 253-261

Chinard, F.P., Tice, C., Bower, V. and Cua, W.O. (1984) Tracer water follows bulk D_2O and H_2O net movement in isolated perfused rat lungs. Microvascular Res. 27, 236 Abstract.

Cua, W.O., Tice, C., Bower, V. and Chinard, F.P. (1986) Effects of flow rates on pulmonary exchange ("capillary") volumes and mean transit times in anesthetized dogs. Fed. Proc. 45, 1150 Abstract.

Cua, W.O., Basset, G., Saumon, G., Garrick, R.A. and Chinard, F.P. (1985) Dependence of tracer water extraction on accessible volume of distribution and independence from temperature changes in isolated perfused rat lungs. Microvascular Res. 29, 213 Abstract.

DeFouw, D.O. and Berendsen, P.B. (1978) Morphologic changes in isolated perfused dog lungs after acute hydrostatic oedema. Circulation Res. 43, 72-82

DeFouw, D.O. and Berendsen, P.B. (1979) A morphometric analysis of isolated perfused dog lungs after acute oncotic oedema. Microvascular Res. 17, 90-103

DeFouw, D.O. and Chinard, F.P. (1983) Morphometric and physiologic studies of alveolar microvessels in dog lungs in vivo after sustained increases in pulmonary microvascular pressures and after sustained decreases in plasma oncotic pressures. Microvascular Res. 25, 56-67

Garrick, R.A. and Chinard, F.P. (1982) Membrane permeability of isolated lung cells to non-electrolytes at different temperatures. Am. J. Physiol. 243, C285-C292

Garrick, R.A., Patel, B.C. and Chinard, F.P. (1982) Erythrocyte permeability to lipophilic solutes changes with temperature. Am. J. Physiol. 242, C74-C80

Garrick, R.A., Polefka, T.G., Cua, W.O. and Chinard, F.P. (1986) Water permeability of alveolar macrophages. Am. J. Physiol. In press.

Garrick, R.A., Ryan, U.S. and Chinard, F.P. (1986) Endothelial cell permeability to water. Biochim. Biophys. Acta. In press.

Goresky, C.A. (1963) A linear model for determining liver sinusoidal and extravascular volumes. Am. J. Physiol. 204, 626-640

Guyton, A.C., Granger, H.J. and Taylor, A.E. (1971) Interstitial fluid pressure. Physiol. Rev. 51, 527-563

Kazemi, H., Hyman, A.L. and Kadowitz, P.J. (1986) Acute Lung Injury. Pathogenesis of Adult Respiratory Distress Syndrome. PSG Publishing: Littleton

Laennec, R.T.H. (1819) De l'Auscultation Médiate. Brosson et Chaude: Paris, vol. II. p. 9.

Parker, J.C., Guyton, A.C. and Taylor, A.E. (1979) Pulmonary transcapillary exchange and pulmonary oedema. Cardiovascular Physiology. Internatl. Rev. Physiol. 18, 261-315

Perl, W., Silverman, F., Delea, A.C. and Chinard, F.P. (1976) Permeability of dog lung endothelium to sodium ions, amides and water. Am. J. Physiol. 230, 1708-1721

Prichard, J.S. (1982) Edema of the Lungs. Thomas: Springfield

Redwood, W.R., Rall, E. and Perl, W. (1974) Red cell permeability deduced from bulk diffusion coefficients. J. Gen. Physiol. 64, 706-729

Said, S.I. ed. (1985) The Pulmonary Circulation and Acute Lung Injury. Futura Publishing: Mount Kisco

Shea, S.M., Karnovsky, M.J. and Bossert, W.H. (1969) Vesicular transport across endothelium: simulation of a diffusion model. J. Theor. Biol. 24, 30-42

Simionescu, M. and Simionescu, N. (1984) Ultrastructure of the microvascular wall: functional correlations. In: Handbook of Physiology. Section 2: The Cardiovascular System. vol. iv. Microcirculation, part 1. American Physiological Society: Bethesda, pp 41-101.

Simionescu, N., Simionescu, M. and Palade, G.G.E. (1975) Permeability of muscle capillaries to small hemepeptides. Evidence for the existence of patent transendothelial channels. J. Cell Biol. 64, 586-607

Staub, N.C. (1984) Pathophysiology of pulmonary oedema. In: Edema, Staub, N.C. and Taylor, A.E., eds. Raven Press: New York

Staub, N.C., Nagano, H. and Pearce, M.L. (1967) Pulmonary oedema in dogs, especially the sequence of fluid accumulation in lungs. J. Appl. Physiol. 22, 227-240

Starling, E.H. (1909) The Fluids of the Body. Keener, Chicago

Tancredi, R.G. and Yipintsoi, T. (1980) Interrelationships of flow, intravascular pressure, and tissue perfusion in the measurement of capillary permeability to sodium in isolated dog lung lobes. Circulation Res. 46, 669-680

Zapol, W.M. and Falke, K.J., eds. (1985) Acute Respiratory Failure. vol. 24 of Lung Biology in Health and Disease. Dekker: New York

In Memoriam

Kjell Johansen: Viking and Physiologist

1932 - 1987

Kjell Johansen, who gave the August Krogh lecture at the XXX International Congress, died suddenly on March 4th, 1987, while on holiday in France. The news of Kjell's unexpected death was rapidly communicated between his friends and colleagues around the world. The world had been Kjell Johansen's laboratory and it was somehow appropriate that physiologists from all corners of the globe were reacting to the tragic shock of his death.

Kjell Johansen was born in Oslo and received his Ph.D. from his home institution. Much of his early career was spent at the University of Washington at Seattle (U.S.A.) which he left in 1971 to become Chairman of the Department of Zoophysiology at Aarhus University in Denmark. Although Kjell Johansen classified himself as a comparative cardiovascular and respiratory physiologist, this designation does not do justice to his wide-ranging interests in the whole field of comparative physiology.

If there is a single theme for which Kjell Johansen will be best remembered it will be the strategy of taking the laboratory to the animal, rather than vice versa, a research strategy that was almost a Kjell Johansen calling card. This approach took him to some of the most interesting (sometimes the most remote!) regions of this planet -- to the Amazon in 1967, and again in 1976; to the Antarctic in 1970; to the Philippine archipelago in 1976 and again in 1979; to Africa in the 1980s, to mention but a few of his more exciting expeditions. Perhaps because of this exposure to animals in such diverse environments, Kjell Johansen displayed an acute awareness of novel physiological problems and the myriad of mechanisms harnessed to solve them.

This essay, based on the August Krogh lecture, is a fitting tribute to both the man and his science. The essay conveys Kjell Johansen's enthusiasm and respect for nature while testifying to his skill and perseverance in studying the evolution and adaptation of physiological systems. Like the man, the essay is lively, interesting, entertaining and bubbling with ideas. Kjell would have liked nothing so much as the thought that it might encourage others to take a broad view of physiological problems and their resolution.

P.W. Hochachka
D.R. Jones

THE AUGUST KROGH LECTURE: THE WORLD AS A LABORATORY
Physiological Insights from Nature's Experiments

Kjell Johansen

Department of Zoophysiology
University of Aarhus
Aarhus, Denmark

August Krogh, the Nobel laureate from 1920 to whom this lecture is dedicated, epitomized the very essence of comparative physiology in his famous statement (Krogh, 1929): "For a large number of problems there will be some animal of choice or a few such animals on which it can be most conveniently studied".

I profess as many before me that the animals' environment and the constraints it offers should also be a paramount consideration in organismic physiology.

In this context I cannot help recalling a discussion I overheard years ago in the Arctic between Larry Irving and some fellow physiologists. The discussion got heated when Larry Irving refused to recognize the white laboratory rat as an animal. He argued that the white rat with food and water ad libitum and a thermostatted cage placed in a regulated light-dark cycle for literally thousands of generations could and should not qualify as an animal. I think Irving won the discussion.

On this basis and these premises I will in the following give a few examples of my research in comparative environmental physiology.

ABOUT COLD FEET

Regulation of skin blood flow

What probably has struck me the most as a biologist in the Arctic and Antarctic has been that the homeotherms there, the birds and mammals, are so strikingly well insulated. This perhaps should not surprise anyone, but in reality this fact shifts the main physiological problem for the polar homeotherm from an easily recognized one of conserving heat in perhaps -40°C to one of getting rid of heat when the polar animal during physical activity increases heat production by a factor of 10 or more. It is easy to reason that polar animals must depend on using selected areas of the skin as dissipating surfaces for heat during increased heat production. Regulation of skin blood flow hence becomes a primary target for study. These surfaces will have to be less insulated by fur, blubber or feathers than the rest of the animal, a fact which makes them vulnerable for costly heat loss at rest or for tissue injuries from freezing when ambient cold becomes severe. Circumvention of these threats must also rest with effective regulation of blood flow. Skin as an organ is not readily isolated from other tissues for direct blood flow measurement without grossly interfering with the integrity of the preparation, and indirect methods for flow measurement will always give uncertainties in interpretation.

For me the ideal preparation for studying skin blood flow became the extremities including the tail of mammals and the feet of birds.

Measuring blood flow to the tail of muskrats at rest and provoked to exercise on a small treadmill in Alaska, using a plethysmographic technique, revealed that tail blood flow could increase 350 times when a heat pad was placed on the trunk or during exercise.

Nerve block of muskrat tails prevented the colossal increase in tail blood flow and it was proposed that skin is endowed with a neurogenic vasodilator mechanism, much as skeletal muscle was known to have (Johansen, 1962).

Next the target animals for studies of skin blood flow and its regulation became Antarctic birds, primarily the giant petrel, *Macronectes giganteus*, and some species of penguins.

The body trunks of these animals have a formidably high insulation but their large naked feet consist mainly if not only of skin. The Macronectes foot with its large swimming web gives easy access to placement of small thermistors inside arteries and veins as well as intracutaneously. Web arterial and venous pressures can be monitored as can blood flow to the leg directly using an electromagnetic flow probe placed on the metatarsal artery.

If these feet were immersed in ice water, an experiment Nature does whenever the bird lands on water, we see a prompt onset of vasodilatation causing a several fold increase in foot blood flow and a temporary rise in arterial blood temperature. No primary vasoconstriction as in the typical homeotherm occurs.

This work (Johansen and Millard, 1973) disclosed that the neurogenic component of the vascular control involved both a tonic constrictor component most likely regulated by an α-adrenergic mechanism and a vasodilator component demonstrated both by nerve section and nerve stimulation. At the time we identified the latter as a cholinergic vasodilator mechanism. Later workers (Murrish and Guard, 1977) concluded that the vasodilatation was dependent on β-adrenergic nerves. Over the years many workers have taken an interest in the cutaneous vasodilatatory mechanism. McGregor (1979) among others has refuted the early suggestion that the neurogenic vasodilatation is cholinergically based, also he has eliminated histamine as a potential neurotransmitter; this amine has been implicated in several attempts to explain cutaneous vasodilatation in mammals. Purinergic as well as peptidergic nervous control of skin vasodilatation also receives scant support from the literature.

The nature of skin neurogenic vasodilatation, which seems to be present in all homeotherms to a different degree, must thus still await its final explanation.

THE AVIAN BROOD PATCH

A remarkable skin vascular organ

Using the world as a laboratory must not be taken to suggest that we never work at home. In the line of skin blood flow and its control recent experiments have been done (Midtgård, Sejrsen and Johansen, 1985) in

Copenhagen on skin circulation in the brood patch which most birds develop during their breeding season. An area on the breast and abdominal skin surface defeathers and develops a profuse vascularization serving to heat the incubated eggs. This vascularization develops with the hormonal changes in the breeding season. Evidence is suggestive that the brood patch occurs as a "seasonal organ". If the vascularization indeed should prove to develop de novo and regress when the breeding season terminates, we have a truly disposable organ such as the placenta, the antlers of reindeer and other ungulates, marvellous model organs for the study of angiogenesis: the formation of blood vessels.

Blood flow through the brood patch is an important factor in the control of egg temperature during incubation. Brood patch blood flow was measured by wash out of ^{133}Xe labelled into the brood patch skin. The egg temperatures could be changed by using artificial metal eggs which could be perfused selectively or collectively with water from a thermostatted system.

When the brood patch was cooled, skin blood flow promptly increased. Importantly, only the area directly cooled showed this cold vasodilatation. Cooling of skin adjacent to the brood patch caused a decreased blood flow and reduced skin temperature. It is certain that the response is strictly local, which will heat individual eggs in a clutch in relation to their respective temperatures and in this way diminish temperature differences between the eggs and have important survival value in synchronizing hatching and successful development of the clutch.

BODY TEMPERATURE REGULATION IN THE TROPICS

Life in a thermostat

We shall now discuss some aspects of body temperature regulation in the tropics. The setting will be in equatorial East Africa, specifically a subterranean habitat about 3 feet underground. There lives a most unusual mammal, the rodent Heterocephalus glaber, the naked mole rat, completely hairless and nearly blind. It lives in an intricately arranged system of burrows, feeding on roots and other plant material.

Unique physiological features of Heterocephalus include its low core temperature of about 30°C, a temperature corresponding very closely to the air temperature in its dark humid habitat where diurnal and seasonal temperature variations are 1 to 2 degrees different from that of the mole rat. This mammal must be one of very few which literally lives in a thermostat set to its own core temperature. Expectedly Heterocephalus has a very low heat production and a high thermal conductivity. The mole rat thus effectively has no resistance to heat flow (Johansen, Maloiy and Kornerup, 1986).

What is it about this animal's ability to thermoregulate? Does it have the sensors and effectors we know as integral parts of homeotherm thermoregulation? If it does, maybe Nature in the naked mole rat has afforded us an animal capable of showing how accurate body thermoregulation may develop when resistance to heat flow develops and becomes dependent on ambient temperature (T_A).

Deep body temperature (T_B) varies rapidly with time when Heterocephalus is taken out of the "thermostat" and reaches near ambient temperature within 2 hours. This results in a nearly linear relation between T_A and T_B when steady state has been reached, usually within 1-2 hours after a change in T_A.

When the mole rat is removed from its "thermostat" and exposed to lower ambient temperatures interesting responses occur. Its strong social behaviour becomes intensified and huddling becomes conspicuous. If T_A is lowered to 15-20°C, shivering particularly of the forepart of the animal is very striking.

Consistently during cold exposure the head skin around the brain case, particularly in the occipital region, was clearly the warmest part of the body, even warmer than the rectal temperature. This higher head temperature correlated with large deposits of brown fat particularly accumulated around the posterior portion of the skull.

It is felt that the experiment Nature has done by placing a mammal in a thermostat set at its core temperature may give us a potential model animal for studies about how regulatory mechanisms may develop. We were strengthened in this view when a colony of mole rats brought from Kenya to Denmark and kept at stable air temperatures of about 24°C rather than their Kenyan habitat temperature of about 30°C, showed a trend for T_B to be kept higher than T_A for longer than observed in the Kenyan animals.

The search for oxygen

We shall now divert your interest to adaptive properties of blood in respiratory gas exchange. In Goethe's famous play Faust, Mefistopheles says to Faust: "Blut ist ein ganz besonderer Saft". (Blood is a remarkable juice). Nothing could be more true. Adaptive relationships between blood O_2 affinity and environmental factors are common and reflect long term as well as acute changes in environmental factors. August Krogh was a pioneer also in this field. Today we know that such adaptive changes in blood respiratory properties can be traced to molecular properties in the haemoglobins. Also modulations of blood O_2 affinity are typically due to ligand or cofactor interaction with the O_2-Hb binding. Short term alterations in O_2 affinity may become manifest within minutes after the environmental or behavioural changes set in, probably reflecting hormonal control of cofactor-Hb interaction. Recent work has also disclosed that the classical Hill plot expressed by the n-value or cooperativity coefficient is not the straight line it used to be but depends on the O_2 saturation level, a phenomenon likely traceable to the state of aggregation of the haemoglobin molecules (Lykkeboe and Johansen, 1978). An increase in n-value with increasing saturation has now been demonstrated in species from all vertebrate classes including a mammalian species (R. Holland, personal communication). The physiological significance of this n-value increase can be most important particularly for low affinity bloods.

We shall first stop in the huge fresh water ocean known as the Amazon. Among the 1500 known species of fish living in the vast region, we shall look at three. Depending on season and location large areas of the Amazon confront the aquatic life with conditions of hypoxia which would require mountains twice as high as Mount Everest to match.

Two of the species I will discuss belong to the relict family Osteoglossidae, which includes only 5 living species of very limited phyletic diversity. These "living fossils" can be found in the same habitats of the Amazon. One, <u>Osteoglossum bicchirosum</u>, (growing to about 30-50 cm) is an exclusive waterbreather. The other, <u>Arapaima gigas</u>, the world's largest fresh water fish, coexists with Osteoglossum, but is an obligatory airbreather and will drown unless it has access to atmospheric air to inhale into its specialized swimbladder lung.

If we compare the O_2-Hb dissociation curves for Osteoglossum and Arapaima at similar temperature and pH we find a much higher affinity in the waterbreathing Osteoglossum expressed by a P_{50} of 6.1 mm Hg compared to 21.0 mm Hg in Arapaima (Johansen, Mangum and Weber, 1978b). When this information is placed in the context of the water P_{O_2} at the site in the Amazon where these studies were done ranging from 15 mm Hg at night to 30 mm Hg during daylight, it becomes clear that the obligatory dependence on airbreathing in Arapaima is an absolute requirement for survival. If the fish had to depend on waterbreathing with gills, its arterial O_2 saturation could not exceed 10-15%. Osteoglossum, however, can rely on gill breathing for saturation of its blood. Interestingly, the O_2 uptake of Arapaima was more than twice that of Osteoglossum in similar sized specimens. The much lower affinity of Arapaima blood is consistent with the much higher O_2 availability in air than water. The low affinity blood will also allow for more efficient O_2 unloading and in this way support a higher level of O_2 uptake.

Notably, Arapaima had a very low gas exchange ratio of its swimbladder lung (about 0.15) suggesting that CO_2 excretion predominantly takes place to the water in the gills or skin as is typical of bimodally breathing fishes. How the fish avoids losing O_2 from blood to water during passage through the gills must depend on a physiological shunt mechanism in the gill and the different diffusion rates and solubility for O_2 and CO_2 in water.

A third species from the Amazon I will discuss is the teleost fish, Synbranchus marmoratus. Synbranchus is a facultative airbreather implying that the entire O_2 requirement can be supported by either water or airbreathing. Synbranchus also voluntarily makes long excursions on land moving in a snake-like fashion. The species can also estivate for long periods in moist soil. Blood respiratory properties in Synbranchus, when in well aerated water and acutely air exposed, show a marked reduction in O_2 affinity with air exposure. This is caused both by reduced pH from retention of CO_2 and from an increase in the red cell concentration of the trinucleoside phosphates ATP and GTP. These data show that blood respiratory properties adapt to the principal mode of gas exchange in a bimodal breather. In distinction to the osteoglossids, the adaptive response of Synbranchus depends on quick onset and reversible ligand changes (H^+ ion and trinucleoside phosphates) and not on intrinsic molecular properties as in Arapaima and Osteoglossum (Johansen, Mangum and Lykkeboe, 1978a).

We may conclude that the time course of adaptation may be influenced by the selection of a particular mechanism to modulate O_2-Hb binding properties. Short-term and often transient changes in environmental oxygen levels appear to induce rapid and reversible changes in oxygen affinity of the blood by alterations in the concentrations of metabolically labile cofactors within the red cell. On the evolutionary time scale, the adoption of airbreathing is accompanied by more profound and genetically fixed changes in oxygen binding, effected by less easily reversible changes in the structure of the haemoglobin molecule.

ESTIVATION AND TORPOR

The lungfish and the hummingbird

Among those of Nature's experiments I have found most rewarding to study are prolonged estivation and daily torpor. These adaptations are indispensable for many animals during drought and starvation. A most striking example is the estivating lungfish. Nocturnal torpor I would like to discuss briefly in the context of the hummingbird. Hummingbirds have such a phenomenally high metabolic rate when active during the day that night-time torpidity for many species when access to food is curtailed becomes obligatory.

The African lungfish, Protopterus, Kamongo in Swahili, is a fish so well known from the fascinating papers and other writings of Homer Smith (Smith, 1930). Since that time many scientists have been interested in the estivating lungfish. The works of A.P. Fishman and his group in Philadelphia among others have been highlighting the subject in recent years (Fishman, Delaney and Laurent, 1985).

It has been my privilege to work on Kamongo with Geoffrey Maloiy in Kenya and with my colleague Lomholt in Denmark on fishes brought there from Kenya.

We were fortunate to find access to an area near Malindi on the Kenya coast where a number of estivating lungfish could be removed intact in the soil in which they had been embedded by Nature about one year earlier. One specimen in particular became unforgettable. A chunk of mud equipped with a Fleisch head connected to a pneumotachograph was sitting on my desk for more than 7 years after removal from the site at Malindi. This was truly a revelation in suspended animation for me, regularly watching the periodic excursion on a recorder reflecting Kamongo's unbelievably small tidal volumes and low breathing effort. One day a foul smell told us that Kamongo had given up waiting for the rains.

Only a few points from our work on estivating lungfish will be brought up here.

With estivation there occurs a most dramatic increase in blood O_2 affinity correlated with a conspicuous reduction in the ratio of red cell organic phosphates to haemoglobin concentration (Johansen, Lykkeboe, Weber and Maloiy, 1976).

We originally reasoned that this dramatic increase in blood O_2 affinity was an adaptive response to an internal hypoxic state such as earlier demonstrated for a number of fishes. When later a needle was placed into the lung space of estivating lungfish, mass spectrometer readings revealed that lung O_2 tension was not at all low as we had surmised it to be. In the meantime, Delaney, Lahiri, Hamilton and Fishman (1974) had also demonstrated that arterial P_{O_2} in Protopterus induced into artificial estivation was high.

Estivation in lungfish is attended by a very large reduction in aerobic metabolism, which can be as low as 5% of the awake non-starved value. Starvation of lungfish in the awake state also reduces O_2 uptake markedly, as it does in ectotherms in general.

In the context of starvation we must of course ask what it is that the tissues have in short supply. Substrate is obviously one commodity although as long as you can fuel metabolism from your own tissues this may not be limiting. Could O_2 be another? If arterial O_2 tension is not reduced in starvation what could be limiting for the tissues in terms of their O_2 supply? Obviously blood flow, which we know will reduce tissue O_2 uptake, for example in diving animals. Also we could speculate that the capillary-to-tissue P_{O_2} gradient could be a limiting factor in tissue O_2 delivery. This gradient will be very much reduced if O_2 is unloaded from

Hb at very low O_2 tensions such as those resulting if blood has a very high O_2 affinity.

We may then perhaps turn around our original and incorrect interpretation of the high O_2 affinity in estivation. We turn cause and effect in reverse, and speculate that some specific factors influencing red cell phosphate metabolism directly may bring about the very high blood O_2 affinity, which in turn will affect the capillary-to-tissue P_{O_2} gradient and cause reduced tissue O_2 uptake.

BEIJA FLOR

The one who kisses the flowers

Hummingbirds, Beija flor in Brazilian, are unique metabolic machines. Their metabolic rate (O_2 uptake) can vary by a factor of more than 300 between torpidity and active flight, a transition which may occur in less than 30 minutes. Nocturnal torpor is a necessity for most hummingbirds, especially species experiencing cold nights, because the high active metabolic rates cannot be sustained when nectar feeding is interrupted and ambient temperature has fallen.

Let us first examine how the respiratory properties of blood may support such high and rapidly changing metabolic rates. In general the O_2 affinity of hummingbird blood is low and the O_2 capacity is high. Three species studied in detail had P_{50} values at pH 7.40 and temperature 39°C between 41.0 and 44.0 mm Hg. The average Bohr factor was -0.39. The thermal sensitivity of the blood O_2 binding was rather low expressed by a ΔH value at P_{50} of -7.65 Kcal.mole^{-1} compared to more than twice that value for the purified haemolysate. A non-linearity of the Hill transformation expressed by increasing n-values at higher O_2 saturations demonstrated that n-values could exceed 7.0 for O_2 saturation 80% or higher (Johansen et al., 1986).

Most authors commenting on such n-value increases, particularly beyond values of 4.0, which raise conflict with conventional views on cooperativity in tetramer vertebrate haemoglobins, do not discuss the physiological implications of this change in O_2-Hb equilibrium. If arterial O_2 tension in hummingbirds ranges between 80.0 and 90.0 mm Hg, values reported typical of other species of resting birds (Kawashiro and Scheid, 1975), a value of 80 mm Hg would give an arterial O_2 saturation of 97.5% in the hummingbird Melanotrochilus. By comparison a linear Hill transformation from the Hill coefficient at P_{50}, would give a saturation of only about 82.0%. Conversely, if the n-value prevailing at P_{50} was applied to the entire equilibrium curve, an arterial P_{O_2} exceeding 140.0 mm Hg would be needed to arterialize the blood to 97.5% at pH 7.40.

Hummingbirds spend nearly all of their awake time in flight, which calls for near maximal efficiency in gas exchange and blood gas transport for extended periods. The saturation dependent increase in n-value may help offset a possible diffusion limitation in the parabronchi during exercise and thus prevent arterial desaturation (Scheid, 1978). Submaximal O_2-Hb loading from reduced residence time of blood in the parabronchial blood capillaries during exercise would also be offset by a steeply rising affinity at the higher saturations.

Summarily, when the Hill plot curves upward in a low affinity blood such as for hummingbird blood at higher levels of saturation, arterial O_2 saturation will be safeguarded while the inherently low affinity of the blood will allow most of the possible O_2 unloading to take place at P_{O_2} values conserving an effective P_{O_2} gradient from capillary blood to cellular sites. It can be calculated that a 40% unloading of O_2 from arterial blood saturated to 90% with O_2 can occur with a P_{O_2} gradient between arterial and venous blood of less than 22.0 mm Hg; an unloading efficiency likely to exceed the maximum possible for any mammalian blood.

The nocturnal torpidity of most hummingbirds may cause deep body temperature to fluctuate by 20 to 25°C within a very short time during entry and arousal from torpor. Such rapidly fluctuating body temperatures will influence pulmonary gas exchange and internal blood gas transport in important ways, including shifting the O_2-Hb and O_2-Mb equilibrium curves. Since arterial P_{O_2} during torpidity is likely to drop due to a marked decrease in ventilatory effort, arterial desaturation can only be prevented if a left shift in O_2-Hb equilibrium attends the reduced body temperature. A ΔH value of about -8.0 $Kcal.mole^{-1}$, such as we measured, may be an ideal compromise between safeguarding arterial saturation and maintaining high venous O_2 tensions at reduced body temperatures. Again the change in n-value at higher saturation will be of major importance.

During regulated torpor at for instance a body temperature (T_B) of 15°C, the O_2 uptake of a hummingbird may be as high as that of the same bird in a homeotherm resting state with a T_B of 40°C. This implies that respiratory gas exchange and blood gas transport must operate with similar efficiency over a body temperature difference of at least 25°C, a situation not matched in any other homeotherm.

In regard to the potential energy saving by torpor in hummingbirds, calculations for one species, Melanotrochilus fuscus, show that 15% of the homeotherm energy expenditure suffices to sustain the energy needs at ambient temperatures between 15 and 20°C (Berger and Johansen, 1986).

Before leaving the Beija flor let us estimate the potential performance of their heart and blood circulation during the maximal energy expenditure of flight. We start with a small species weighing 2.5 grams, have an O_2 uptake rate in flight of 40 ml $O_2 \cdot g \cdot hr^{-1}$. Blood O_2 capacity is set to 22 vol% and blood volume to 8% of body weight. If now the utilization of O_2 from the circulating blood is 60% and we use the Fick principle to calculate cardiac output, we arrive at the figure of 12.6 $ml \cdot min^{-1}$. This is a staggering 5 times the body weight of the bird per minute at a heart rate of up to 1400 beats per minute. Blood volume for the bird calculated as 0.2 ml, will give an average circulation time of about 1 second for the blood volume. This figure raises the obvious question if the residence time of circulating blood in the lung and tissue capillaries is long enough for the blood to serve its functions. The answer is, of course, yes; the small hummingbirds have long demonstrated that.

If, however, we seek information from the literature about the kinetics of the reactions in red cells between haemoglobin and its ligands and consider the diffusion rates and boundary conditions of red cell gas exchanges, we will find that the smallest hummingbirds are at a size approaching limitation by cardiovascular performance. The rate of blood flow should thus be added to other well known factors which set the lower limit for the size of homeotherms (Schmidt-Nielsen, 1984).

ON BEING TALL

Cardiovascular function in the giraffe

The physiology of scaling has rightfully enjoyed widespread interest among physiologists, certainly among comparative physiologists. The hummingbirds being so small have understandably defended an important role in scaling discussions. Scaling refers to correlations with weight, volume and linear dimensions. One dimension which to my mind has been neglected in this interesting discussion is that of length. Both horizontal and vertical lengths of animals are of great interest in physiology, particularly in cardiovascular physiology. Length in the vertical direction is height. If this dimension is large the animal is tall. I want to allude briefly to the physiology of being tall.

In a museum in Berlin, there is a skeleton of a Brachiosaurus, an extinct sauropod that lived 65 million years ago. The skeleton reaches nearly 12 metres high and gives any cardiovascular physiologist awesome thoughts about the hydraulic engineering of such a creature.

Today we can watch living giraffes reaching nearly 6 metre heights as survivors of Nature's experiments on being tall.

Alan Burton's wonderful reading on the Physiology and Biophysics of the Circulation (Burton, 1965), chapter 9, starts with the quotation (Lyly, 1580): "The truth, the whole truth and nothing but the truth". Burton goes on and now I quote him: "It is no harder, in the circulation, for blood to flow uphill than downhill." If this is nothing but the truth, it is also the truth that a full-grown giraffe, 5-6 metres tall, has a systemic blood pressure at heart level about double that of normal man and most mammals. Since the days of Poiseuille and perhaps before, our understanding of haemodynamics has been conceived by clear minds working with rigid glass tubes and coloured water. Much later a pump in lieu of the heart was placed in the glass models, but this pump was typically not pulsatile and if it was, flow in the large major tubes leaving the pump did not come to a halt or reverse between beats such as in the living organism (Spencer and Greiss, 1962).

With collapsible tubes and passive valves between the intermittently beating pump and the tube system, and flow records at the base of the aorta showing reversed and later zero flow in diastole (which does not imply that flow in the lesser arteries and capillaries is halted), "the whole truth and nothing but the truth" must be that when systole starts, the heart must overcome the pressure keeping the valves shut, which must include the height of the fluid column resting on them, a possible back pressure from a more peripheral windkessel function, the resistance of the microcirculation and possibly collapsed veins, before blood returns to the heart by a waterfall effect.

If we next turn to the other end of the giraffe, to the legs where arterial, venous and probably microvascular pressures must far exceed those at heart level and use this to reason about tissue fluid balance, we find that August Krogh did it before us.

What Krogh asked the giraffe is an example of his so-called "visual thinking". Krogh wrote: "It would be extremely interesting to know just how the giraffe avoids filtration oedema in its feet". Krogh tried to find means to study the subject, but only came as far as watching giraffes in the Copenhagen Zoo and verified that they did not have oedema in their feet, except in one instance when he recalled that a giraffe after standing still for a considerable length of time did show some swelling of its feet.

Fifty years later our group asked the same question (Hargens, Millard, Johansen, Gershuni, Pettersson, Burroughs, Meltzer and Van Hoven, 1986). We were able in 1985 to work with the giraffe in Nature's laboratory.

First let us briefly comment on morphological specializations of interest to the question of fluid balance. Peripheral arteries in the extremities have a phenomenally high muscular component giving the arteries a pinhole lumen and a very thick muscular wall. This picture extends down to the microvasculature.

Williamson, Vogler and Kilo (1971) demonstrated that the capillary membrane thickness in giraffe legs is more than twice that typical of mammals. Importantly, this may reduce capillary permeability when going from the neck to the leg. These morphological features may be important in reducing capillary pressures, reducing the effective surface area, the hydraulic conductivity and increase the reflection coefficient for proteins, thus impeding excessive filtration.

In addition, there is a conspicuous subcutaneous connective tissue fascia along the extremities and neck, presumably having an important G-suit effect. The G-suit was thus not at all invented by aviation medicine but by the giraffe. The connective tissue G-suit may also serve to reduce filtration by increasing interstitial tissue pressure and reduce venous capacitance. A well developed lymphatic system will also serve to prevent oedema formation.

Transcapillary fluid balance depends on the four "Starling pressures". Most importantly these are all dynamic and change with time, some more than others. We were able to measure the arterial, venous, and interstitial tissue pressures dynamically using telemetry techniques when giraffes were walking in their corral. Pressures on the neck were simultaneously recorded for comparison.

When these data are plugged into the Starling equation they reveal that in the neck region there should be no problem with oedema formation because a net reabsorption pressure prevails. In the legs, however, using data from a standing giraffe there is a gross imbalance of transcapillary pressures which would favour filtration. When oedema does not appear to happen it must relate to highly variable pressures with movement of the animal.

CEPHALOPODS

Very old members of the jet set

I surmise that more than 90% of all physiological research concerns mammals and more than 95% deals with the vertebrates. Perhaps more than 20% deals with the laboratory rat which Larry Irving refused to consider an animal. The vertebrates represent one animal phylum out of more than twenty. The Director of Research in Nature's laboratory must find this a highly biased choice of experimental subjects by us physiologists.

This striking imbalance must have its basis in the misconception that the mammal and man in particular has a special and important role to play as experimental subject in physiology. I take exception to this and again I would like to quote August Krogh from a lecture he gave at the International Physiology Congress in 1929. The title of the lecture was: "Progress in Physiology". Krogh said: "I want to say a word for the study of comparative physiology also for its own sake. You will find in the lower animals mechanisms and adaptations of exquisite beauty and the most surprising character (and I think nothing can be more fascinating than the senses and instincts of insects as revealed by modern investigations)". Another pertinent quotation goes as follows: "It is virtually a truism that the simplest organism that carries out the phenomenon in question will give the basic answer first".

I would like to deal briefly with some physiological aspects of the cephalopods - the most awesome and agile animals in the oceans. Some squids swim faster than most fishes and travel by jet. Cephalopods apparently represent the largest biomass in water. It is their blood that I want to discuss, blue blood it is, when its respiratory protein haemocyanin (Hcy) is oxygenated.

The O_2 affinity of cephalopod blood is like most bloods pH sensitive as expressed by the Bohr coefficient. No other group of animals, however, exhibits such a large variability in the Bohr factor between species. It may range from -1.80 as Redfield and Goodkind demonstrated 56 years ago for the squid Loligo, down to a modest -0.20 recently shown for Nautilus (Lykkeboe and Johansen, 1983).

The Bohr effect is generally accredited important physiological significance in blood gas transport in that metabolically produced CO_2 will promote unloading of bound O_2 in the capillaries. It is common to assume that this facilitation should be proportional to the size of the Bohr coefficient. However, closer scrutiny puts a large question mark by this generalization, because the Bohr effect is tantamount to an O_2-linked binding of protons (H^+ ions) to the respiratory pigment. This process is referred to as the Haldane effect. According to the linkage equation described by Wyman (1964), the Bohr coefficient and the Haldane coefficients are identical, i.e.

$$\frac{\Delta \log P_{50}}{\Delta pH} = \left\{\frac{\Delta cRPH}{\Delta cRPO_2}\right\} pH$$

(RP = respiratory protein)

If now, as is the case in many cephalopods, the Bohr coefficient is numerically greater than -1.0 (i.e. the slope is steeper than -1.0), it implies that more than 1 mmol of hydrogen ions will be bound to the pigment per mmol of O_2 unloaded. This is most important when our discussion is limited to hydrogen ions resulting from aerobic metabolism, since the CO_2 metabolically produced from aerobic metabolism can maximally yield 1 mmol of hydrogen ions per mmol of O_2 consumed, if the gas exchange ratio is unity. In fact if the Bohr coefficient (slope) is steeper than -1.0, perhaps -1.8 as in the case of Loligo, a pH increase rather than a decrease should accompany deoxygenation; i.e. venous blood should have a higher pH than arterial. This would align poorly with the usual reasoning that a large Bohr factor will aid O_2 unloading from the blood. A large Bohr factor may then be a detriment rather than an asset for O_2 unloading.

Or could it be that there exists another source of CO_2 or H^+ ions not anaerobically produced? If it is CO_2, it has to be O_2 linked, in other words be set free on deoxygenation, but independent of pH!!!

We have probed into this problem and can confirm that such a CO_2 binding exists (Lykkeboe, Brix and Johansen, 1980).

The question of whether pH can actually decrease during O_2 unloading even if a high ($>$-1.0) Bohr factor exists, may be answered affirmatively after our experiments. A prerequisite for this must, however, be the presence of an oxygen-linked CO_2 component large enough to generate a functional Haldane coefficient (mM CO_2/mM O_2 at constant pH) well below unity. This condition is likely to be species-dependent and relate to metabolic rate and metabolic scope. In *Sepia latimanus*, a moderately active cephalopod, the condition is marginally present, but in the case of *Loligo pealei*, a far more active species, it may be more developed.

In the light of these ideas and results it is suggested that we re-evaluate the physiological consequences and significance of the Bohr and Haldane effects. We must expand the traditional and still prevalent viewpoint that metabolically produced CO_2 promotes dissociation of O_2 bound to the respiratory pigment and thus facilitates tissue O_2 transport in proportion to the magnitude of the Bohr coefficient. If large Bohr coefficients are present deoxygenations of the pigment will primarily promote the removal of CO_2 with a minimal change in hydrogen ion concentration (isohydric CO_2 transport). Thirdly, the magnitude of the Bohr effect will be decisive for the influence of a pH change on the O_2 affinity (Lykkeboe and Johansen, 1983).

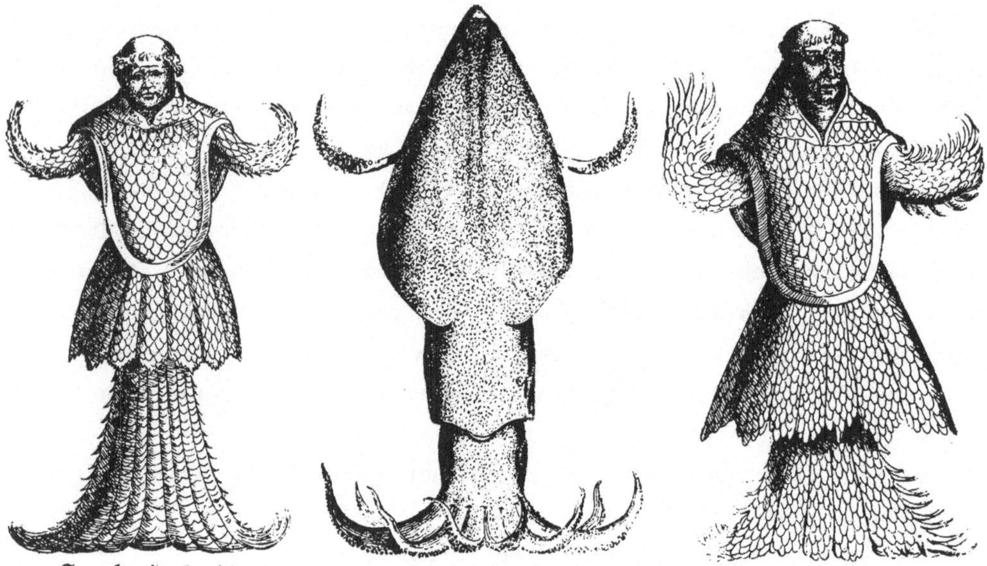

Sømunken efter Rondelet. Loligo-Blæksprutte fra Kattegattet. Sømunken efter Belon.

Fig. 1. After J. Steenstrup, 1855.

EPILOGUE

It is my impression that in our day and age the domain of politics dominates everything, also science unfortunately. It has not always been like this. Let me use my fascination with cephalopods to illustrate my point.

In the year 1550 a strange creature was caught in Øresund, the narrow strait separating present day Sweden and Denmark. The creature, nearly dead, was brought before the King (Fredrik II). He had a head like a man and a monk's crown, a bald head with a ring of hair. He also wore a coat like monks did. The King buried the creature with full royal honours. The famous author Holberg (born and raised in Norway, but claimed by Danes as theirs) wrote in his book: The History of Denmark, that the sea monks were the result of the Lutheran reformation when the monks became unwanted and had to take refuge in the oceans. The Danish King Fredrik II must have had similar thoughts because he had sent a drawing of a sea monk to the Spanish Emperor Karl and reassured him that the Catholic faith still blossomed in northern Europe in spite of Luther's reformation. Historians now tell us that because of this reassurance a treaty was signed between Emperor Karl and the Kings of Scotland and Denmark. The treaty is said to have given important stability and peace in Europe at the time.

In our time we as biologists have no influence on political decisions. Nature will go on making experiments, however, and as scientists in that laboratory we will always be behind and will never be out of a job. I wonder what a mermaid is?

ACKNOWLEDGEMENT

I would like to express my gratitude to the Nordic Insulin Laboratory in Copenhagen for their support and funding of the August Krogh Lecture. It is no happenstance that the Insulin Laboratory wished to make this dedication to August Krogh who played such an important role in promoting the understanding of insulin research and the production of insulin.

REFERENCES

Berger, M. and Johansen, K. (1986) The stages of torpor in hummingbirds. In manuscript.

Burton, A.C. (1965) Physiology and Biophysics of the Circulation. Year Book Medical Publ.: Chicago, 217p.

Delaney, R.G., Lahiri, S., Hamilton, R., and Fishman, A.P. (1974) Aestivation of the African lungfish *Protopterus aethiopicus*: cardiovascular and pulmonary function. J. exp. Biol. 61, 111-128.

Fishman, A.P., Delaney, R.G. and Laurent, P. (1985) Circulatory adaptation to bimodal respiration in the dipnoan lungfish. J. Appl. Physiol. 59, 285-294.

Hargans, A.R., Millard, R.W., Johansen, K., Gershuni, D.H., Pettersson, K., Burroughs, R., Meltzer, D.G.A., and Van Hoven, W. (1986) Blood and interstitial fluid pressures in feet and neck of the giraffe. Fed. Proc. 45, 758.

Johansen, K. (1962) Heat exchange in the muskrat tail. Evidence for vasodilator nerves to the skin. Acta Physiol. Scand. 55, 160-169.

Johansen, K., and Millard, R.W. (1973) Vascular responses to temperature in the foot of the giant fulmar, *Macronectes giganteus*. J. Comp. Physiol. 85, 47-64.

Johansen, K., Lykkeboe, G., Weber, R.E. and Maloiy, G.M.O. (1976) Respiratory properties of blood in awake and estivating lungfish, *Protopterus amphibius*. Respir. Physiol. 27, 335-345.

Johansen, K., Mangum, C.P. and Lykkeboe, G. (1978a) Respiratory properties of the blood of Amazon fishes. Can J. Zool. 56, 898-906.

Johansen, K., Mangum, C.P. and Weber, R.E. (1978b) Reduced blood O_2

affinity associated with air breathing in osteoglosside fishes. Can. J. Zool. 56, 891-897.

Johansen, K., Berger, M., Bicudo, J.E.P.W., Ruschi, A., and De Almeida, P.J. (1986) Respiratory properties of blood and myoglobin in hummingbirds. Physiol. Zool., in press.

Johansen, K., Maloiy, G.M.O., and Kornerup, S. (1986) Metabolic rate and body temperature regulation in the naked mole rat, Heterocephalus glaber. In preparation.

Kawashiro, T., and Scheid, P. (1975) Arterial blood gases in undisturbed resting birds; measurement in chicken and duck. Respir. Physiol. 23, 337-342.

Krogh, A. (1929) The progress in Physiology. Am. J. Physiol. 90, 243-251.

Lykkeboe, G., and Johansen, K. (1978) An O_2-Hb "paradox" in frog blood? (n-values exceeding 4.0). Respir. Physiol. 35, 119-127.

Lykkeboe, G., Brix, O., and Johansen, K. (1980) Oxygen linked CO_2 binding independent of pH in cephalopod blood. Nature 287, 330-331.

Lykkeboe, G., and Johansen, K. (1983) A cephalopod approach to rethinking about the importance of the Bohr and Haldane effects. Pacific Science 36, 305-313.

McGregor, D.D. (1979) Noncholinergic vasodilator innervation in the feet of ducks and chickens. Am. J. Physiol. 237, H112-117.

Midtgård, U., Sejrsen, P., and Johansen, K. (1985) Blood flow in the brood patch of Bantam hens: evidence of cold vasodilatation. J. Comp. Physiol. 155, 703-709.

Murrish, D.E. and Guard, G.L. (1977) Cardiovascular adaptations of the giant petrel, Macronectes giganteus, to the Antarctic ecosystems. Llano, G.A., ed. Smithsonian Inst.: Washington. pp 511-530

Redfield, A.C., and Goodkind, R. (1929) The significance of the Bohr effect in the respiration and asphyxiation of the squid, Loligo pealei. J. exp. Biol. 6, 240-349.

Schmidt-Nielsen, K. (1984) Scaling. Why is animal size so important? University Press: Cambridge, 241 p.

Smith, H. (1930) Metabolism of the lungfish Protopterus aethiopicus. J. Biol. Chem. 88, 97-130.

Spencer, M.P., and Greiss, F.C. (1962) Dynamics of ventricular ejection. Circulation Res. 10, 274-279.

Williamson, J.R., Vogler, N.J., and Kilo, C. (1971) Regional variations in the width of the basement membrane of muscle capillaries in man and the giraffe. Am. J. Pathol. 63, 359-370.

Wyman, J., Jr. (1964) Linked functions and reciprocal effects in hemoglobin: a second look. Adv. Prot. Chem. 19, 224-286.

THE DUCK AS A DIVER

David R. Jones

Department of Zoology
6270 University Blvd.
University of British Columbia
Vancouver, B.C. V6T 2A9 Canada

SUMMARY

During forced submergence adult dabbling ducks (Anas platyrhynchos) can remain submerged for times only matched, among endotherms, by the larger marine mammals. Blood flow is restricted to two tissues, the heart and brain, and it is possible to show that it is the O_2 consumption by these tissues that limits underwater survival. In dabblers circulatory adjustments occur slowly and are due to arterial chemoreceptor stimulation as blood O_2 falls and CO_2 rises, whereas in divers adjustments occur rapidly and are set in train by water contacting nasal receptors.

Ducks submerge voluntarily for very short periods of time. Dabbling ducks show no heart rate changes during submergence. In divers, however, heart rate increases just before a dive and falls precipitously on submergence to stabilize within 2-5 sec submergence in the range of 100-200 beats.min^{-1}. Also, blood flow distribution is different from that in forced dives; flow is preferentially directed to the exercising muscles of the hind limbs.

Heart rates between 2 and 5 sec submergence are linearly related to the logarithm of heart rate in the immediate pre-dive period, for unrestrained ducks making voluntary dives and dabbles, for restrained ducks forcibly submerged while at rest and when submergence terminates an exercise bout. There is an increase in vagal activity of about half maximum in all these manoeuvres which is caused by stimulation of nasal receptors in forced dives by restrained ducks while naso-, baro- and chemoreceptors have little influence on the cardiac response in all other types of dive. This suggests that only in forced dives, by restrained animals, is cardiac control largely reflexogenic.

The natural diving abilities of ducks were widely appreciated after the publication of Dewar's (1924) monograph, although their capacity to withstand remarkably long periods of forced submergence, under laboratory conditions, had been known for many years previously. Richet (1899) claimed that dabbling ducks of less than 2 kg could be submerged for almost 20 min before the last "gasp", heart beat or twitch. Recently more sophisticated end-points, such as time to a flat EEG have been used to delimit underwater endurance. Hudson and Jones (1986) found that for dabbling ducks (A. platyrhynchos) maximum dive time (MDT, min) was related to body mass (M, kg) as follows

$$MDT = 6.6\ M^{0.64}$$

Although this is somewhat less than Richet found, underwater endurance is still impressive in that it falls in the same range as that of Harbour seals some 10-20 times heavier and is about 6 times longer than the maximum dive time of a Canada goose (Branta canadensis) of the same size.

The reason that ducks can dive for so long is that the blood is redistributed to two tissues, the heart and the brain, thereby restricting oxygen use from the blood and pulmonary stores. This has been shown both directly and indirectly over the years (Scholander, 1940; Eliassen, 1960; Jones, Bryan, West, Lord and Clark, 1979) and most recently by using macro-aggregated albumin (MAA) labelled with Technetium 99m (Heieis and Jones, 1986). MAA lodge in the capillaries and the animal can be scanned using a gamma camera to find out where the flow was going. Since MAA breaks down within a few hours and Technetium 99m has a half-life of 6.02 hours, the same duck can be used to determine blood flow distribution at rest, and during forced and voluntary diving over a period of several days (Fig. 1). The blood flow redistribution is essential for survival because duck brains do not appear to have any special biochemical mechanisms to preserve their physiological integrity in apnoeic asphyxia compared with chicken brains. Bryan and Jones (1980a,b) measured NADH accumulation by microspectrofluoroscopy as an indicator of cellular redox balance and found that the EEG went flat in both ducks and chickens when NADH reached 30-40% of the death-induced maximum. If heart rate adjustments are prevented by

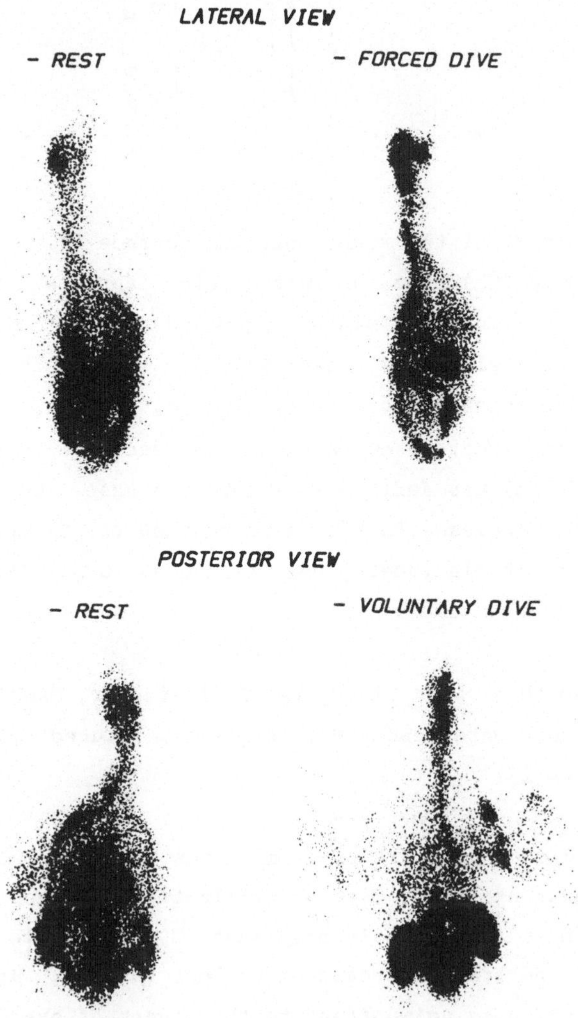

Fig. 1. Blood flow distribution during forced and voluntary submergence by restrained and free lesser scaup (A. affinis) respectively. Images from gamma-camera scans. Technetium 99m labelled macro-aggregated albumin was injected into the cardiovascular system of the birds through a cannula, the tip of which was located just outside the left ventricle. Upper panels - lateral views - resting (left) and after 2 min of forced submergence (right). Lower panels - posterior views - resting (left) and during a voluntary dive (right). In all panels the head is uppermost. (Heieis and Jones, 1986)

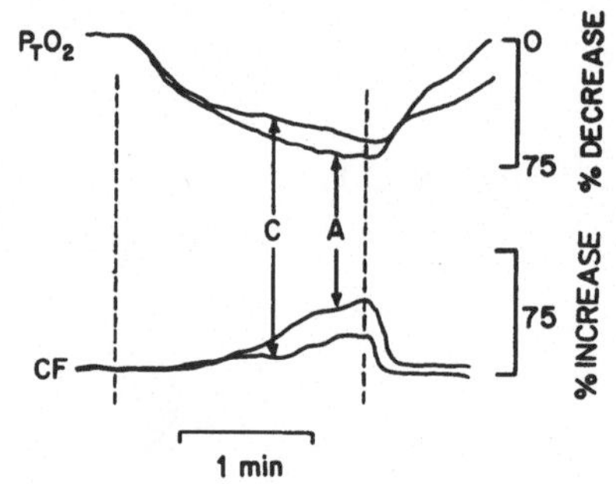

Fig. 2. Oxygen tension of the right cortical surface ($P_{T_{O_2}}$) and corrected fluorescence (CF) from the left cortical surface, indicating NADH accumulation, during 2 min of apnoeic asphyxia ("simulated dive" between vertical dashed lines) in a paralyzed duck before (C) and after an atropine injection (A) $P_{T_{O_2}}$ is expressed as the percent decrease in electrode current when the decrease from normoxia to anoxia (death) was defined as a 100% decrease. CF is expressed as the percent increase in fluorescence when the CF change from normoxia to anoxia (death) was defined as 100%. (Modified from Bryan and Jones, 1980b)

injecting atropine then brain tissue P_{O_2} falls faster, NADH accumulates more rapidly and underwater endurance is greatly reduced compared with control ducks (Fig. 2).

Hence it should be possible to predict maximum dive time from measures of the oxygen stores and their rate of utilization. Oxygen stores increase more in proportion to increases in body mass ($M^{1.19}$), while the combined mass of the heart and brain increases much less ($M^{0.65}$). However, the mass of the heart and brain is unimportant to the length of the dive, what is required is the metabolic rate of these two organs. Metabolic rate can be obtained by multiplying the mass exponent by the metabolic exponent (0.75) so that

$$MDT = \frac{\text{Oxygen stores}}{\text{Oxygen uptake by Heart and Brain}}$$

$$\propto \frac{M^{1.19}}{(M^{0.65})^{0.75}}$$

$$\propto M^{0.7}$$

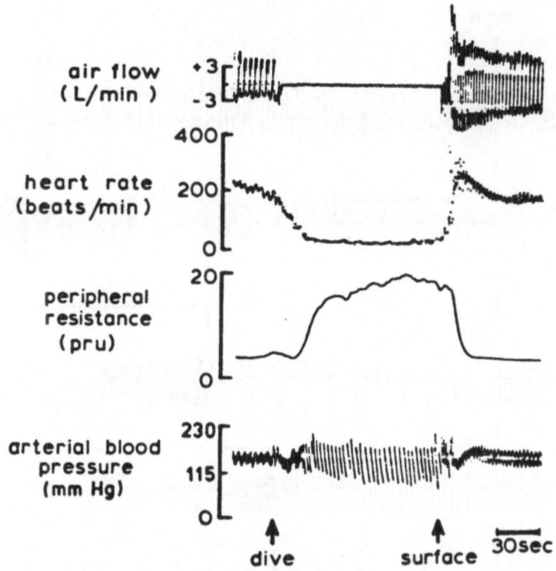

Fig. 3. Respiratory and cardiovascular responses to forced submergence of a domestic duck (A. platyrhynchos). Air flow was obtained from a pneumotachograph attached to a tracheal cannula. "Peripheral resistance" refers to the resistance to blood flow in a constant flow perfused, vascularly isolated, hind limb. (Gabbott and Jones, 1985, unpublished observations).

This calculated value is quite close to the measured value of $M^{0.64}$, and suggests that the changing relationship of available oxygen to the aerobic metabolic rate of heart and brain accounts for the increasing tolerance to asphyxia with body mass and age (Hudson and Jones, 1986).

During breath-hold diving, heart rate falls to about 1/10th the pre-dive rate (Fig. 3) and, since stroke volume is largely unchanged, the fall in heart rate reflects a similar reduction in cardiac output. Mean blood pressure is little altered so if cardiac output has fallen to 1/10th of pre-dive then total peripheral resistance must go up around 10 times. In fact, measurements in the hind limb show that the resistance of that bed (HLVR) increases by 7 to 8 times (Fig. 3). In dabbling ducks these adjustments take 30 sec before they are fully effective and during that period blood oxygen content falls rapidly. When diving adjustments are complete then blood oxygen content falls extremely slowly since blood oxygen is now principally being utilized by the heart and brain.

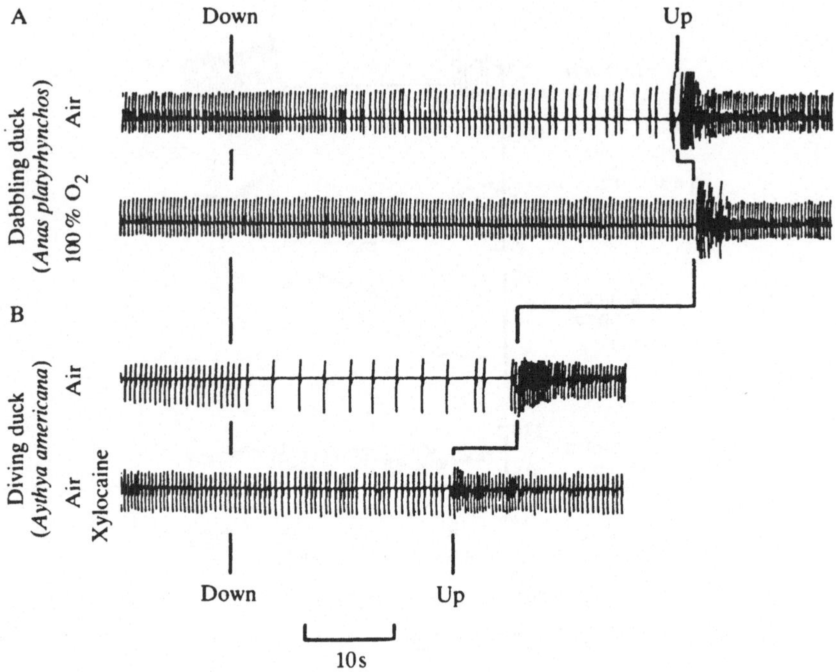

Fig. 4. Traces showing heart rates of restrained ducks before, during, and after forced submergence. (A) A dabbling duck having breathed air (upper trace) and oxygen (lower trace) before submergence. (B) A diving duck having breathed air before submergence (upper trace). The lower trace shows the cardiac response to submergence after the application of a local anaesthetic to the nares. "Down" refers to the time at which the beak entered the water and "up" refers to the time when the animal surfaced. (Furilla and Jones, 1986a, by permission).

In dabbling ducks these cardiovascular adjustments are initiated, and maintained, by stimulation of carotid body chemoreceptors by the progressively more hypoxic and hypercapnic blood. Denervation of the carotid bodies virtually eliminates diving bradycardia (Jones and Purves, 1970) although about 1/3 of the change in HLVR is caused by excitation of central chemoreceptors (Jones, Milsom and Gabbott, 1982). An easier way of illustrating the chemoreceptor contribution to diving responses is to let dabbling ducks breath oxygen before short dives (Fig. 4) but, surprisingly, breathing oxygen does not retard onset of cardiovascular adjustments in true diving ducks. In divers, such as the redhead (Aythya americana), bradycardia develops immediately on submergence and can be inhibited only by spraying a local anaesthetic over the external and internal nares

(Fig. 4, Furilla and Jones, 1986a). Hence nasal receptors, which contribute only to apnoea in dabbling ducks (Bamford and Jones, 1974), have an important cardiovascular influence in diving ducks.

Although ducks can withstand many minutes of forced submergence, in nature most voluntary dabbles and dives are extremely short. Dabbles seldom exceed 5 sec duration and dives are usually less than 20 sec. Voluntary head submersion by dabbling ducks causes little or no change in heart rate (Furilla and Jones, 1987). This is not surprising given that the periods of submergence are too short to cause sufficient changes in blood gas tensions to stimulate systemic chemoreceptors. On the other hand, heart rate changes would be expected in divers submerging voluntarily as soon as the nasal receptors contact water. While this appears to be the case close correlation of heart rates broadcast from ducks by telemetry with cinefilms of diving behaviour shows that some seconds before the dive both breathing and heart rates increase from around 20 breaths.min^{-1} and 120 beats.min^{-1} to about 60 breaths.min^{-1} and at least 300 beats.min^{-1}; heart rate then falls just before or actually on submergence (Butler and Woakes, 1979; 1982; Woakes and Butler, 1983).

The first cardiac interval in voluntary dives is usually the longest and heart rate then rises to stabilize in the range of 100-200 beats.min^{-1} within 2 to 5 sec submergence (Furilla and Jones, 1986a). This is very different from the response in forced dives when heart rate falls to about 20 beats.min^{-1} in a similar time period. Blood flow redistribution is also quite different in voluntary compared with forced dives. Lesser scaup (Aythya affinis) have been trained to carry a motorized injection syringe as a "back-pack". A wing artery is occlusively cannulated and the tip advanced to lie just outside the aortic valves. The syringe is filled with MAA labelled with Technetium 99m and during a voluntary dive the motor is activated by a light pulse from a flash unit. Blood flow during voluntary dives is preferentially directed to the leg muscles which are active during submergence (Fig. 1).

It has been argued that in forced dives the cardiovascular responses are accentuated by "fear" or "stress" which are not components of the voluntary dive response (Kanwisher, Gabrielsen and Kanwisher, 1981). Certainly, if circulating catecholamine levels are indicative of "fear" or "stress" then dabbling ducks (A. platyrhynchos) forced to dive must be very

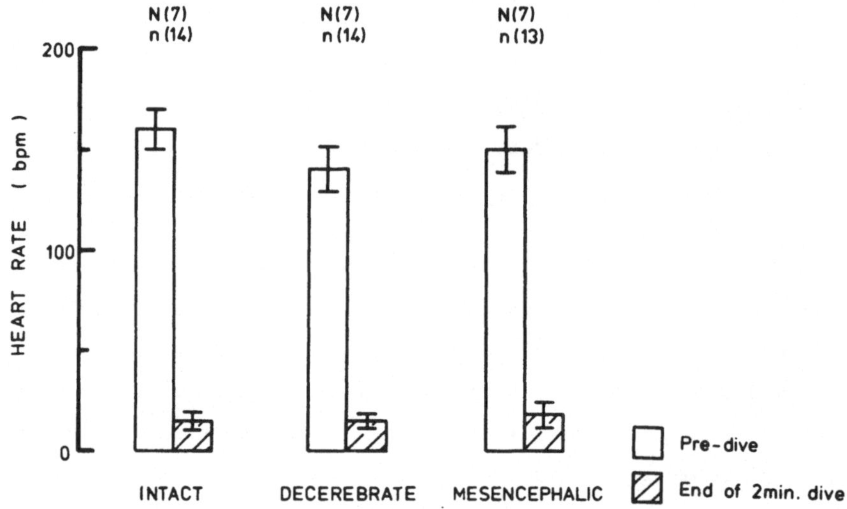

Fig. 5. Effect of decerebration and rostral mesencephalic transection in decerebrate ducks (A. platyrhynchos) on the cardiac response to forced submergence. n = number of observations on N animals. (Gabbott and Jones, 1985, unpublished observations).

frightened, since both circulating adrenaline and noradrenaline increase by 3 orders of magnitude yielding plasma catecholamine levels which are the highest ever recorded (Hudson and Jones, 1982). However, brain transection across the rostral mesencephalon in chronically decerebrated dabbling ducks has no effect on the cardiac response to forced submergence (Fig. 5.). Although a fear response may be excited reflexly by neural inputs to the ponto-medullary region there is obviously no cognitive or even hypothalamic component involved in the cardiac response to forced diving (Gabbott and Jones, 1985).

Heart rates in the same range as those seen in forced dives can be provoked in diving ducks by preventing them from surfacing at the end of a voluntary dive (Furilla and Jones 1986b). In fact, by manipulating the diving behaviour it is possible to obtain a wide range of pre-dive and dive heart rates in diving ducks submerging voluntarily. For instance, reducing the water depth obliges ducks to dabble instead of dive resulting in pre-dive and dive heart rates in dabbles that are considerably below those seen in dives. In voluntary dives, the pre-dive tachycardia can be reduced by β-blockade whereas for forced dives pre-dive heart rate can be increased by exercising ducks on a treadmill up to the moment of submersion. A scatter plot of all pre-dive and dive heart rates, after 2-5 sec submergence,

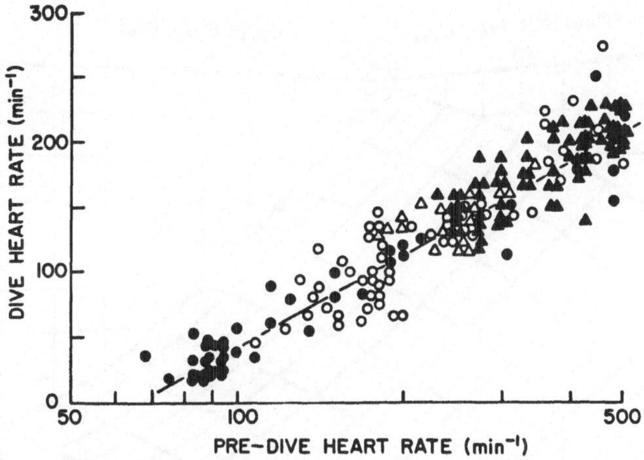

Fig. 6. The relationship between dive (or trapped) heart rate and the logarithm of pre-dive (or pre-trap) heart rate for the following dives by diving ducks (Aythya sp.). ● , restrained dives including those after exercise; O, trapped dives; ▲ , voluntary dives including β-blocked dives; Δ , dabbles and voluntary face immersion in a beaker of water. The equation for the line is y = -451 + 246 log x, where y = dive (or trapped) and x = pre-dive (or pre-trap) heart rate with a coefficient of determination (r^2) of 0.98. (Furilla and Jones, 1986b, by permission).

reveals that a linear relation can be established between dive heart rate and the logarithm of pre-dive heart rate with a coefficient of determination of 0.98 (Fig. 6; Furilla and Jones, 1986b).

This relationship raises the question of whether it can be supported by current knowledge of either afferent or efferent neural mechanisms which affect cardiac control. Afferent mechanisms could include neural input from naso-, chemo- and cardiovascular mechanoreceptors. However, anaesthetisation of nasal receptors with a local anaesthetic has only a slight effect on the dive/pre-dive heart rate relation, which contrasts markedly with the effect of anaesthetisation on the cardiac response to forced diving (Fig. 4). Allowing ducks to breathe oxygen before voluntary dives to reduce input from cardiovascular chemoreceptors in dives has no effect on the dive/pre-dive relation. The same lack of any effect is also seen in chronic barodenervated ducks diving voluntarily. Consequently, it seems unlikely that the explanation of this relation lies in similar afferent neural control in all types of dives.

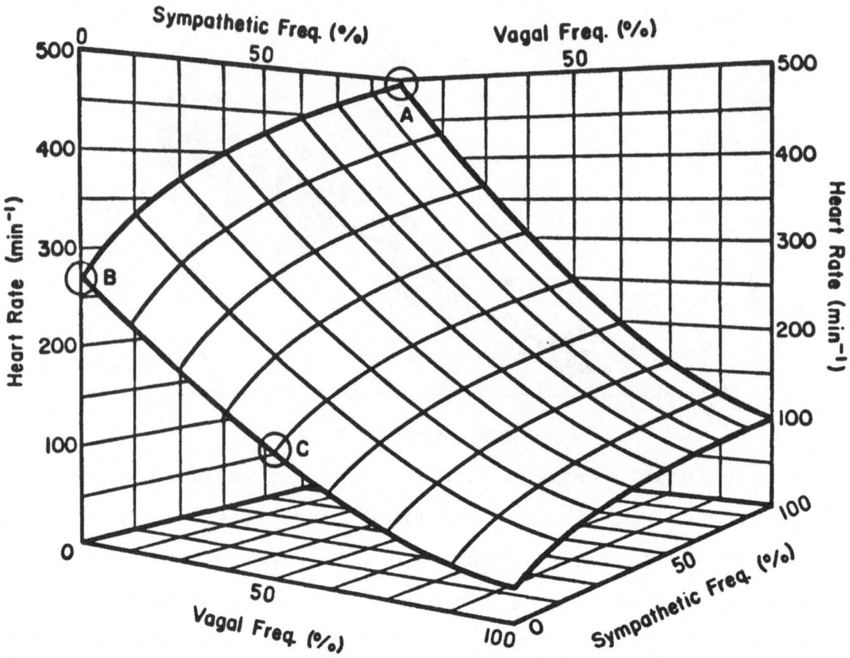

Fig. 7. The relationship of heart rate to bilateral stimulation of the distal cut ends of the vagus and cardiac sympathetic nerves. 100% represents the frequency of stimulation above which no further changes in heart rate occurred with increases in frequency. The heart rate caused by a given level of vagal and sympathetic stimulation was plotted on "perspective" graph paper and the surface was drawn, by eye, to encompass all heart rates obtained in the stimulation experiments. See the text for an explanation of points A, B, and C. (Furilla and Jones, 1986b, by permission)

Efferent control of the heart has been investigated by implanting stimulating electrodes bilaterally, on cut distal ends of the vagi and cardiac sympathetic nerves. Vagal stimulation results in rapid heart rate changes while sympathetic stimulation causes much slower effects, thus it is doubtful if changes in sympathetic activity would have any marked effect up to 5 sec submergence. Heart rates versus normalised stimulation frequencies of vagal and sympathetic nerves are shown in Fig. 7, and the surface in Fig. 7 illustrates all heart rates that can be produced by any combination of sympathetic and vagal stimulation. Regions which appear to pertain to intact animals can be identified on this surface. For instance, maximal sympathetic activation in the absence of vagal stimulation gives heart rates of 500 beats.min^{-1} (point A in Fig. 7) which is the highest rate observed before voluntary dives. β-blockade with propranolol yields heart rates around 300 beats.min^{-1} just before voluntary dives (point B). Finally, before forced dives heart rate is between 90 and 140 beats.min^{-1} and since β-blockade does not lower heart rate further (tested in two ducks), cardiac efferent control in these animals is probably described by the region around point C on Fig. 7. If sympathetic activity does not decrease in a dive (it cannot when pre-dive rates are represented by B and C) then the relative increase of vagal activity required to cause the diving heart rates observed must be similar in voluntary dives, represented by points A and B, and in forced dives represented by point C. Specifically, an increase in vagal activity of about half maximum will give the dive heart rates observed between 2 and 5 sec in voluntary dives, with and without propranolol, and in forced dives. Further, heart rates in all other dabbles and dives, including heart rate changes when ducks are trapped under water compared with pre-trapped rates, appear to be the result of a similar increase in vagal activity. The curvilinear relation between sympathetic and vagal activity and heart rate, alone or in combination (Fig. 7), is obviously linearised by taking the logarithm of the pre-dive heart rate in the dive/pre-dive heart rate relation. A similar linearising effect results from taking the logarithm of heart rates existing before and plotting these against heart rates resulting from any 50% of maximum increase in vagal activity (Furilla and Jones, 1986b).

The fact that we have been unable to show any major reflexogenic influences on heart rate in voluntary dives leads us to propose that psychogenic influences are much more likely to be expressed in free than in restrained dives in diving ducks. For instance, psychogenic influences affect dive heart rate both initially, and after 2-5 sec submergence, through their influence on the pre-dive rate. Anticipation of the dive response, described in voluntary dives by Butler and Woakes (1982), also implies a profound psychogenic influence. Finally, the fact that heart rate can immediately drop when the normal diving pattern is disrupted in unrestrained animals, suggests that integrative processes occur which are well above the brainstem level in free dives. These conclusions appear to be in conflict with those obtained from studying restrained and free head submersions by dabbling ducks suggesting that psychogenic influences affect the former and not the latter (Kanwisher et al., 1981; Blix, 1985). Obviously these earlier data were insufficient really to decipher any fundamental cause and effect relationships.

ACKNOWLEDGEMENTS

This research was made possible by grants from N.S.E.R.C.C., the Canadian National Sportsmen's Funds, and the B.C. Heart Foundation. I am grateful to my colleagues Graham Shelton, Pat Butler, Owen Bamford and Dennis Hudson and my students Lowell Langille, Nigel West, Bill Milsom, Bob Bryan, Geoff Gabbott, Bob Furilla, Frank Smith, Agnes Lacombe, Harry Mangalam, Manabu Shimizu and Mark Heieis for maintaining my interest and enthusiasm for this subject for more than 20 years. Finally, I would like to thank Peter Hochachka, David Randall and Mike Hedrick for their comments on an earlier draft of this manuscript.

REFERENCES

Bamford, O.S. and Jones, D.R. (1974) On the initiation of apnoea and some cardiovascular responses to submergence in ducks. Resp. Physiol. 22, 199-216.

Blix, A.S. (1985) The emotional components of the diving responses in mammals and birds. In: Arctic Underwater Operations Rey, L., ed., Graham and Trotman: London

Bryan, R.M. and Jones, D.R. (1980a) Cerebral energy metabolism in diving and non-diving birds during hypoxia and apnoeic asphyxia. J. Physiol. (Lond.) 299, 323-336.

Bryan, R.M. and Jones, D.R. (1980b) Cerebral metabolism in mallard ducks during apneic asphyxia: the role of oxygen conservation. Am. J. Physiol. 239, R352-357.

Butler, P.J. and Woakes, A.J. (1979) Changes in heart rate and respiratory frequency during natural behaviour of ducks, with particular reference to diving. J. exp. Biol. 79, 283-300.

Butler, P.J. and Woakes, A.J. (1982) Telemetry of physiological variables from diving and flying birds. Symp. Zool. Soc. Lond. 49, 106-128.

Dewar, J.M. (1924) The Bird as a Diver. Witherby: London

Eliassen, E. (1960) Cardiovascular responses to submersion asphyxia in avian divers. Årbok. Univer. Bergen 2, 1-100.

Furilla, R.A. and Jones, D.R. (1986a) The contribution of nasal receptors to the cardiac response to diving in restrained and unrestrained redhead ducks (Aythya americana) J. exp. Biol. 121, 227-238.

Furilla, R.A. and Jones, D.R. (1986b) The relationship between dive and pre-dive heart rates in restrained and free dives by diving ducks. J. exp. Biol. in press.

Furilla, R.A. and Jones, D.R. (1987) Heart rate response to diving and dabbling in the Mallard (Anas platyrhynchos) Physiol. Zool. in press.

Gabbott, G.R.J. and Jones, D.R. (1985) Psychogenic influences on the cardiac response of the duck (Anas platyrhynchos) to forced submission. J. Physiol. (Lond.) 371, 71 P.

Heieis, M. and Jones, D.R. (1986) Unpublished observations.

Hudson, D.M. and Jones, D.R. (1982) Remarkable blood catecholamine levels in forced dived ducks. J. exp. Zool. 224, 451-456.

Hudson, D.M. and Jones, D.R. (1986) The influence of body mass on the endurance to restrained submergence in the pekin duck. J. exp. Biol. 120, 351-367.

Jones, D.R., Bryan, R.M., West, N.H., Lord, R.H. and Clark, B. (1979) Regional distribution of blood flow during diving in the duck (Anas platyrhynchos) Can. J. Zool. 57, 995-1002.

Jones, D.R., Milsom, W.K. and Gabbott, G.R.J. (1982) Role of central and peripheral chemoreceptors in diving responses of ducks. Am. J. Physiol. 243, R537-R545.

Jones, D.R. and Purves, M.J. (1970) The carotid body in the duck and the consequences of its denervation upon the cariac responses to immersion. J. Physiol. (Lond.) 211, 279-294.

Kanwisher, J.W., Gabrielsen, G. and Kanwisher, N. (1981) Free and forced diving in birds. Science 211, 717-719.

Richet, Ch. (1899) De la résistance des canards à l'asphyxie. J. physiol. pathol. gén. 1, 641-650.

Scholander, P.F. (1940) Experimental investigations on the respiratory function in diving mammals and birds. Hvalråd. Skrift. 22, 1-131.

Woakes, A.J. and Butler, P.J. (1983) Swimming and diving in tufted ducks, Aythya fuligula, with particular reference to heart rate and gas exchange. J. exp. Biol. 107, 311-329.

THE ROLE OF MITOCHONDRIA IN THE CONTROL OF CELLULAR CALCIUM HOMEOSTASIS

Jeanie B. McMillin

Division of Cardiovascular Disease
Department of Medicine
University of Alabama at Birmingham
Birmingham, AL, U.S.A.

INTRODUCTION

Demonstration of an active Ca^{2+} transport system in the mitochondrial inner membrane has led to investigation of the physiological function of this activity. Control of intracellular Ca^{2+} ion homeostasis and/or modulation of mitochondrial metabolism may be important consequences of Ca^{2+} flux across the mitochondrial membrane. Distribution of Ca^{2+} between the cytosol and the two major intracellular Ca^{2+} transport organelles (the mitochondria and the reticular network) will be determined by the environmental levels of Ca^{2+} and the kinetic constants of each system. The high affinity (but low capacity) of the reticulum for Ca^{2+} is within the range of cytosolic Ca^{2+} buffering, i.e., from 150-200 nM, and is dealt with elsewhere in this symposium. The ability of the mitochondria to respond to changing Ca^{2+} concentrations in order to activate metabolism or to buffer cytosolic Ca^{2+} is dependent upon intracellular conditions which may affect the transporters. Mitochondrial Ca^{2+} flux is controlled by both an import pathway and an export pathway which are thought to operate in parallel. Any alteration in the balance between the two pathways can affect cytosolic Ca^{2+} buffering and/or modulate energy production via Ca^{2+} sensitive dehydrogenase activities in the mitochondrial matrix. Mitochondrial Ca^{2+} uptake is a low affinity, high capacity process, linked to the fundamental energy producing mechanism across the inner membrane. While Ca^{2+} uptake proceeds by a potential-dependent uniport pathway, Ca^{2+} efflux may occur by a variety of mechanisms which are tissue specific. These include a) Na^+-Ca^{2+} exchange, b) a Na^+-independent pathway and c) a phospholipase A_2-

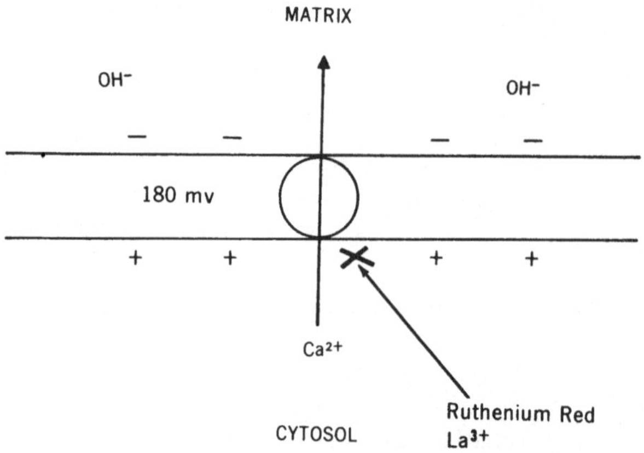

Fig. 1. Mitochondrial calcium uniport

dependent permeability pathway. In order to understand factors which may regulate the expression of Ca^{2+} flux across the inner mitochondrial membrane, characteristics of the Ca^{2+} uptake and export pathways will be considered separately.

THE MITOCHONDRIAL UNIPORTER: Ca^{2+} UPTAKE BY MITOCHONDRIA AND REGULATION OF Ca^{2+} HOMEOSTASIS

The transport of Ca^{2+} across the inner membrane into the mitochondrial matrix takes place via an electrophoretic uniport mechanism driven by the potential gradient across the membrane (about 180 mV, negative inside) (Fig. 1). The electrogenic nature of the mitochondrial Ca^{2+} transporter was established by experiments in which distributions of Ca^{2+} and Rb^+ (in the presence of valinomycin) were correlated with the Nernstian membrane potential. Approximately the same values for the membrane potential could be obtained by assuming a net charge transfer of 2 associated with the movement of Ca^{2+} (Rottenberg and Scarpa, 1974). Inhibition of Ca^{2+} uptake by lanthanides (Mela, 1968) and the glycoprotein stain, ruthenium red (Moore, 1971) gave further support to the concept of a carrier-mediated mechanism. Because of the link with the protonmotive force generated by the electron transport chain, Ca^{2+} transport into the matrix is accompanied by respiratory stimulation and a redox shift in cytochrome b (Chance, 1965). The total capacity of mitochondria for Ca^{2+} is large, up to 2 μmol Ca^{2+}/mg protein in the presence of proton donating anions, e.g., phosphate (Lehninger, 1974). In suspensions of isolated mitochondria exposed to substrate and to relatively high concentrations of Ca^{2+} (0.2 mM), the maximal velocity of Ca^{2+} uptake is rapid (over 1 μmol/min/mg in cardiac mitochondria) (McMillin-Wood, Wolkowicz, Chu, Tate, Goldstein and Entman, 1980). However, in the intact cell, the realized velocities of mitochondrial Ca^{2+} transport are likely quite low, due to the sigmoidal kinetics of the uniporter and the high K_{app} for Ca^{2+}, in the presence of physiologically prevailing Mg^{2+} concentrations (Scarpa and Graziotti, 1973).

It is now generally agreed that the kinetics of the mitochondrial transporter are not sufficient to compensate for rapid changes in cytosolic free Ca^{2+}, in particular, regulation of the beat-to-beat Ca^{2+} cycle in mammalian hearts (Scarpa and Graziotti, 1973). However, much attention has been given to the possibility that mitochondria may play a role in control of steady-state Ca^{2+} concentrations in the cell. In suspensions of isolated mitochondria, the level to which extramitochondrial Ca^{2+} is buffered depends upon the presence of Mg^{2+} in the medium. Since Mg^{2+} increases the sigmoidal response of the mitochondrial uniporter to Ca^{2+}, Mg^{2+} should raise the free extramitochondrial Ca^{2+} at steady state (Fig. 2). Indeed, increasing pMg^{2+} to 2.5 (3 mM) increases the pCa^{2+} from 6.1 (0.8 μM) in the absence of added Mg^{2+} to 5.75 (1.8 μM) (Becker, 1980). The lower extramitochondrial Ca^{2+} values achieved at steady state in the presence of Mg^{2+} in similar experiments from other laboratories, e.g. Becker (1980), may reflect the presence of phosphate as the permeant ion or to the presence of Mg^{2+} chelates.

The relative contributions of mitochondria and microsomes to the regulation of Ca^{2+} at levels approximating those found in the cytosol have been explored in isolated, reconstituted organellar systems as well as in digitonin-permeabilized hepatocytes (Becker, 1980). Suspensions of mitochondria in an assay environment resembling the cellular cytosol are

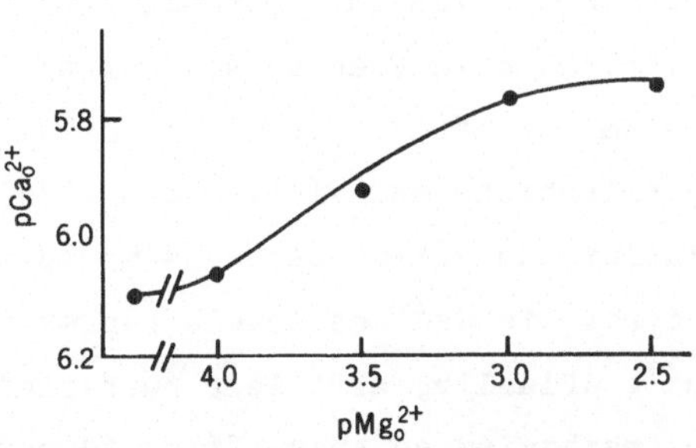

Fig. 2. Effect of Mg^{2+} on steady-state pCa^{2+} maintained by mitochondria (Nicholls, 1978)

able to maintain a steady state free Ca^{2+} level of 0.5 μM. Further additions of Ca^{2+} produce an initial rapid rise in extramitochondrial Ca^{2+} followed by a decay to the steady state value, indicating the high capacity of mitochondria for Ca^{2+}. The presence of microsomes further lowers the steady state Ca^{2+} concentration to 0.2 μM which rises back to 0.5 μM when the microsomal stores are saturated. Identical patterns are observed in permeabilized hepatocytes, suggesting that similar pathways of control may be operative in vivo (Becker, 1980). The net release of Ca^{2+} from mitochondria upon the addition of microsomes or EGTA to the assay also points to a physiological role for a Ca^{2+} efflux mechanism in the mitochondrial inner membrane (see below for discussion of the Ca^{2+} exporter). These experiments suggest that operation of the mitochondrial uniporter provides a kinetically controlled model which is able to maintain precise regulation of extramitochondrial pCa^{2+}.

Differences in the mitochondrial capacity for intracellular Ca^{2+} buffering are found between intact cells and saponin permeabilized cells. In intact cells, approximately 60-80% of the total cellular Ca^{2+} is in the mitochondria, as demonstrated by uncoupling agents (Joseph, Coll, Cooper, Marks and Williamson, 1983). However, in saponin-permeabilized cells where the prevailing free Ca^{2+} is maintained at 0.2 μM, mitochondria appear unable to accumulate Ca^{2+}, until cytosolic Ca^{2+} levels are increased above 3μM (Burgess, McKinney, Fabiato, Leslie and Putney, 1983). The suggestion (Burgess et al., 1983) that decreased mitochondrial affinity of the uniporter for Ca^{2+} is a consequence of cellular permeabilization and loss of some soluble effector molecule was explored in experiments by Nicchitta and Williamson (1984). At probable free physiological levels of the polyamine, spermine, the mitochondria demonstrate the capacity to transport Ca^{2+} at concentrations below 0.5 μM, suggesting a change in the affinity of the uniporter. Most significantly, with spermine addition, the cytosolic free Ca^{2+} decreases from 0.8 to 0.25 μM. Concomitant effects of spermine on efflux of Ca^{2+} from mitochondria (see subsequent section) are seen as an apparent decrease in the Km for efflux (from 9.6 to 2.3 μM) with no change in the maximal velocity (Nicchitta and Williamson, 1984). The combined effect of spermine on influx and efflux pathways would allow the mitochondria to buffer cytosolic Ca^{2+} at levels representative of prevailing cell free Ca^{2+}. Finally, efficient buffering requires saturation of the efflux mechanism so that efflux is constant at any change in mitochondrial Ca^{2+} content (Joseph et al., 1983; Hansford and Castro, 1982). Therefore, any change in the steady state distribution of Ca^{2+}

across the inner mitochondrial membrane must result from a change in the affinity of the uniporter for $Ca2+$. This may represent the unique action of modifiers such as steroid hormones and growth factors (which alter ornithine decarboxylase activity) on the control of cytosolic free Ca^{2+} concentrations. Whether or not the efflux pathway is indeed saturated in vivo depends upon accurate assessment of total mitochondrial Ca^{2+} content in situ. This consideration is also important to matrix dehydrogenase control by Ca^{2+} (see discussion on metabolic control).

MITOCHONDRIAL CALCIUM EFFLUX

Na^{+}-independent pathway

The existence of a separate mitochondrial efflux pathway for Ca^{2+} was first suggested by experiments to quantify the activity of Ca^{2+} (and Mn^{2+}) in the mitochondrial matrix (Puskin and Gunter, 1973). Distribution of free Ca^{2+} is significantly less than would be predicted from the Nernst equation, assuming equilibration of Ca^{2+} via the Ca^{2+} uniporter. It was proposed that existing Ca^{2+} steady state is a net result of uniport activity, which acts to accumulate Ca^{2+}, and an export pathway (Puskin, Gunter, Gunter and Russell, 1976). The efflux pathway is detectable even when the membrane potential is sufficiently high to exclude Ca^{2+} loss by reversal of the uniporter. Ca^{2+} release via the electroneutral efflux pathway is also unaffected by addition of ruthenium red. Mitochondria from excitable tissues (heart and brain) and nonexcitable tissues (kidney and liver) apparently possess differing pathways of efflux. In liver and kidney mitochondria, this process appears to be Na^{+}-independent, although a small component of Na^{+}-stimulated efflux may also be present. The Na^{+}-dependent exchange present in heart and brain mitochondria will be discussed separately in a later section. Thus, in liver (and kidney) mitochondria, release of Ca^{2+} to the cytosol occurs in the absence of Na^{+} by a mechanism independent of the transmembrane electrochemical gradient. In early studies designed to characterize the Na^{+}-independent pathway, evidence was presented which suggested that Ca^{2+} efflux proceeds by an electroneutral $Ca^{2+}/2H^{+}$ antiport process (Puskin et al., 1976; Akerman, 1978; Fiskum and Cockrell, 1978).

The stoichiometry of the Ca^{2+}/H^{+} antiport was determined in mitochondria treated to eliminate the possibility of peripheral proton fluxes due to phosphate movements (Fig. 3 and Fiskum and Lehninger, 1979).

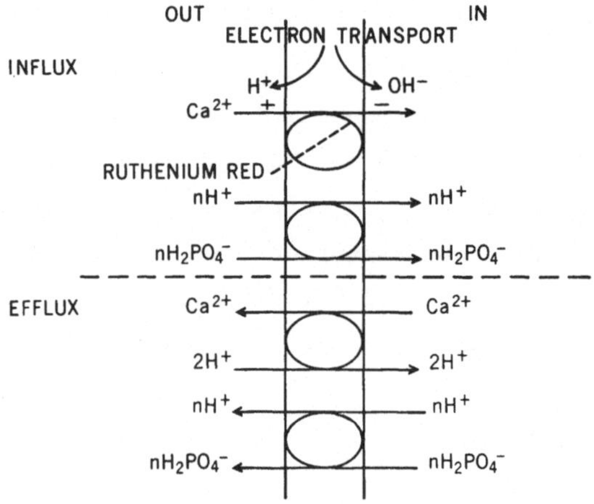

Fig. 3. Ca^{2+} uptake and release pathways in liver mitochondria (Fiskum and Lehninger, 1979)

Upon addition of succinate, Ca^{2+} uptake is coupled via the electrochemical gradient to proton ejection by a 1:2 stoichiometry, consistent with the inward movement of a net charge of 2. The average $Ca^{2+}:H^+$ stoichiometry for Ca^{2+} efflux is also 1:2, which would be anticipated for electroneutral exchange (Fiskum and Lehninger, 1979).

More recently, the nature of the Ca^{2+} efflux pathway in liver mitochondria has been re-examined. Since acid-induced release of mitochondrial Ca^{2+} may result from localized damage by H^+ to a susceptible population of mitochondria in the vicinity of the acid addition, experiments were carried out in which the pH was lowered by introduction of buffer subsequent to ruthenium red in order to observe the efflux pathway (Gunter, Chace, Puskin and Gunter, 1983). No additional Ca^{2+} efflux was observed at any pH in the range of 7.2 to 6.6. Furthermore, direct measurement of media alkalinization coupled to Ca^{2+} efflux at a variety of media pHs produced $H^+:Ca^{2+}$ stoichiometries ranging from +7 (pH 6.5) to -3.5 (pH 8.4). The data are inconsistent with the expected properties of an electroneutral antiport process where the stoichiometry of exchange should remain constant. These data do not support the hypothesis that the Ca^{2+} efflux mechanism present in liver mitochondria is a passive Ca^{2+}-$2H^+$ exchanger. It seems more likely that the Ca^{2+} efflux previously observed, results from de-energization of the mitochondria with subsequent Ca^{2+} release via a permeability pathway when the chemical potential gradient for Ca^{2+} is high (see below). However, in support of a mediated export mechanism for Ca^{2+}, recent results suggest that Ca^{2+} release from liver mitochondria may take place by a sodium-independent pathway (Gunter et al., 1983) which operates exclusive of any obligatory proton exchange. Kinetically, the mechanism of the Na^{+}-independent pathway is second order, characterized by the flat saturation dependence of the transport curve. By gentle Ca^{2+} depletion procedures, it is possible to distinguish between saturation of mediated transport and the velocity "runaway" conditions which signal nonspecific increases in mitochondrial permeability (Gunter, unpublished findings).

Permeability Pathway

Studies carried out on cardiac mitochondria demonstrate that H^{+}-induced Ca^{2+} efflux may be the result of increases in the permeability of the mitochondrial membrane. In support of permeability alterations, Jurkowitz and Brierley (1982) found that the rate of H^{+}-induced Ca^{2+} release increases with increasing Ca^{2+} loading. Preparations of mitochondria demonstrating enhanced susceptibility to spontaneous Ca^{2+} release contain lower total adenine nucleotide than do more intact preparations. In mitochondria systematically depleted of adenine nucleotide by PP_i treatment, the rate of H^{+}-induced Ca^{2+} efflux is inversely related to the adenine nucleotide content. Finally, H^{+}-dependent Ca^{2+} efflux in cardiac mitochondria incubated in the presence of acetoacetate is accompanied by loss of endogenous Mg^{2+} and by mitochondrial swelling (Jurkowitz and Brierley, 1982). The similarity between the swelling change and ion loss suggests that nonspecific changes in membrane permeability are associated with H^{+}-induced Ca^{2+} release.

The physiological significance of mitochondrial Ca^{2+} overload and changes in the properties of the proton-conducting inner membrane may be relevant to pathological states such as ischaemic heart disease. Low flow ischaemia is associated with contracture of the heart with partial mechanical recovery upon reperfusion, although end diastolic pressure remains elevated (Henry, Schuchleib, Davis, Weiss and Sobel, 1977). These mechanical changes are directly correlated with accumulation of Ca^{2+} in the mitochondrial fractions isolated from the ischaemic and reperfused hearts and imply a net increase in total myocardial calcium. It should also be noted that at the levels of Ca^{2+} accumulated during cardiac ischaemia, the mitochondria can no longer maintain efficient buffering of the cytosol (Becker, 1980; Jurkowitz and Brierley, 1982). Thus, elevated matrix Ca^{2+}

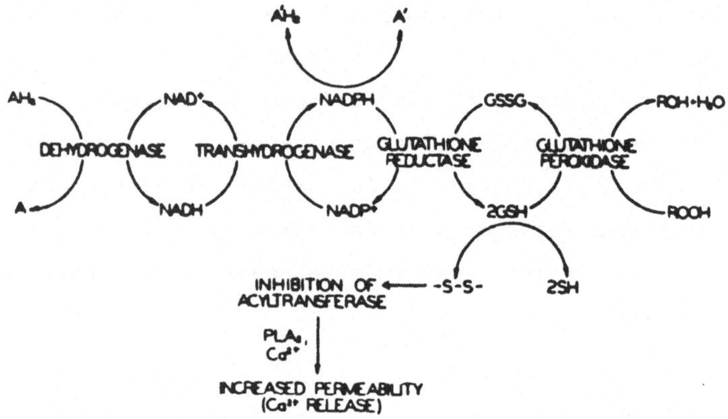

Fig. 4. Calcium efflux from mitochondria: increased mitochondrial permeability due to altered gluthathione redox (Beatrice, Stiers and Pfeiffer, 1984)

in pathology may be associated with permeability alterations which eventually determine reversibility of the injury.

In contrast to heart and liver, tumor mitochondria demonstrate remarkable resistance to damage by high levels of accumulated Ca^{2+} (Fiskum and Cockrell, 1983). Approximately three times the Ca^{2+} load is required to uncouple tumor mitochondrial respiration compared to liver mitochondria. This difference has been attributed to elevations in matrix Mg^{2+} in the matrix of tumor mitochondria. The mechanism of this effect may be related to inhibition of a ruthenium red-insensitive efflux pathway, which is induced by phospholipase A_2 activation, leading to changes in membrane permeability.

The nonspecific permeability increase in mitochondrial membranes associated with Ca^{2+} releasing agents (and decreases in mitochondrial ions and adenine nucleotides) has been proposed to arise from changes in thiol group reduction by a reaction coupled to the energy-linked transhydrogenase (Harris, Al-Shaikhaly and Baum, 1979). Demonstration of losses in NAD^+-dependent respiration in mitochondria exposed to the disulfide cystamine were related to a breakdown in mitochondrial permeability barriers and loss of matrix co-factors, e.g., Mg^{2+} (Skrede, 1966). Beatrice, Stiers and Pfeiffer (1984) proposed that accumulation of phospholipid degradation products as a result of Ca^{2+} activated phospholipase A_2 and inhibition of lysophosphatide reacylation are causative factors in the increased permeability (Fig. 4). The loss of reacylating activity may in turn be

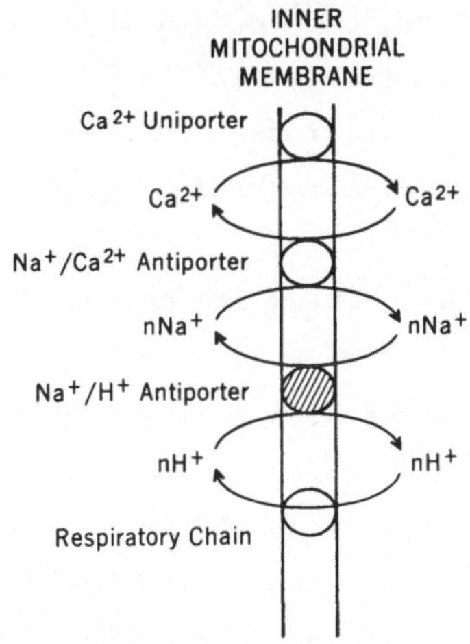

Fig. 5. The Na^+/Ca^{2+} cycle (Crompton, 1980)

controlled by decreases in reduced matrix glutathione which correlates with mitochondrial swelling and sensitivity to Ca^{2+}. In experiments on Ca^{2+}-dependent mitochondrial swelling where NAD^+ and $NADP^+$ redox states can be manipulated independent of glutathione there is a convincing correlation between mitochondrial permeability and of reduced glutathione levels (Beatrice et al., 1984). While this permeability pathway is significant *in vitro*, any *in vivo* role of these membrane changes in the control of cellular Ca^{2+} remains to be established.

Na^+-Dependent Pathway

While controversy has arisen concerning the Na^+-independent Ca^{2+} efflux pathway, most investigators agree that Na^+-dependent Ca^{2+} exchange is predominant in a variety of tissues. According to this mechanism, the direct exchange of Na^+-in for Ca^{2+}-out is electroneutral, ruthenium red-insensitive and operates independently of any change in the transmembrane potential gradient (Affolter and Carafoli, 1980). The Na^+-induced efflux of Ca^{2+} may also be linked to a Na^+/H^+ antiporter which allows Na^+ to exit from the mitochondria in exchange for protons (Fig. 5). Evidence for a mitochondrial carrier that catalyzes an exchange between Na^+ and H^+ was provided by matrix swelling experiments in the presence of sodium acetate or potassium acetate (Vaghy, Johnson, Matlib, Wang and Schwartz, 1982).

Respiration-inhibited mitochondria swell rapidly in iso-osmotic sodium acetate, but not in potassium acetate. This swelling is the apparent result of acetate entry as the undissociated acid with subsequent Na^+ entry in exchange for H^+. Therefore, Ca^{2+} efflux in exchange for Na^+ will result in a dissipation of pH via successive $Ca^{2+}:Na^+$ and $Na^+:H^+$ exchanges. However, kinetic control of this process is probably governed by factors other than the transmembrane electrochemical gradient.

The initial rates of Ca^{2+} efflux are dependent upon the Na^+ concentration in the medium. A sigmoidal relationship between Na^+ and rate of Ca^{2+} efflux exists, where the Hill coefficient is approximately 3 (Crompton, Capano and Carafoli, 1976). The 2 Na^+/Ca^{2+} antiport, but not the sodium-independent pathway, is inhibited by plasma membrane Ca^{2+} channel blockers, the most potent inhibition being observed with diltiazem (Vaghy et al., 1982). The diltiazem interaction with the exchange is specific and does not affect either Ca^{2+} influx on the uniporter or oxidative phosphorylation. The kinetics of the 2 Na^+/Ca^{2+} exchange are complex, so that in addition to the cooperative behavior with respect to Na^+, increasing concentrations of extramitochondrial Ca^{2+} also markedly inhibit Na^+-dependent Ca^{2+} efflux (Hayat and Crompton, 1982). This inhibition is consistent with the presence of external regulatory sites sensitive to changes in Ca^{2+} over the physiological range (0-2 μM). The maximal velocity of Ca^{2+} release on the exchange is about 10 nmol/mg/min at 30° with an apparent Km of 5.7 μM for matrix-free Ca^{2+} (Coll, Joseph, Corkey and Williamson, 1982).

Although the 2 Na^+/Ca^{2+} exchange is mainly associated with cardiac, skeletal muscle and brain mitochondria, there is evidence that this carrier is also present in liver mitochondria, albeit at low activities (Heffron and Harris, 1981). In perfused liver, Na^+-dependent Ca^{2+} efflux from isolated liver mitochondria is activated by beta-adrenergic agonists (Goldstone and Crompton, 1982), further supporting the presence of a second Ca^{2+} efflux process in liver. Na^+-induced Ca^{2+} release also may act to stimulate Ca^{2+} efflux by changes in inner membrane permeability. In addition to attenuation of Na^+-dependent Ca^{2+} release by dibucaine and tetracaine, which are known inhibitors of phopholipase A_2 (Harris and Heffron, 1982), loss of Ca^{2+} in the presence of ruthenium red is also accompanied by efflux of Mg^{2+} and adenine nucleotides, again suggesting phospholipase A_2 activation (Harris and Cooper, 1981). However, the extent to which the amount of Ca^{2+} loaded into the matrix plays a role in the observed permeability changes has not been systematically studied. In fact, matrix concentrations of Ca^{2+} in vivo as determined by electron-probe analysis are likely to be less than one-tenth of the levels loaded in experiments _in vitro_. Recent estimates suggest that muscle mitochondrial Ca^{2+} content _in vivo_ is less than 3-5 nmoles/mg (McCormack and Denton, 1986). This finding implies that the Ca^{2+} efflux pathway may not be saturated under normal physiological conditions. This low content of total Ca^{2+}, i.e., below 3 nmoles/mg, is inconsistent with the ability of mitochondria to buffer cytosolic Ca^{2+} accurately (Nicchitta and Williamson, 1984; Hansford and Castro, 1982). However, the possibility still remains that mitochondria may act to modify changes in cytoplasmic Ca^{2+} if these concentrations rise above 1 μM.

REGULATION OF MITOCHONDRIAL METABOLISM BY Ca^{2+}: ROLE OF Ca^{2+} IN DEHYDROGENASE ACTIVATION

In addition to increasing evidence of low Ca^{2+} content _in situ_ in mitochondria from a variety of tissues, low endogenous Ca^{2+} levels in isolated mitochondria can also be achieved if care is taken to prevent mitochondrial Ca^{2+} transport during tissue homogenization and centrifugation. With these precautions, activation of Ca^{2+}-sensitive dehydrogenases is observed at a matrix-free Ca^{2+} concentration that is an order of magnitude less than is required to saturate the Ca^{2+} efflux pathway.

These considerations give strength to the view that the primary function of Ca^{2+} transporters in the mitochondrial inner membrane is to relay changes in cytosolic Ca^{2+} to mitochondrial matrix enzymes whose activity is stimulated by Ca^{2+}. The principles behind this regulatory process are illustrated in Fig. 6, and have been described in detail by McCormack and Denton (1985) and Hansford (1985) in excellent review articles. Increases in cell membrane Ca^{2+} permeability, e.g., via hormonal activation of receptors, are proposed to increase cytoplasmic Ca^{2+}. In addition to stimulating cytoplasmic Ca^{2+}-dependent processes such as glycogen phophorylase kinase and muscle contraction, the increased Ca^{2+} concentration in the cytosol is communicated to the mitochondrial matrix via transport on the electrophoretic uniporter. The elevation in matrix Ca^{2+} transmits the Ca^{2+}-dependent increases in cellular energy demand to three key regulatory enzymes in oxidative metabolism: pyruvate dehydrogenase, NAD^{+}-isocitrate dehydrogenase and α-ketoglutarate dehydrogenase. These dehydrogenases are activated several-fold by increasing concentrations of Ca^{2+} in the range of 0.1-10 μM. Pyruvate dehydrogenase is activated by a Ca^{2+}-stimulated phosphatase which increases

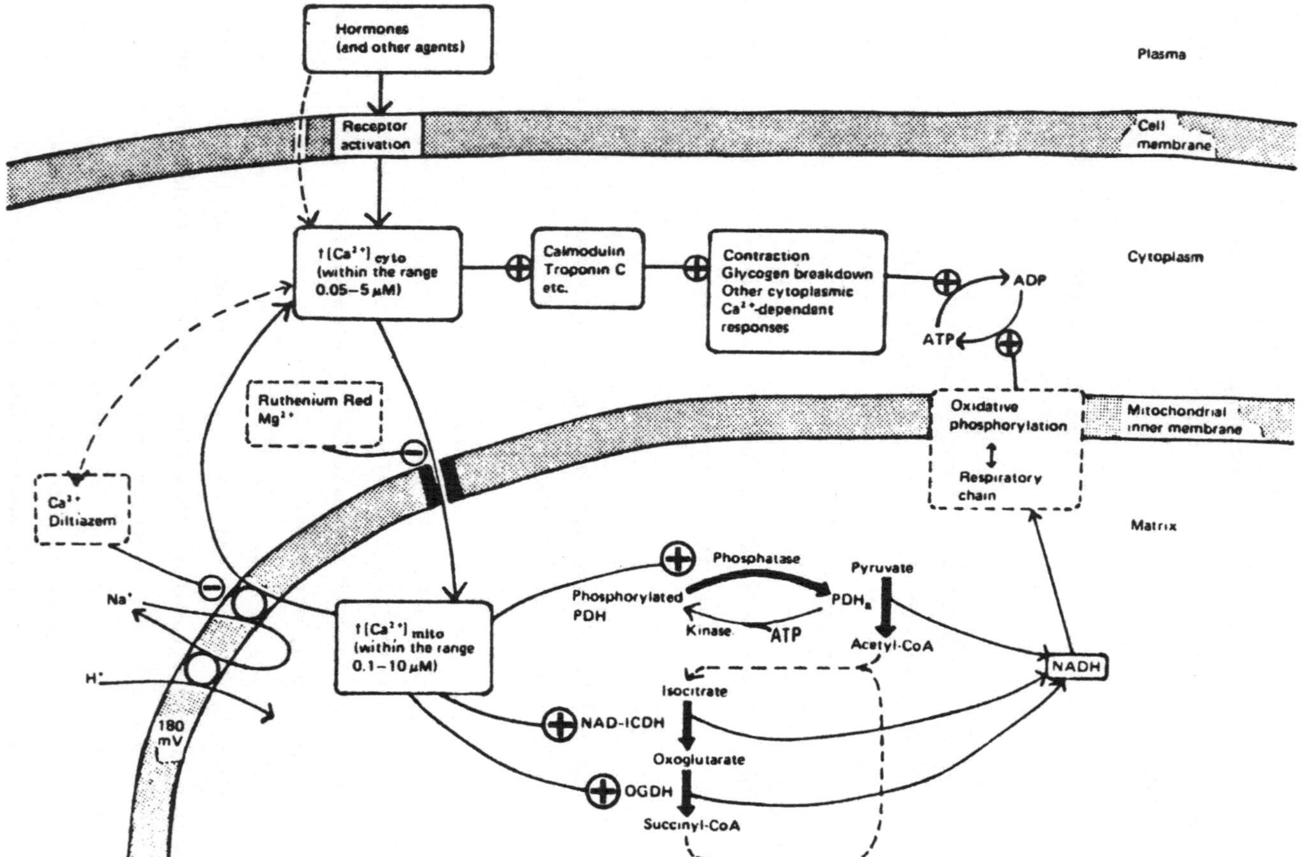

Fig. 6. Hormonal control of intramitochondrial Ca^{2+}-sensitive enzymes (McCormack and Denton, 1985)

the amount of the nonphosphorylated form of the enzyme. Both isocitrate and α-ketoglutarate dehydrogenases are directly activated by Ca^{2+} which decreases the Km values for their substrates. A ten-fold increase in extramitochondrial Ca^{2+} from 50 to 500 nM stimulates ketoglutarate-supported respiration and reduction of citric acid cycle $NADP^+$. This stimulation occurs in the presence of physiological concentrations of Na^+ (to discharge matrix Ca^{2+}) and Mg^{2+} (to limit uptake on the uniporter).

In studies with isolated, intact mitochondria, the proportion of pyruvate dehydrogenase in its active, dephosphorylated form was monitored as a function of extramitochondrial Ca^{2+} in media containing Na^+ and Mg^{2+}(Denton and McCormack, 1985). At a concentration of 0.4 μM Ca^{2+} outside, the enzyme is half maximally activated. Under these conditions, the matrix concentration of free Ca^{2+} is just 2-3 times the outside concentration. Therefore, when this concentration gradient is abolished by addition of uncoupler, half-maximal activation of pyruvate dehydrogenase occurs at 1 μM which is comparable to the value for the separated enzyme. In the absence of Na^+ and Mg^{2+}, the extramitochondrial Ca^{2+} which gives half maximal activation is significantly less than the value observed in the presence of these ions. This is especially true in heart mitochondria which have a more active Na^+:Ca^{2+} antiport. If care is taken to preserve intramitochondrial Ca^{2+} during isolation following hormonal stimuli, then the activation of pyruvate dehydrogenase persists and can be discharged by extramitochondrial Na^+ (Denton and McCormack, 1985). Thus, these experiments validate the mechanism of respiratory stimulation in vivo.

Similar types of activation can be seen in isolated cardiac myocytes in response to agents which produce elevated cytosolic Ca^{2+}, e.g., KCl, which opens the voltage-dependent Ca^{2+} channels; ouabain, which inhibits Na^+, K^+-ATPase and thus increases Na^+:Ca^{2+} exchange; and veratridine which opens the Na^+ channels, leading to reversal of Na^+:Ca^{2+} exchange (Hansford, 1986). When the intracellular Ca^{2+} dye, Quin 2, is loaded into myocytes to determine qualitative changes in cytosolic free Ca^{2+}, the high degree of intracellular Ca^{2+} buffering slows the response of pyruvate dehydrogenase to increased cytosolic Ca^{2+}. The magnitude of both the depolarization-induced increase and the veratridine-induced increase in cytosolic Ca^{2+} and pyruvate dehydrogenase activation changes in parallel with the concentration of the effectors added (Hansford, 1986). The mitochondrial uniport inhibitor ruthenium red, blocks the response of pyruvate dehydrogenase to KCl and veratridine plus ouabain, whereas ryanodine, which appears to inhibit Ca^{2+} release from the sarcoplasmic reticulum, has no effect. The importance of the Ca^{2+} signal to mitochondrial energy production is thus quantitatively significant, even in the face of the consequences of increased contractile work, where mitochondrial ATP/ADPxP_i and NADH/NAD^+ decrease. The latter are also known to be important signals for dehydrogenase activation. Since ruthenium red does not abolish the increase in cytosolic Ca^{2+} nor does it alter the frequency of contractile waves in the myocyte, its direct effect in these cells is on mitochondrial Ca^{2+} uptake and subsequent elimination of Ca^{2+} activation of pyruvate dehydrogenase.

These experiments, as well as additional bodies of data from Hansford, and from Denton and McCormack, provide compelling evidence for Ca^{2+} activation of energy metabolism at the dehydrogenase level. In addition to metabolic control of oxidative phosphorylation by ADP, Ca^{2+} stimulation of ketoglutarate and NAD^+ isocitrate dehydrogenase activities may provide a biological advantage in excitable tissues (Hanford, 1985). When there is no dehydrogenase activation, phosphorylation of ADP produces pronounced oxidation of the NAD^+/NADH couple. In contrast, dehydrogenase activation increases the reduction of this couple which then serves as a driving force for oxidative phosphorylation. Thus, high rates of oxidative metabolism can proceed in tissues which require high rates of energy transduction, e.g., heart. Under these conditions, there will be little or no change in the NAD^+/NADH couple or in the cytosolic phosphorylation potential.

Future experiments may be directed toward the resolution of several issues. First, the physiological relevance of Ca^{2+}-activated mitochondrial dehydrogenase activity in liver mitochondria is still unknown. Furthermore, it will be important to demonstrate actual elevations in mitochondrial Ca^{2+} content in vivo within the range needed for dehydrogenase stimulation following the appropriate hormonal or trans sarcolemmal ionic stimulus. Secondly, it is not known if tissue specificity contributes to the mode of mitochondrial Ca^{2+} handling, i.e., metabolic activation versus cytosolic Ca^{2+} buffering, or if any organ can switch between these modes. What are the factors or signals which determine the tissue response?

In conclusion, in spite of the controversies surrounding operation of the mitochondrial systems governing Ca^{2+} efflux and steady state values of matrix free Ca^{2+}, the physiological meaning of modulation of matrix Ca^{2+} is compelling with respect to the control of catabolic metabolism and energy production in the cell. Information should be forthcoming which will provide a detailed understanding of mitochondrial Ca^{2+} regulation in vivo as well as the significance of this regulation in the cellular integration of metabolism.

REFERENCES

Affolter, H. and Carafoli, E. (1980) The Ca^{2+}-Na^{+} antiporter of heart mitochondria operates electroneutrally. Biochem, Biophys. Res. Commun. 95, 193-196.

Akerman, K.E.O. (1978) Effect of pH and Ca^{2+} on the retention of Ca^{2+} by liver mitochondria. Arch. Biochem. Biophys. 189, 256-262.

Beatrice, M.C., Stiers, D.L. and Pfeiffer, D.R. (1984) The role of glutathione in the retention of Ca^{2+} by liver mitochondria. J. Biol. Chem. 259, 1279-1287.

Becker, G.L., Fiskum, G. and Lehninger, A.L. (1980) Regulation of free Ca^{2+} by liver mitochondria and endoplasmic reticulum. J. Biol. Chem. 255, 9009-9012.

Burgess, G.M., McKinney, J.S., Fabiato, A., Leslie, B.A. and Putney, J.W. (1983) Calcium pools in saponin-permeabilized guinea pig hepatocytes. J. Biol. Chem. 258, 15336-15345.

Chance, B. (1965) The energy-linked reaction of calcium with mitochondria. J. Biol. Chem. 240, 2729-2748.

Coll, K.E., Joseph, S.K., Corkey, B.E. and Williamson, J.R. (1982) Determination of the matrix free Ca^{2+} concentration and kinetics of Ca^{2+} efflux in liver and heart metochondria. J. Biol. Chem. 257, 8696-8704.

Crompton, M. (1980) The sodium ion/calcium ion cycle of cardiac mitochondria. Trans. Biochem. Soc. 8, 261-262.

Crompton, M., Capano, M. and Carafoli, E. (1976) The sodium-induced efflux of calcium from heart mitochondria. A possible mechanism for the regulation of mitochondrial calcium. Eur. J. Biochem. 69, 453-462.

Denton, R.M. and McCormack, J.G. (1985) Ca^{2+} transport by mammalian mitochondria and its role in hormone action. Am. J. Physiol. 249, E543-E554.

Fiskum, G. and Cockrell, R.S. (1978) Ruthenium red sensitive and insensitive calcium transport in rat liver and Ehrlich ascites tumor cell mitochondria. FEBS Lett. 92, 125-128.

Fiskum, G. and Cockrell, R.S. (1983) Uncoupler-stimulated release of Ca^{2+}

from Ehrlich ascites tumor cell mitochondria. Arch. Biochem. Biophys. 240, 723-733.

Fiskum, G. and Lehninger, A.L. (1979) Regulated release of Ca^{2+} from respiring mitochondria by $Ca^{2+}/2H^{+}$ antiport. J. Biol. Chem. 254, 6236-6239.

Goldstone, T.P. and Crompton, M. (1982) Evidence for β-adrenergic activation of Na^{+}-dependent efflux of Ca^{2+} from isolated liver mitochondria. Biochem. J. 204, 369-371.

Gunter, T.E., Chace, J.H., Puskin, J.S. and Gunter, K.K. (1983) Mechanism of sodium independent calcium efflux from rat liver mitochondria. Biochemistry 22, 6341-6351.

Hansford, R.G. (1985) Relation between mitochondrial calcium transport and control of energy metabolism. Rev. Physiol. Biochem. Pharmacol. 102, 1-72.

Hansford, R.G. Biochem. J. (1986), in press.

Hansford, R.G. and Castro, F. (1982) Intramitochondrial and extramitochondrial free calcium ion concentrations of suspensions of heart mitochondria with very low, plausibly physiological, contents of total calcium. J. Bioenerg. Biomembr. 14, 361-376.

Harris, E.J., Al-Shaikhaly, M. and Baum, H. (1979) Stimulation of mitochondrial calcium ion efflux by thiol-specific reagents or by thyroxine. Biochem. J. 182, 455-464.

Harris, E.J. and Cooper, M.B. (1981) Calcium and magnesium ion losses in response to stimulants of efflux applied to heart, liver and kidney mitochondria. Biochem. Biophys. Res. Comm. 103, 788-796.

Harris, E.J. and Heffron, J.J.A. (1982) The stimulation of the release of Ca^{2+} from mitochondria by sodium ions and its inhibition. Arch. Biochem. Biophys. 218, 531-539.

Hayat, L.H. and Crompton, M. (1982) Evidence for the existence of regulatory sites for Ca^{2+} on the Na^{+}/Ca^{2+} carrier of cardiac mitochondria. Biochem. J. 202, 509-518.

Heffron, J.J.A. and Harris, E.J. (1981) Stimulation of calcium-ion efflux from liver mitochondria by sodium ions and its response to ADP and energy state. Biochem. J. 194, 925-929.

Henry, P.D., Shuchleib, R., Davis, J., Weiss, E.S. and Sobel, B.E. (1977) Myocardial contracture and accumulation of mitochondrial calcium in ischemic rabbit heart. Am. J. Physiol. 233, H677-H684.

Joseph, S.K., Coll, K.E., Cooper, R.H., Marks, J.S. and Williamson, J.R. (1983) Mechanisms underlying calcium homeostasis in isolated hepatocytes. J. Biol. Chem. 258, 740-741.

Jurkowitz, M.S. and Brierley, G.P. (1982) H^+-dependent efflux of Ca^{2+} from heart mitochondria. J. Bioenerg. Biomembr. 14, 435-449.

Lehninger, A.L. (1974) Role of phosphate and other proton-donating anions in respiration-coupled transport of Ca^{2+} by mitochondria. Proc. Nat. Acad. Sci., USA 71, 1520-1524.

McCormack, J.G. and Denton, R.M. (1985) Hormonal control of intramitochondrial Ca^{2+}-sensitive enzymes in heart, liver and adipose tissue. Trans. Biochem. Soc. 13, 664-667.

McCormack, J.G. and Denton, R.M. (1986) Ca^{2+} as a second messenger within mitochondria. TIBS, 11, 258-262

McMillin-Wood, J., Wolkowicz, P.E., Chu, A., Tate, C.A., Goldstein, M.A. and Entman, M.L. (1980) Calcium uptake by two preparations of mitochondria from heart. Biochim. Biophys. Acta. 591, 251-265.

Mela, L. (1968) Interactions of La^{3+} and local anesthetic drugs with mitochondrial Ca^{++} and Mn^{++} uptake. Arch. Biochem, Biophys. 123, 286-293.

Moore, C.L. (1971) Specific inhibition of mitochondrial Ca^{++} transport by ruthenium red. Biochem, Biophys. Res. Comm. 42, 298-305.

Nicchitta, C.V. and Williamson, J.R. (1984) Spermine. A regulator of mitochondrial calcium cycling. J. Biol. Chem. 259, 12978-12983.

Nicholls, D.G. (1978) The regulation of extramitochondrial free calcium ion concentration by rat liver mitochondria. Biochem. J. 176, 463-474

Puskin, J.S. and Gunter, T.E. (1973) Ion and pH gradients across the transport membrane of mitochondria following Mn^{++} uptake in the presence of acetate. Biochem. Biophys. Res. Comm. 51, 797-803.

Puskin, J.S., Gunter, T.E., Gunter, K.K. and Russell, P.R. (1976) Evidence for more than one Ca^{2+} transport mechanism in mitochondria. Biochemistry 15, 3834-3842.

Rottenberg, H. and Scarpa, A. (1974) Calcium uptake and membrane potential in mitochondria. Biochemistry 13, 4811-4817.

Scarpa, A. and Graziotti, P. (1973) Mechanisms for intracellular calcium regulation in heart. I. Stopped-flow measurements of Ca^{++} uptake by cardiac mitochondria. J. Gen. Physiol. 62, 756-772.

Skrede, S. (1966) Effects of cystamine and cysteamine on the adenosine-triphosphatase activity and oxidative phosphorylation of rat-liver mitochondria. Biochem. J. 98, 702-708.

Vaghy, P.L., Johnson, J.D., Matlib, M.A., Wang, T. and Schwartz, A. (1982) Selective inhibition of Na^+-induced Ca^{2+} release from heart mitochondria by diltiazem and certain other Ca^{2+} antagonist drugs. J. Biol. Chem. 257, 6000-6002.

ON EVOLUTION OF RENAL FUNCTION AND WATER-SALT HOMEOSTASIS

Yu. V. Natochin

Sechenov Institute of Evolutionary Physiology
and Biochemistry
Leningrad, U.S.S.R.

Evolutionary physiology is a term which is not unique in meaning. On one hand, it conveys the idea of functional development in the period of growth and development of any individual from its birth to death, on the other, the development of function in animals of various systematic groups since the first forms of life appeared on the Earth. The term biological evolution embraces irreversible historical development of all living things. There is, however, some difference in what the words "development" and "evolution" are taken to mean by Russian and foreign authors. In my opinion, to avoid any possible misunderstanding it should be made quite clear from the very start what we put in the term.

Developmental physiology does not cover to the full the range of problems belonging to evolutionary physiology. The latter term was suggested by Severtsov (1914) as an alternative to evolutionary morphology. What it came to mean being applied to physiology is given in the works of Orbeli (1933, 1961), Koshtoyants (1932) and Ginetsinsky (1961). In their and their collaborators' works it is pointed out as necessary to analyse the problems of development, evolution of functions by a combined study of onto- and phylogenesis of functions. With respect to kidney, a very close view can be found in the works of Smith (1951, 1953).

In the present review an attempt is made to give a general picture of the development of water-salt homeostasis and the main effectory organ responsible for it in vertebrates - kidney. The data on the ontogenesis of function in question would not be enough to make the picture complete. For the approach to the studies of development, evolution of function to be more extended and reliable, some other approaches should be added. First, it is a comparative study on a wide scale of the given function in animals belonging to different levels of phylogenetic development. Haeckel (1874), Severtsov (1914), Koshtoyants (1932) and Orbeli (1961) pointed out that it is very important to compare in onto- and phylogeny not only structures, but functions as well. Orbeli (1961) attracted attention to the third and fourth lines of investigation in the study of evolution of function; this problem should be approached through stages in the disturbance of functional relations in pathology, when examining patients, and when disturbance in functional activity is induced experimentally, e.g. as a result of denervation of the organ. The next, fifth line of research concerning evolution of function is to find out the boundaries of a change of the function, which would allow the animal to adapt to extreme conditions. With respect to the water-salt homeostasis, of particular interest is adaptation to life in desert and arctic zones, highly freshened or hyperhaline water basins. This line of research covers also the development or acute reaction of organisms in such extreme conditions as a space flight or high and low pressure. The sixth, and last line of research deals with a specific reaction of organisms belonging to different levels of evolutionary and individual development to toxic agents. This direction, which could be rightly referred to as evolutionary toxicology, allows the comparison of the kind of dysfunction in representatives of different classes of vertebrates and of various stages of development, the response to the action of harmful factors.

In formulating general principles of evolution of function, in order to avoid a mistake which is quite common when comparing individual peculiarities of species or certain peculiarities of the ontogenesis of animals belonging to the same species, all or some of the above approaches should be used together. This would allow one to suggest possible grounds for evolution of function.

In this review an attempt is taken to put forward on the basis of the results obtained in our laboratory and the data available from the literature, some principles underlying the development of water-salt homeostasis and evolution of renal function.

Nowadays in physiology there is a strong tendency to study the membrane and molecular bases of function of organisms. No doubt, this makes a good contribution to physiological science, but at the same time a study of the organism as a whole, of the place and role these molecular processes have in it has become a real necessity. No less important, as it seems, would be the analysis of tendencies and of principles of functional development.

The concept of physiology as it is seen nowadays, encompasses much more than a specific function of a separate organ, even if it is the only performer of this function. The physiology of water-salt homeostasis may serve as a rather good example that a stable level of some constants of fluids in the milieu intérieur such as their volume, osmotic concentration, ion content, are provided by a number of elements, though in higher vertebrates quite often kidney is the main effector organ. When studying this problem, we came to the conclusion that for a better understanding of the principles of evolution of function it is necessary to understand first the principles of evolution of the given functional system which includes specific receptors (e.g. osmo- or volumoreceptors), central mechanisms of regulation of this function, afferent pathways of regulation, and the effector organs.

In mammals the principal organ is kidney, in some other vertebrates urinary bladder, skin, salt glands, gills can also be involved. To achieve our goal, it is not enough as it came out, to trace the evolution of function of the effector organ in question. Some other criteria, and in some cases quite different principles are to be formulated when the main morpho-functional elements of the organ are concerned, in our case it is the nephron. And at last, the development of specialized functions is concerned first of all with rearrangement, with specialization of the organ cells, renal cells. Thus, the update approach to the problem of evolution of function would cover events belonging to four different categories which are determined by the level of organization of the living systems. These are: 1) evolution of functional systems, 2) evolution of function of organ, 3) evolution of functional units, and 4) evolution of specialized cell.

EVOLUTION OF THE SYSTEM OF WATER-SALT BALANCE

The unicellular animals had only intracellular fluid. As time went on, cells combined to form colonies, and the first multicellular organisms appeared; this was accompanied by formation of extracellular gel, and later of extracellular fluids. The extracellular gel is likely to differ from the surrounding medium in its ion content. Evidently, there was a special physiological system responsible for maintaining a constant volume of fluid within gel. A study we carried out of jellyfish Tiaropsis multicirrata which lives in ocean water showed that animals with no kidney or any other excretory organs have special physiological mechanisms responsible for stabilizing the volume and ion content of their body (Natochin, Lavrova and Gusev, 1971). Very important for a progressive development of the water-salt metabolism was the formation of coelom and circulation systems which provided organs with nutrients, oxygen, and allowed the access of substances to the excretory organs. The evolution of animals was accompanied by the appearance of a number of extracellular fluids, among which as very important should be mentioned blood, lymph, coelomic fluid, and such specialized fluids as cerebrospinal, intraocular and others.

In addition to Myxine, representatives of the other classes of vertebrates are known to have a similar ion content of the blood serum. For the function of physiological systems, however, it is necessary that the concentration of each ion in blood should be stable. In our laboratory this value was determined by calculating the coefficient of variation of osmolality and some ions in the blood serum of different animals and man. Taken several times during a usual work day, these values in man showed osmolality of concentration of active Ca^{2+} and Na^{+} to be most stabilized. The variation coefficient for these values is as little as 0.6 - 1.4%, whereas for potassium, magnesium, and some other ions it is much higher. A very close tendency is revealed for the values of water-salt homeostasis when comparing vertebrates from cyclostomata to mammals, the constants being less stabilized, however. The results of a comparative study of osmoregulation in birds whose blood plasma osmolality, unlike that of mammals, was found to be very liable to changes give another proof for reliability of the above data. It is advisable that osmolality and ion concentration of the blood plasma should be determined on animals in usual conditions, and under a standard load as well. Thus, under dehydration equal to 10% of body weight shifts in ion and osmotic homeostasis were found to be small in mammals, more pronounced in amphibians, reptiles, and birds.

To find the answer concerning the water and salt content of the body in the process of individual development the following must be determined. First, ion concentration of fluids in the milieu intérieur; second, ion balance in the developing individual; third, the role of placenta and kidney in ion homeostasis; fourth, the role of regulatory factors in stabilization of water and ion metabolism; fifth, maintenance of ion homeostasis in extreme conditions.

Beginning from the first days after birth and during his whole life the ion composition of blood serum in man remains, as a rule, stable (Natochin, 1976). In experiments on sheep it was shown that the value of Na^+ and K^+ concentrations in the blood plasma of embryos was rather similar to that of the adult animal (Lumbers and Stevens, 1983). This means that concentration of electrolytes in the blood of mammals from the last stages of embryogenesis and during their life is maintained on a stable level. In the past, especially in the 60's, there appeared many publications devoted to the problem of electrolytes in the organs of mammals during their development. The available data, however, did not give a complete picture of the development of electrolyte composition in the foetus. In experiments on rats we studied the electrolyte content in the foetal body from the 15th day until birth. Beginning from the 15th to 21st day of pregnancy the foetal weight increased 23.5-fold, the amount of water calculated per dry weight constantly decreased (from 10.9 ± 0.36 to 6.63 ± 0.47 g H_2O/g of dry weight), and what is very significant with electrolytes being constantly increased in number, their ratio was maintained on the same level (Natochin, Dolgopolova, Lavrova, Shakhmatova, Serova, Denivsova and Iliushko, 1984), in spite of the fact that cells of many organs and tissues started developing fast, and that there was some difference in qualitative characteristics of ion composition and percentage of cells, and in the amount of extracellular fluid.

The data obtained suggest that in mammals the system responsible for stabilization of water-salt content in the developing foetus is very effective indeed. Neither a water load (Natochin et al., 1984), nor NaCl solution, or increased (39-fold) consumption of Na^+ during pregnancy (Kirksey, Pike and Callahan, 1962) led to any statistically significant difference in Na^+ and K^+ concentration in the blood serum and in the foetus. At the same time, ion deficiency is rather dangerous, because a lower content of magnesium and sodium in the daily food of pregnant rats led to more cases of foetal resorption.

Thus, it can be said that in mammals, i.e. placental organisms, whose development requires regular uptake of water, salts, and organic

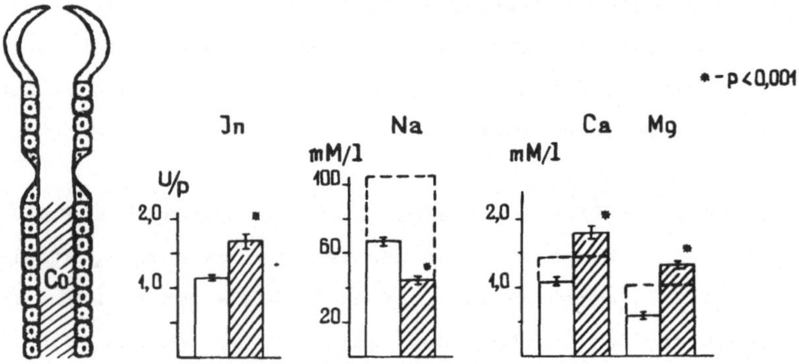

Fig. 1: A change in ion and fluid reabsorption after microinjection of 5.10^{-4}M $CoCl_2$ in the early distal tubule of the newt kidney (according to O.A. Goncharevskaya, Yu.G. Monin and Yu.V. Natochin). The white columns - control, dashed columns - injection of $CoCl_2$, dotted line - ion concentration in the solution for tubular microperfusion.

substances, the ratio of electrolytes in the developing individuals is very stable.

Our attention was also focused on the formation of electrolytes in embryogenesis of freshwater and marine teleosts. This choice was made because embryonic development of these animals takes place not within the female, but beyond it, in the medium poor (fresh water) or rich (sea water) in different salts. The electrolyte composition of embryos was determined before the larvae passed on to active feeding, which allowed us to find out how much their development depended on the external sources of ions and on the ratio of electrolytes that makes this development possible. According to the results obtained, in each egg there are enough electrolytes for the individual to develop before it takes to active feeding, and their ratio in the larvae remains without change (Natochin, Lukianenko, Lavrova and Metallov, 1976). A considerable difference in the content of electrolytes in some stages of fish development is connected with formation of the perivitelline space and has nothing to do with the embryo.

The problem of functional regulation of different organs responsible for water-salt homeostasis has come to be more complicated than was seen not long ago. In addition to hormones which regulate the cellular function, a high complexity of the intracellular system responsible for hormonal signals, interaction of nerve and humoural stimuli, and hormone modulators such as prostaglandins were also found to be very important. Until quite recently, the basal plasma membrane of nephron cells was considered to be the centre of the action of hormones. Some substances, however, some diuretics in particular (Burg, Stoner, Cardinal and Green, 1973; Natochin, 1982), a number of trace elements act on the outer surface of the apical plasma membrane (Goncharevskaya, Monin and Natochin, 1986; Natochin, Goncharevskaya, Monin and Shakhmatova, 1986) (Fig. 1). Taking into account such great complexity of regulation of cellular function, we can at present speak only about the effect some substances produce on the function of nephron cell, but it is too early - as we have no reliable data - to speak about the origin and perfection of responses in the evolution of renal function.

As a tendency of evolution of regulatory systems, an increased number of physiologically active substances exerting an action on function and a change of their physiological role could probably be mentioned. Consider two examples. Most important regulators of Ca^{2+} balance are known to be active metabolites of vitamin D_3, especially $1,25(OH)_2$-vit.D_3. Specific regulators of Ca^{2+} metabolism are parathormone and calcitonin. Ca^{2+} balance as well as hydroxylation of vitamin D_3 is likely also to be affected by some other hormones, including prolactin. There are data according to which in vertebrates the hormones which take an active part in Ca^{2+} homeostasis gradually increase in number (Dacke, 1979). This conclusion can be extended to hormones involved in regulation of water and sodium metabolism. Vasotocin, and later vasopressin are capable of regulating excretion of water in terrestrial vertebrates; in cyclostomata and teleosts vasotocin lacks this ability (Bentley, 1971).

To sum up, the conclusion can be made that the principles underlying the progressive evolution of a system of water-salt homeostasis are the following: first, increase of the stability of homeostasis and second, increase of regulatory factors.

EVOLUTION OF RENAL FUNCTION

The comparison of morphological, functional, and biochemical characteristics of kidney in representatives of different classes of vertebrates, a comparative study of excretory organs in vertebrates and invertebrates, as well as other organs involved in water and salt metabolism which appear in some groups of animals in the evolutionary process, the analysis of data on the ontogenesis of kidney, its state in extreme conditions and in pathology -- all this allows the formulation of some principles of evolution of renal function.

Increase of multi-functionality

It would not be correct to regard kidney as an excretory organ only. In fact, in higher animals and man it performs many homeostatic functions. In mammals kidney is involved in regulation of water-salt homeostasis (regulation of the volume of extracellular fluid, concentration of osmotically active substances in the blood serum, its ionic composition and pH), excretion of the end metabolic products, metabolism of proteins, carbohydrates, lipids; it is the site where physiologically active substances are produced which regulate the arterial pressure, blood coagulation, modulate hormone action, here the active form of vitamin D_3 is synthesized, catabolism of peptide hormones and some other functions are carried out (Brenner and Rector, 1976). The extension of the functional importance of kidney and its multifunctionality which reaches higher levels are revealed in the evolution of vertebrates. For example, while in *Myxine* the kidney cannot form hypotonic urine, in lamprey this process is carried out rather effectively (Dantzler, 1985). The data on the action of such blockers of chloride and sodium transport as furosemide and amiloride on the kidney of animals belonging to all classes of vertebrates, beginning from lamprey (Natochin, 1976), give evidence of a rather great similarity in membrane processes which underlie the mechanism of urine dilution. In *Agnatha* and *Osteichthyes* there is no mechanism to enable the kidney to save water. First this function appears in Anuran amphibians, warm-blooded vertebrates are the first animals whose kidney is capable of maintaining water-salt metabolism, in spite of a high water deficiency, as a result of osmotic concentration of urine.

Intensification of processes responsible for renal function

According to A.N. Severtsov, intensification of function should be regarded as an indication of progressive development. The study of evolution of renal function serves as a good proof of this. The ratio of the kidney weight to the body surface, or body weight in vertebrates of similar weight is a rather stable figure (in %): in lamprey 0.55, frog 0.38, pigeon 0.52, rat 0.67, newborn child 0.66. At the same time, compared to cold-blooded, in warm-blooded vertebrates the glomerular

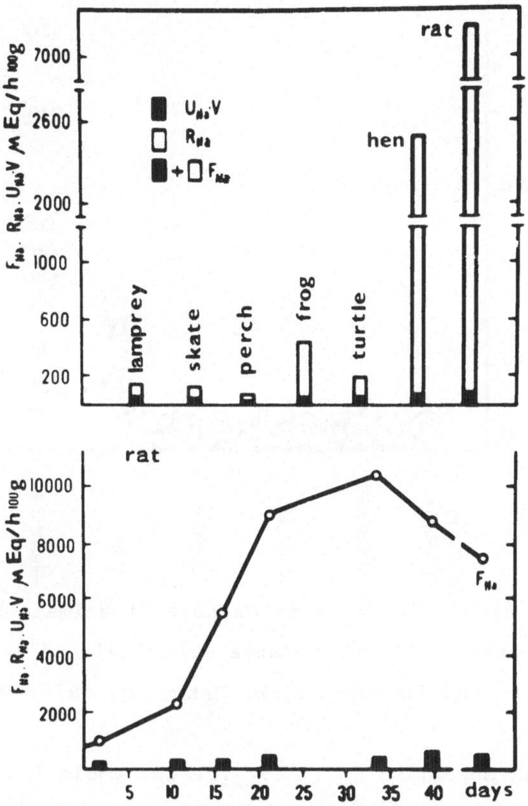

Fig. 2: Sodium glomerular filtration reabsorption and excretion by the kidney of different vertebrates and rats of various ages.

filtration rate (and as a result, the amount of reabsorbed substances, sodium for example) is much higher (Natochin, 1976; Dantzler, 1985). In lamprey and fish the glomerular filtration rate is 20 to 100 times lower than in mammals, this figure holds for reabsorption of ions from the tubular lumen into blood (Figs. 2,3). Though such a sharp increase in the intensity of transport processes in the kidney of warm-blooded animals creates favourable conditions for regulation so that the milieu intérieur would be more stable, it requires much energy. The mammalian kidney is an organ with the highest level of blood circulation, at rest up to 20% of blood from the heart comes to the kidney, and its consumption of oxygen is 10% of the total amount the organism needs.

A reliable proof that intensification of renal function is a reflection of progressive evolution of the organ rather than a mere consequence of warm-bloodedness, compared to cold-blooded vertebrates, are the data obtained in the study of renal function of mammals in post-natal

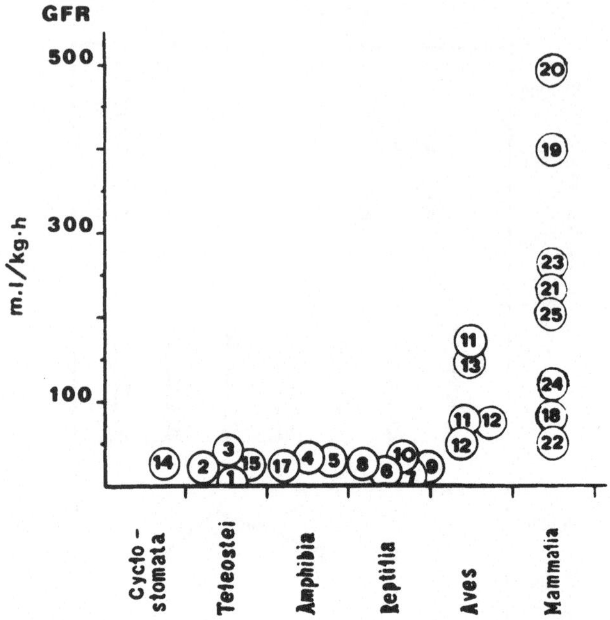

Fig. 3: Glomerular filtration rate in kidneys of animals belonging to different classes of vertebrates (recalculated according to Smith, 1951; Renkin and Gilmore, 1972; Natochin, 1976; Dantzler, 1985).

1. Anguilla anguilla
2. Salmo gairdneri
3. Fundulus kansae
4. Rana clamitans
5. Xenopus laevis
6. Gopherus agassizii
7. Pseudemys scripta
8. Crocodylus johnsoni
9. Crocodylus acutus
10. Varanus gouldii
11. Gallus domesticus
12. Callipedla gambelii
13. Sturnus vulgaris
14. Lampetra fluviatilis
15. Squalus acanthias
16. Mustelus canis
17. Necturus maculosus
18. Tachyglossus aculeatus
19. Mouse
20. Rat
21. Guinea pig
22. Horse
23. Dog
24. Man
25. Cat

ontogeny. In the first weeks after birth there is a marked increase in the glomerular filtration rate, a rise in the intensity of tubular transport. The fact that such substances as divalent ions in marine teleosts, uric acid in reptiles and birds must be removed from blood at a low glomerular filtration rate due to a relatively small arterial blood supply of the kidney (per minute) leads to the mode of blood supply of kidney being changed in evolution. Many vertebrates, except Agnatha, fresh-water fish and mammals, have a renoportal system by which venous blood from the back part of the body comes to the kidney. This greatly contributes to the possibility of kidney to excrete a number of substances by secreting them into a tubular lumen. What has been said allows the suggestion that an increase of the glomerular filtration rate and intensification of renal tubular reabsorption may be regarded as one of the characteristic features of progressive evolution of this organ.

The principle of superlayer

The study of brain function led L.A. Orbeli to the idea that new functions emerge not as a mere substitution for the previous, but are added to and suppress the latter; nevertheless each organ, no matter how high its specialization may be, possesses the features which reflect the history of its development. The kidney of birds and mammals can be a good example of the validity of this principle. In addition to osmotic dilution, their organ can also perform the function of osmotic concentration of urine; this became possible due to a new structure which is absent in other vertebrates, i.e. the kidney consisting of two layers, namely cortex and medulla. As was thought earlier, osmotic concentration of urine was connected with Henle's loop which the warm-blooded animals developed in the course of evolution. However, structurally similar nephron loops were found in the kidney of lamprey (Goncharevskaya, 1976) which gives good grounds to suggest that subdivision of kidney into cortex and medulla, high level of renal arterial blood supply, and availability of nephron loops made it possible to develop a new structure (medulla), and a new function (osmotic concentration of urine). In mammals the function of osmoregulation gradually develops in ontogenesis (Dicker, 1970), rather often it is the function to be disturbed in pathology (Brenner and Rector, 1976) and after a space flight (Natochin, Kozyrevskaya and Grigoryev, 1975).

Oligomerization of organs and polymerization of functional units

Dogel (1954) gave grounds to the law of oligomerization of organs in the process of evolutionary development of multicellular animals. The law holds with respect to evolution of excretory organs, numerous metamerically located metanephridia are substituted by a pair of excretory organs in molluscs or maxillary glands in Crustacea. All vertebrates have a pair of kidneys. However, the data obtained from the anatomical study of kidneys give evidence that they consist of a great number of relatively similar elementary units - nephrons. The comparison of the quantity of nephrons per g of kidney weight in some representatives of different classes of vertebrates shows it to vary from 7600 in frog to 73680 in Psammomys, in man it is 7000. Thus, the principle of oligomerization of organ should be added by that of polymerization of functional units; this would make the function of organ more reliable, as a result the number of erroneous reactions would be minimized and the stability of the physiological system increased.

Change of the organ function

The principle of functional modification, change of the organ function was formulated by Dohrn (1875) who believed that a step-like change in function of the same organ brings about a change of this organ. Renal function may be regarded as a resultant of several items one of which stands for the main function, whereas others are additional functions. When the main function dies away, and one of the auxiliary functions gains something in its importance, the resultant function undergoes some modification, and as a result the organ would change too. The suggestion of Dohrn based on the results of morphological studies has been proved by the new data available on comparative physiology of kidney. In teleosts, in addition to being involved in water-salt metabolism, excretion of a number of the end metabolic products, kidney is also a haemopoietic organ. In the kidney of terrestrial vertebrates there was found no lymphoid tissue, but according to numerous findings, the kidney of mammals has an important role in regulation of erythropoiesis and blood coagulation. It seems likely that earlier the kidney was involved in production of blood cells and their release into blood, but later it could perform only a regulatory function with respect to erythropoiesis.

Substitution of an organ or its function

Adaptation to another environment, to new conditions may lead to the substitution of an organ or make it function in quite another way. Great attention was paid by evolutionary morphologists to the analysis of how organs underwent changes in natural selection. In an approach to the problem of substitution of organs and functions through the data on water-salt metabolism, it would seem worth mentioning that a change in the function of an organ, or substitution of one organ by another, the latter being capable of performing a new function, depends not only on the potential capability of the initial structure, but on the stage of development of other functional systems of the organism. All vertebrates, except mammals, are adapted to live in the sea because some of their organs take part in hypo-osmotic regulation. For osmoregulation animals drink sea water and freshen it by secreting Na^+ and Cl^-; this is done in teleosts by gills, in elasmobranchs by rectal gland, in reptiles and birds by salt glands. The kidneys of these animals are not able yet to provide a rather high level of osmotic concentration and save water, as this would require a high level of kidney arterial blood supply, which is the case only with mammals. Thus in adaptation to an hyperosmotic environment, instead of kidney gill chloride cells and salt glands are involved in osmoregulating reactions, and only in mammals is this function performed by kidneys.

Regress of function

The cause of transformation of kidney in fish was their migration from fresh water to sea. As a matter of fact, in marine fish kidneys are not involved in osmoregulation, as a rule they secrete urine which is iso-osmotic to the blood serum; according to the literature, only in a few species was urine found to be slightly hypotonic. In these fish the role of kidney was, in particular, to excrete divalent ions which came in abundance into the fluid of the internal medium from sea water. What is significant in the structure of their kidney is the fact that the glomeruli in many teleosts are reduced, in some cases there was a complete lack of glomeruli, and some species had lost even distal tubules (Smith, 1953). A well-developed capacity to excrete magnesium, sulphates and some other divalent ions was accompanied by disappearance of other functions. The kidney of these fish was not able to secrete osmotically diluted urine, which is another proof of its regression. Regression of the kidney function contributed, however, to morpho-functional progress of this group of vertebrates since it had led to such modification of their kidney which increased its capability to secrete divalent ions, and a loss of water with urine was lowered. This modification was evidently based on the initial forms of kidney activity, secretion of magnesium and calcium in particular, which in turn, is due to the ion-exchange processes such as Na^+/Mg^{2+} and Na^+/Ca^{2+} (Natochin and Gusev, 1969; Natochin, 1976; Renfro and Shustock, 1985; Cliff and Beyenbach, 1986).

Irreversibility of regressive evolution of function

Continuously increasing differentiation and adaptation to a narrow ecological area result in a decreased plasticity. Some new progressive features may appear, accompanied by a lack of others, which would never reappear again unchanged. In paleontology the idea of irreversibility of the evolutionary process was put forward by Dollo. This does not suggest that the degradation of structure in an animal group is doomed to be fatal, the course of evolution may change and go along a new progressive line but this direction will always differ from the previous. A very close connection between structure and function can be seen in that in animals with their increasing specialization to living conditions, better adaptation to the environment can be accompanied by a loss of certain functions, which makes the capability to adaptation weaker. Consider some examples illustrating evolution of kidney adaptation in rodents. Compared to the white rat which has no problem in excreting and saving water in case of water insufficiency, the kidney of the water vole Arvicola terrestris L. and big gerbil Rhombomys opimus L. not only acquired some new features that contribute to their adaptation to the environment but, in addition, lost some of their functions. The water vole spends its life near water, the structure of its kidney makes it possible to increase the efficiency of water excretion, the inner medulla is reduced, nephrons with a long Henle's loop have disappeared, the ion content of renal medullary tissue is changed. The kidneys of these animals are not able any longer to form urine with a high osmotic concentration. Contrary to them, the big gerbil that lives in deserts can stay without drinking water. This is possible due to the fact that its kidney contains a great number of Henle's loops with a thin long loop and the inner medulla and long papilla, and so is capable of forming urine with osmotic concentration 15 times that of blood. The above changes belong to the osmoregulating function of the kidney, whereas its other functions, such as volume- and ion-regulating, are equally effective (Ivanova, Natochin, Serebryakov, Goncharevskaya, Knyazkova, Lavrova, Nasledova, Pechurkina, Podsekaeva and Shakhmatova, 1980; Natochin, Ivanova, Serebryakov, Podsekaeva, Pechurkina, Lavrova, Lavrinenko, Melidi, Nasledova and Shakhmatova, 1982).

EVOLUTION OF FUNCTIONAL UNITS

The morpho-functional unit of kidney in all vertebrates is the nephron. It is common knowledge that the structure of nephrons is not always the same. Nephrons can differ sufficiently by the presence or absence of glomeruli and some parts of tubules, by how their characteristics change in representatives of different classes of vertebrates in ontogenesis and pathology. Heterogeneity of nephrons is

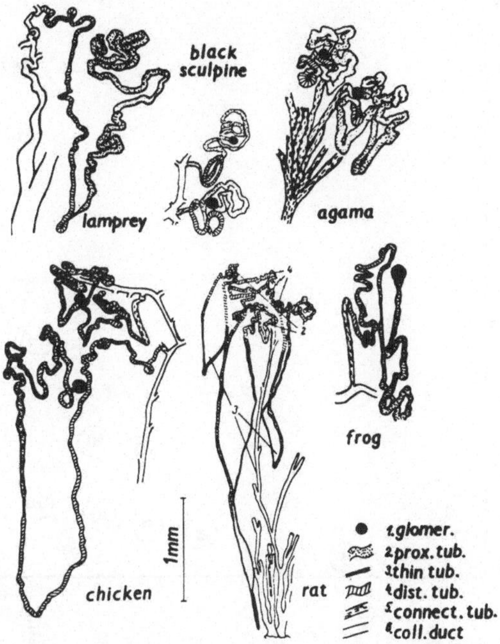

Fig. 4. Microanatomy of nephrons in the kidney of various vertebrates (Goncharevskaya, 1976).

found to be a regular renal feature of representatives of all classes of vertebrates (Fig. 4), its functional role being revealed quite clearly in the case of birds and mammals. In the latter the role of different populations of nephrons is found not to be similar.

Of particular interest are the new data on the role of ADH in increasing heterogeneity of nephrons. In adult Brattleboro rats with inherited diabetes insipidus which can secrete no ADH, superficial nephrons prevail in number over juxtamedullary nephrons. This does not agree with the situation in the case of normal rats. Two-week old Brattleboro rats injected with ADH develop a typical nephron heterogeneity (Trinh-Trang-Tan, Diaz, Grunfeld and Bankir, 1981). This indicates the role of genetic and phenotypical factors in heterogeneity of nephron populations. Variations in nephron heterogeneity are clearly seen in the adaptive evolution of the related forms. When studying the kidney of rodents, we found that the water vole has, in fact, no population of nephrons with a long Henle's loop, while in the desert big gerbil all nephrons have a long thin Henle's loop (Ivanova et al., 1980). In the mammalian kidney with its long thin Henle's loop, osmotic concentration is made possible by accumulation of urea and sodium, in the medulla of the avian kidney and in the outer

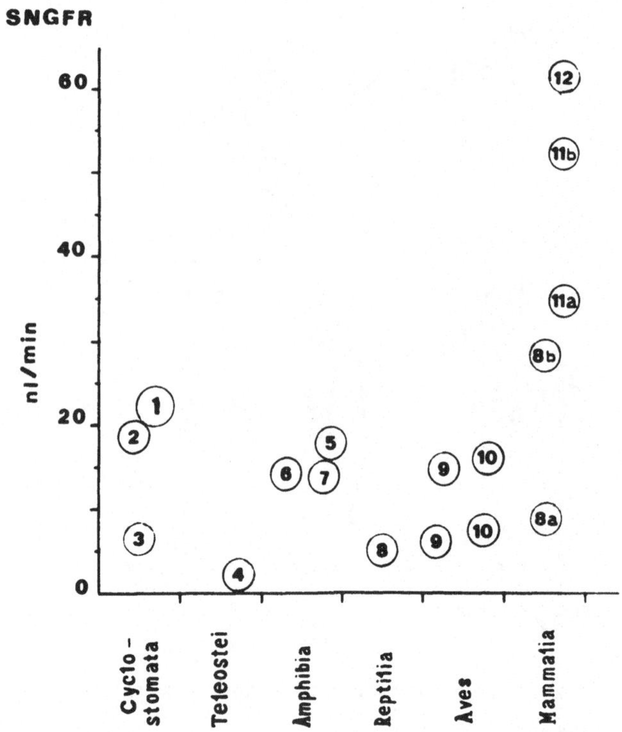

Fig. 5. Single nephron glomerular filtration rate in kidneys of animals belonging to different classes of vertebrates (according to Renkin Gilmore, 1973; Arendhorst, Gottschalk, 1985; Dantzler, 1985).

1. Myxine glutinosa	8. Thamnopis sirtalis
2. Eptatretus stouti	8. a,b - Psammomys obesus
3. Lampetra fluviatilis	9. Callipepla gambelii
4. Salmo gairdneri	10. Surnulus vulgaris
5. Amphiuma meana	11. Rat, a - euvolemia
6. Necturus maculosus	b - volume expansion
7. Rana pipiens	12. Dog

medulla of mammals it is maintained only by sodium chloride. It follows that structural heterogeneity is closely connected with functional and biochemical peculiarities.

Increase of differentiation

The kidney of vertebrates from cyclostomata to mammals, with no exception, has the glomerular apparatus, proximal and distal segments of nephron. There is an increase of differentiation to some other parts of tubular distal segments in warm-blooded animals. To a very great extent, however, there are also differentiated nephrons in elasmobranch fish in whose kidney the nephron loops are likely to be involved in reabsorption of urea. Differentiation is extended both to the structural elements of nephron, namely cells, zones of cellular contacts and to the connective tissue around them, and to the arrangement of tubules and blood vessels. In some cases, e.g. in some forms of teleosts, there is a smaller number of nephron segments and a glomerulus or distal tubule can be absent (Smith, 1951).

Intensification of reabsorption and secretion

One of the characteristic features of evolution of kidney - intensified reabsorption of ultrafiltrate - extends to its functional units, namely nephrons (Fig. 5). In mammals the rate of fluid reabsorption, which is gradually becoming higher in the course of ontogenesis (Horster, 1985), exceeds per unit of tubule length that of the cold-blooded vertebrates (Dantzler, 1985). Quite an opposite picture is observed in pathological cases, e.g. chronic renal insufficiency.

What has been said above is concerned with a total volume of fluids and substances being absorbed from the ultrafiltrate. In case it becomes vitally important for some substances or a group of substances to be excreted (e.g. in teleosts), a special mechanism responsible for intensified secretion of divalent ions is formed.

Formation of morpho-functional complexes

There is a close connection between the functioning of nephrons and the vessel system of kidney, of great importance for some renal functions is the interstitium. Osmotic concentration of urine can be regarded as one of the most important gains of the progressive evolution of kidney. The main structural component of the nephron responsible for this function is the Henle's loop with a vasa recta. Together with the well developed interstitium they form the renal medulla, countercurrent multiplication. A complete picture of evolution of functional units of kidney needs good understanding of how blood and lymph vessels, nerve elements and the connective tissue were interrelated in the development of kidney.

Extension of regulation of functional activity

Further differentiation of nephrons, their close connection with vessels, a higher level of the regulation of function of some nephron parts due to physiologically active substances open the way for kidney to be better adapted to the requirements of the organism. This situation arises when the involvement of distal parts of the nephron in reabsorption of some ions would compensate for the changes in the proximal segment. This is most evident in the case of mammals where a decrease of proximal reabsorption of Na^+ and Cl^- can be followed by an increased compensatory reabsorption of these substances in the ascending part of Henle's loop. It would be quite logical to suppose that what the nephron really owes to evolution, i.e. increased capability of glomerular filtration and proximal reabsorption, is accompanied by the formation in the thick parts of Henle's loop and distal tubules of a very powerful system of reabsorption of some ions, which is capable of absorbing a great variety of substances and compensating for their altered transport in the upper parts of the nephron.

EVOLUTION OF FUNCTION OF NEPHRON CELLS

The comparison of structural and functional characteristics of cells from different parts of the nephron in vertebrates in adult animals, and in their ontogenesis, gives grounds for some suggestions concerning the principles of their evolution, and in this particular case allows the comparison of the homologous structures in representatives of the same phylum.

Of great importance for evolution of the nephron cell as well as other osmoregulatory and excretory organs is specialization of its apical and basal parts, which is connected with the formation of asymmetrical cell, and in this way is responsible for a directional transport. The ion channels and pumps on the plasma membranes of the opposite parts of the cell are arranged so as to make it possible for the substances in the process of reabsorption to be absorbed from the tubular lumen and be reabsorbed into blood; in case of secretion the process goes in quite the opposite direction. In the course of evolution, as a result of cell specialization, the molecular mechanisms of transport are arranged so that they appear to be located in one of the plasma membranes and are almost absent in the other. This is true with respect to the cells which are responsible for the directional transport of glucose and different amino acids, for protein and ion absorption. Besides endogenous substances, specialization of transport systems can also be responsible for secretion from blood into the tubule of foreign organic acids and bases.

Increase of some forms of the initial activity of cell provides the basis for its modification in the course of evolution of kidney. Thus, for its function to be performed, for the components to be built and to be provided with enough energy each cell needs amino acids, glucose, nonesterified fatty acids which it takes from the extracellular fluid. In the majority of tubule cells this process is similar to that in other cells of the organism, but there is one nephron segment (proximal) where this form of activity is transferred into a specialized function responsible for reabsorption of glucose, amino acids, etc. This intensified cellular activity reaches its peak in the kidney of mammals where the rate of glomerular filtration is very high, and so is the rate of reabsorption of glucose, amino acids, vitamins, and other substances.

Compared to lower vertebrates, the nephrons of mammals and birds contain, as a rule, more cells with different structure and function. First of all, this concerns their form, size, ultrastructure, and function. The basis for the above is the altered shape of the plasma and variations in the infoldings of the basolateral membranes. Cells differ also in the number of mitochondria and granuli, and in the ability to perform pinocytosis and exocytosis. In addition, a great contribution to the nephron cell are specialized organelles which cover the tubulocisternal system, the tubular apparatus of the cells involved in secretion of organic acids, and the vacuolar apparatus. The above examples seem to be enough for the conclusion to be made that one very important feature of evolution of any specialized cell is further differentiation of its structure. Though the same ions are absorbed in different parts of the nephron, ionic channels in the apical membranes of their cells are known to vary to a great extent. This follows from the action of ion transport inhibitors. Furosemide on the side of nephron lumen acts as a blocker of Cl^- transport in the thick ascending part of Henle's loop in mammals and in the early segments of the distal tubule in amphibians. Amiloride inhibits Na^+ transport of the end parts of the distal tubule in amphibians and mammals. The fact that furosemide and other diuretics are so effective in the kidney of animals belonging to almost all classes of vertebrates, including aglomerular fishes, speaks in favour of a qualitative molecular similarity of the nephron ion transport in all vertebrates. So far there is no kinetic data whether or not the properties of ion channels undergo any change in the process of kidney evolution; of no less interest would be the answer to the question why the action of such blockers of Na^+ and Cl^- transport as furosemide and amiloride is limited to one part of the nephron only, though both ions are absorbed in all other parts as well. It should be mentioned at this point that with cellular specialization in any organ of a multicellular organism, the evolution of cell would not obligatorily proceed by way of intensification of metabolism. Differentiation of cells in the nephron of mammals, which goes in parallel with the formation of the Henle's loop thin part responsible for counter-current multiplication (Braun, 1985), may serve as an example of the formation of flat epithelial cells not rich in mitochondria, with no brush border and a small number of infoldings in the basolateral membranes. As these cells must function in the kidney medulla where under osmotic concentration the osmotic pressure in some species can exceed 200 atmospheres, the level of glycolysis in the cells is high. This means that progressive evolution of any organ may be accompanied by such changes in some types of its cells that to be performed, new functions would require simplification of the cell structure, oligomerization of their organelles, less intensive metabolism.

Differentiation of cells gives grounds for intensification of their activity. In the kidney of mammals in ontogenesis the surface of basolateral membranes in the cells of a proximal convoluted tubule is increased 2.2-fold for the same cellular volume ($\mu m^2/\mu m^3$), the surface of the luminal membrane changes from 0.08 to 0.18 $\mu m^2/\mu m^3$ (Horster, 1985). Mitochondria grow in volume and are located quite close to the plasma membranes. Similar tendencies are revealed in the cells of the proximal convoluted tubules in vertebrates. Compared to mammals, lamprey, fresh-water fishes, amphibians and reptiles have fewer infoldings in their kidney, their crypts are arranged more chaotically, and their connection with the infoldings of the basal plasma membrane is not so pronounced (Natochin, 1976) (Fig. 6).

These peculiarities in the cellular ultrastructure of the proximal tubule correlate very well with intensification of succinic dehydrogenase activity that takes place in the process of development, intensified reabsorption of Na^+ and other components of the ultrafiltrate.

In recent times in the physiology of the kidney an important role in the process of absorption of substances has been ascribed to specialization

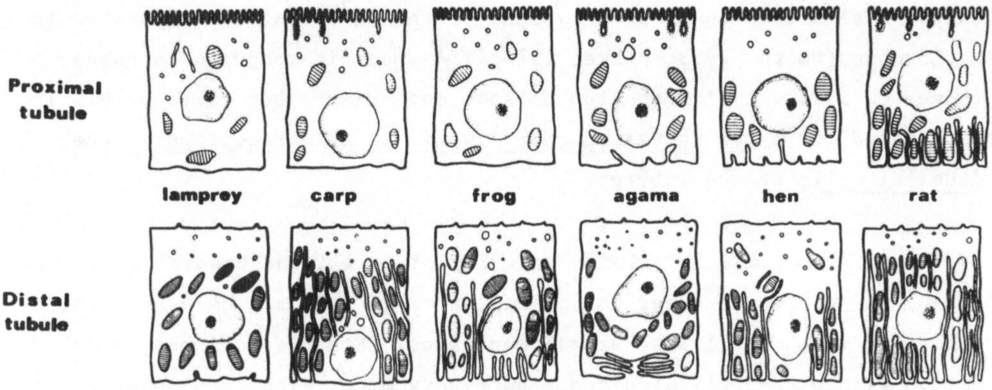

Fig. 6.

of cellular contact areas, paracellular transport. Specialization of cellular contact areas is the main factor to determine how many and which substances would move into blood from the tubule lumen by gradient. Electrical resistance of the proximal tubule wall of the rat kidney is about 10 times lower than in amphibians (Boulpaep, 1979), which correlates quite well with intensified reabsorption of fluid in this part of nephron in the animals under study. Successive stages of differentiation of the junctional complexes are also found in the proximal renal tubule of mammals in post-natal development. It follows that variability of some components of intercellular junctions, including the area of tight junctions, desmosomes and gap junctions, their structure and properties responsible for permeability of nonelectrolytes, cations, anions, water, specialization of different parts of tubules and their change both in onto- and phylogenesis of vertebrates show the very important role they have in evolution of the kidney.

The organism does not merely need some particular forms of cellular activity, but it seeks for the ways to regulate this activity, depending on its own state and, as a result, to maintain stability of the inner medium. Efferent nerves, different hormones, and other physiologically active substances are involved in regulation of cellular activity of nephron. In spite of the fact that many of the mediators and hormones of higher animals have been found in living animals with most primitive organization, according to the available data in the course of evolution of the kidney the number and kind of regulatory influences both grow and change. This refers to the effect of arginine-vasotocin on tubular permeability to water, this effect is found to take place beginning with the amphibian kidney. A similar situation is typical of the Ca^{2+}-reabsorbing system to parathormone (Smith, 1953; Dacke, 1979). A specific response to these hormones is a result of formation of some mechanisms, but first of all it is accounted for by an increased quantity of specific receptors in the epithelial cells of the tubules.

Regulation of the functional activity of cell is made possible by the receptor of peptide hormones and of mediators of the nervous system which are located on the cell basolateral membranes. Steroid hormones which function within the cell also contribute very much. The data we obtained in stimulating Na reabsorption by Co^{2+} from the tubule lumen (Fig. 1), i.e. from the outer surface of the apical membrane, shows this way of regulation to be quite possible in natural conditions (Natochin et al., 1986). In the kidney of mammals, as is known, the concentration of non-reabsorbed substances is greatly increased, which can be accounted for by a big volume of fluid being absorbed by distal tubules. This gives a local (in the nephron lumen) rise in concentration of substances capable of acting on the nephron cell from the side of the luminal membrane, but having no effect on the other cells. It cannot be excluded at the same time that some endogenous physiologically active substances, with a number of trace elements among them, would exert action on the nephron cells. This is the exact way to interfere with the action on the kidney when diuretics such as furosemide, amiloride and some others are applied in clinics. According to the available data however, concerning different sensitivity of the kidney cells to hormones in the process of their development, it is not possible yet to give quantitative characteristics of this discrepancy.

Of great interest also is the problem of secondary messengers which are responsible for functional adaptation in the nephron cells. There are data on the role of cyclic nucleotides, inositoltriphosphate, calcium in the intracellular reactions under the action of vasopressin, parathormone, mediators of regulation of renal function. Data are being accumulated on the role of these substances in the regulating function of kidney cells, and on the mechanism of their action; so far however, there are no grounds for evaluating the contribution of secondary messengers to the adaptive reactions on different stages of kidney evolution.

In addition to being capable of changing ion and water permeability of the epithelial cell plasma membrane, cAMP has also an effect on the permeability of tight junction and regulates the transepithelial resistance

(Duffey, Hainau, Ho and Bentzel, 1980). These data present a great interest, for variations in the properties of cellular contacts in different parts of the nephron, and a likely possibility of these junctions in the nephron wall to be regulated, may be regarded as important factors in the evolution of the kidney and its adaptive reactions.

Thus, evolution of structure and function will undoubtedly look even more interdependent if approached on the molecular level, and the ways of transformation of functions of different organs and systems in the course of evolution should be accounted for on the basis of regularities underlying evolution of function of some particular specialized cell. On the cellular level, however, to a greater degree can be revealed similarity rather than difference among animals under study, be it concerned with individual development, animals belonging to different classes, or closely related groups which have been adapted to different environments. On the cellular level potency of the kidney can be quite clearly seen, and the role of interrelation of cells comes to be understandable, as well as populations of nephrons, zones of kidney in its progressive evolution.

What has been said above allows the conclusion that however far from each other forms and ways of adaptation to different conditions may be, the latter is governed by a very strict logic, a limited number of principles which underlie all possible variations in the water-salt metabolism, structure and function of the kidney. These principles were followed by a growing complexity of organisms as evolution proceeded. This is the view shared by A. Einstein who strongly believed that evolution proceeds in the direction of increasing simplicity of logical bases.

In this review an attempt has been undertaken to trace some of the main lines of evolution of renal function. We wish to express hope that further progress in the study of the physiology of water-salt metabolism and physiology of the kidney not only will prove our suggestions to be right, but will give grounds to formulate the principal tendencies in the evolution of this function and answer the question what made it go just this way.

REFERENCES

Arendshorst, W.J. and Gottschalk, C.W. (1985) Glomerular ultrafiltration dynamics: historical perspective. Am. J. Physiol. 248, F163-F174.

Brenner, B.M. and Rector, F.C. (Eds.) (1976) The Kidney. Saunders: Philadelphia

Bentley, P.J. (1971) Endocrines and Osmoregulation. A Comparative Account of the Regulation of Water and Salt in Vertebrates. Springer: Berlin.

Boulpaep, E.L. (1979) Electrophysiology of the kidney. In: Transport Organs. Giebisch, G., ed., Springer: Berlin, pp 97-144

Braun, E.J. (1985) Comparative aspects of the urinary concentrating process. Renal Physiol. 8, 249-260.

Burg, M., Stoner, L., Cardinal, J. and Green, N. (1973) Furosemide effect on isolated perfused tubules. Am. J. Physiol. 225, 119-124.

Cliff, W.H. and Beyenbach, K.W. (1986) Transepithelial MgCl2 secretion and NaCl secretion drive fluid secretion in flounder renal proximal tubules. Renal Physiol. 9, 75.

Dacke, C.G. (1979) Calcium Regulation in Sub-Mammalian Vertebrates. Academic Press: London.

Dantzler, W.H. (1985) Comparative aspects of renal function. In: The Kidney Physiology and Pathophysiology. Seldin, W.D. and Giebisch, G., eds., Raven Press: New York, pp 333-364

Dicker, S.E. (1970) Mechanisms of Urine Concentration and Dilution in Mammals. Arnold: London.

Dogel, V.A. (1954) Oligomerization of Homologous Organs as one of the Main Ways of Animal Evolution. University Press: Leningrad.

Dohrn, A. (1875) Der Ursprung der Wirbelthiere und das Prinzip des Functionswechsels. Engelmann: Leipzig.

Duffey, M.E., Hainau, B., Ho, S. and Bentzel, C.J. (1981) Regulation of epithelial tight junction permeability of cyclic AMP. Nature 294, 451-453.

Ginetsinsky, A.G. (1961) On evolution of function and functional evolution. USSR Acad. Sci., Moscow.

Goncharevskaya, O.A. (1976) Nephron organization of the kidney and proximal reabsorption in vertebrates (microdissection and micropuncture studies). J. Evol. Biochem. Physiol. 12, 113-119.

Goncharevskaya, O.A., Monin Yu.G. and Natochin Yu.V. (1986) The influence of furosemide and Co^{2+} on electrolyte and water transport in newt distal tubule and frog skin. Pflügers Arch. 406, 557-562.

Haeckel, E. (1940) The main principle of organic development. In: F. Muller, E. Haeckel, The Principle Law of Nature. Collected works, pp. 169-186, USSR Acad. Sci., Moscow.

Horster, M. (1985) Ontogenetic processes in nephron epithelia: function, enzymes, and structures. In: The Kidney, Physiology and

Pathophysiology. Seldin, D.W. and Giebisch, G., eds., Raven Press: New York, pp 317-332

Ivanova, L.N., Natochin, Yu.V., Serebryakov, E.P., Goncharevskaya, O.A., Knyazkova, L.G., Lavrova, E.A., Nasledova, N.I., Pechurkina, N.I., Podsekaeva, G.V. and Shakhmatova, E.I. (1980) Comparative study of the concentrating mechanism in the kidney of the big gerbil (Rhombomys opimus L.) and the water vole (Arvicola terrestris L.). Comp. Biochem. Physiol. 662, 499-505.

Kirksey, A., Pike, L. and Callahan, J.A. (1962) Some effects of high and low sodium intakes during pregnancy in the rat. II. Electrolyte concentrations of maternal plasma, muscle, bone and brain and of placenta, amniotic fluid, fetal plasma and total fetus in normal pregnancy. J. Nutrition 77, 43-51.

Koshtoyants, Kh.S. (1932) Some Problems of Physiology and Theory of Development. Medical Publishers: Moscow.

Lumbers, E.R. and Stevens, A.D. (1983) Changes in fetal renal function in response to infusions of a hyperosmotic solution of mannitol to the ewe. J. Physiol. (Lond.) 343, 439-446.

Natochin, Yu.V. (1976) The Kidney: Regulation of Ionic Balance. Nauka: Leningrad.

Natochin, Yu.V. (1982) Mechanism of drugs action on ion and water transport in renal tubular cells. Progr. in Drug Res. 26, 87-142.

Natochin, Yu.V., Dolgopolova, G.V., Lavrova, E.A., Shakhmatova, E.I., Serova, L.V., Denisova, L.A. and Iliushko, N.A. (1984) Dynamics of water, electrolytes and trace elements content in placenta, foetus and some female organs during pregnancy. Sechenov Physiol. J. 70, 206-212.

Natochin, Yu.V., Goncharevskaya, O.A., Monin, Yu.G. and Shakhmatova, E.I. (1986) Regulation of Na and water transport in tight amphibian epithelia by extracellular apical Ca^{2+} and Co^{2+}. Renal Physiol. 9, 99-100.

Natochin, Yu.V. and Gusev, G.P. (1970) The coupling of magnesium secretion and sodium reabsorption in the kidney of teleost. Comp. Biochem. Physiol. 37, 107-111.

Natochin, Yu.V., Ivanova, L.N., Serebryakov, E.P., Podsekaeva, G.V., Pechurkina, N.I., Lavrova, E.A., Lavrinenko, V.A., Melidi, N.N., Nasledova, N.I. and Shakhmatova, E.I. (1982) Volume- and ion-regulating renal functions in the big gerbil (Rhombomys opimus L.) and the water vole (Arvicola terrestris L.). Comp. Biochem. Physiol. 72A, 535-539.

Natochin, Yu.V., Kozyrevskaya, G.I. and Grigoryev, A.I. (1975) Functional tests in the study of water-salt metabolism and renal function in cosmonauts. Acta astron. 2, 175-188.

Natochin, Yu.V., Lavrova, E.A. and Gusev, G.P. (1971) A study of ionic, osmotic and volume regulation in the medusa Tiaropsis multicirrata. J. Evol. Biochem. Physiol. 7, 138-144.

Natochin, Yu.V., Lukianenko, V.I., Lavrova, E.A. and Metallov, G.F. (1976) Electrolyte composition of embryo and larva of the Russian sturgeon in the process of development. Comp. Biochem. Physiol. 55A, 57-59.

Orbeli, L.A. (1933) On the evolutionary principle in physiology. Priroda, 3-4, 77-86.

Orbeli, L.A. (1961) Collected works. USSR Acad. Sci.: Moscow

Renfro, J.L. and Shustock, E. (1985) Peritubular uptake and brush border transport of ^{28}Mg by flounder renal tubules. Am. J. Physiol. 249, F497-F506.

Renkin, E.M. and Gilmore, J.P. (1973) Glomerular filtration. In: Handb. Physiol., Sect. 8. Renal Physiology Am. Physiol. Soc.: Washington, pp 185-248.

Severtsov, A.N. (1914) Modern Problems of Evolutionary Theory. Moscow.

Smith, H.W. (1951) The Kidney. Structure and Function in Health and Disease. Oxford: New York.

Smith, H. (1953) From Fish to Philosopher. Little, Brown: Boston.

Trinh-Trang-Tan, M.M., Diaz, M., Grunfeld, J.P. and Bankir, L. (1981) ADH-dependent nephron heterogeneity in rats with hereditary hypothalamic diabetes insipidus. Am. J. Physiol. 240, F372-F380.

MECHANISMS OF WATER TRANSPORT ACROSS TUBULAR EPITHELIA: ROUTES FOR MOVEMENT

Guillermo Whittembury and Paola Carpi-Medina

Centro de Bíofisica y Bíoquimica
Instituto Venezolano de Investigaciones Científicas
P.O. Box 21827
Caracas 1020A, Venezuela

ABSTRACT

Continuous water pathways pierce both apical and basolateral cell membranes of the proximal kidney tubule to be used by water during osmotic equilibration between cells and luminal and peritubular media, because (a) the water osmotic permeability coefficient of apical and basolateral plasma membranes, P_{os}^{ca} and P_{os}^{cb}, respectively is high; (b) their activation energy, E_a, is as that of free water movement; (c) the sulfhydryl reagent pCMBS inhibits markedly (but reversibly) P_{os}^{ca} and P_{os}^{cb} increasing their E_a to values similar to those observed in lipid bilayers without pores; (d) measurements of P_d, the water diffusive permeability coefficient using proton relaxation NMR indicate that $(P_{os}^{ca} + P_{os}^{cb})/P_d$ is near 23 in controls and 3 with pCMBS. Scatchard or Hill plots of the degree of inhibition of P_{os}^{cb} and of P_d as a function of [pCMBS] gives values for N of 4 and 2, respectively, indicating that more than 1 pCMBS molecule binds to each water pathway to block it. In addition to these transcellular pathways, the following observations indicate that paracellular pathways for water flow exist in leaky epithelia: (a) some large extracellular solutes are dragged by water in four leaky epithelia: gall bladder, *Necturus* proximal tubule, rat proximal tubule and *Rhodnius* malpighian tubule; (b) the transcellular water osmotic permeability coefficient is smaller than the transepithelial (P_{os}^{te}) values measured in the rabbit proximal straight tubule; (c) pCMBS inhibits P_{os}^{te} by 60% under conditions in which P_{os}^{cb} is inhibited much more. This requires a significant paracellular permeability.

INTRODUCTION

The distal segments of the kidney pose no problem to the understanding of water transport. Under physiological conditions their paracellular permeability is zero. Therefore ions and water must move through the cells. They transport an hyperosmotic solution, their water permeability is small and large osmotic gradients exist across them. Thus water movement across these epithelia is thought to be due to osmotic forces. On the other hand, leaky epithelia absorb or secrete nearly isosmotic solutions (Shafer, 1984). Although cells are kept together by junctional complexes, a paracellular pathway for ion flow has been shown to exist in these epithelia, based on three independent observations. First the transepithelial resistance is much lower than the cell membrane resistance indicating that there exists a low paracellular resistance which effectively constitutes a pathway for current and therefore for ion flow across these epithelia. Then the demonstration that the electrondense ion La^{3+} crosses these epithelia at the level of the junctional complexes (hydrated La^{3+} is at least as big as Na^{+}) (Whittembury and Rawlins, 1971). Finally the direct demonstration that the highest current flow occurs indeed where cells are joined together (Frömter, 1972; see also Berry, 1983; Weinstein and Windhager, 1985; Windhager, 1979 for references).

In these tissues it has been shown that water moves passively, secondary to ion flow (Weinstein and Windhager, 1985; Windhager, 1979); but the pathway that water may follow is still under controversy (Berry, 1983). This secondary movement of water may only occur across the cells, or part of the water may also move between cells (Berry, 1983; Whittembury, Verde-Martínez, Linares and Paz-Aliaga, 1980). We will review here first the nature of the transcellular pathways; and the evidence showing that water flows between cells via the junctional complexes in addition to the known transcellular water flow.

PATHWAYS FOR WATER FLOW ACROSS THE APICAL AND THE BASOLATERAL CELL MEMBRANE IN THE RABBIT PROXIMAL STRAIGHT TUBULE (PST)

Water may cross the cell membrane by diffusion through the lipid matrix, and/or by flow through continuous pathways that may be opened at a given moment piercing the cell membrane from cytoplasm to extracellular space. Information concerning the nature of these pathways in kidney cells may be obtained by taking advantage of previous experience with artificial lipid bilayers and with red blood cells (Finkelstein, 1984; Moura, Macey, Chien, Karan and Santos, 1984; Solomon, Chasan, Dix, Lukacovic, Toon and Verkman, 1983). (A) In artificial bilayers doped with pore-forming antibiotics and in untreated red blood cells, (1) the water osmotic permeability coefficient, P_{os}, is high; (2) its Arrhenius activation energy, E_a, is as that of free water movement, i.e. 4.2 Kcal/mole; (3) the water diffusive permeability, P_d, is much lower than P_{os}, so that the ratio of P_{os}/P_d is clearly greater than 1. (B) In non-doped artificial bilayers and in red blood cells which have been treated with the sulfhydryl reagent para-chloromercuribenzenesulfonate, pCMBS, (1) P_{os} is low; (2) its E_a is higher than 10 Kcal/mole; (3) P_d has values similar to P_{os} so that P_{os}/P_d is about 1. These results are taken to indicate that in condition A there are continuous pathways piercing the artificial membranes and the red cell membranes. Thus P_{os} is high, water flow is relatively free of great interactions with the membrane fabric, therefore E_a is low. Current views hold that the ratio of P_{os}/P_d indicates the number of water molecules occupying the pore in a single file (rather than the pore radius (Levitt, 1984)). In the case of the non-doped membranes and of the red cells treated with pCMBS, water crosses these membranes essentially by diffusion through the membrane fabric; thus the P_{os} observed is mainly due to diffusion secondary to the gradient in water activity created by the osmotic difference. The degree of interaction between water and the lipid membrane fabric is high, thus E_a is high.

Water osmotic permeabilities

To explore whether the apical and the basolateral cell membranes of the proximal tubule adhere to any of these alternatives, we have measured in the proximal straight tubules, PST, the water osmotic permeabilities, P_{os}^{ca} and P_{os}^{cb} across the apical and across the contraluminal cell membrane and their temperature dependence in order to obtain the Arrhenius E_a, under control conditions. Then the tubules were treated with pCMBS and measurements of P_{os}^{ca}, of P_{os}^{cb} and of E_a were repeated. For this purpose PST were isolated from the rabbit kidney, they were mounted, between pipettes, in a chamber that could be thermostatted from 4 to 37°C. The tubules were observed with an inverted microscope and a TV camera. According to whether the measurement was of P_{os}^{ca} or of P_{os}^{cb}, changes in lumen diameter and cell height (González, Carpi-Medina, Linares and Whittembury, 1984) and of cell height alone (Carpi-Medina, 1986; Carpi-Medina, Lindemann, González and Whittembury, 1984) were recorded with a special processor which continuously records these parameters vs. time with precision better than 0.03 μm and 17 msec (Carpi-Medina et al., 1984). P_{os}^{ca} and P_{os}^{cb} were estimated from the rate of change of lumen diameter and cell height with time per unit osmotic concentration difference. Osmotic steps were set up across the basolateral membrane in less than 100 msec. After control measurements of P_{os}^{ca} and of P_{os}^{cb} at two temperatures, 2.5 mM pCMBS was added and P_{os}^{ca} and P_{os}^{cb} were determined again at the same temperatures. Control values for P_{os}^{ca} and P_{os}^{cb} were, respectively 46 ± 7 and 90 ± 13 in units of 10^{-4} cm^3/s.Osmolar./cm^2 of basal membranes. E_a for P_{os}^{cb} was 3.2 ± 1.4 Kcal/mole. In the presence of pCMBS, P_{os}^{ca} and P_{os}^{cb} decreased to 23 ± 15 and $26 \pm 17\%$ of the control figures, respectively, and E_a for P_{os}^{cb} was 9.2 ± 2.2 Kcal/mole. E_a for P_{os}^{ca} could not be estimated due to difficulties in keeping a viable perfused preparation for the long time required to perform the experiment (Carpi-Medina, 1986). Thus, in the control experiments P_{os}^{ca} and P_{os}^{cb} are high and the E_a is not different from a value of 4.2 expected from free movement. Treatment with pCMBS markedly lowers P_{os}^{ca} (see also Pratz, Ripoche and Corman, 1986) and P_{os}^{cb}, and increases E_a. It must be pointed out that this effect of pCMBS readily reverses upon addition of dithiothreitol (DTT) to the preparation (Carpi-Medina, 1986; Whittembury et al., 1984). If one bears in mind the possibility that during measurement of P_{os}^{ca} not only the apical, but also part of the basolateral membrane could have contributed to osmotic equilibration through the junctions, the values obtained for P_{os}^{ca} are probably somewhat of an overestimate of the true permeability. The same argument may hold for the values of P_{os}^{cb}.

Water diffusive permeability (P_d)

P_d was measured in isolated proximal tubule cells using proton NMR (Carpi-Medina, León, Espidel and Whittembury, 1986). Cells isolated by mechanical means were incubated in artificial plasma doped with $MnCl_2$ (Steward and Garson, 1985). The relaxation time T_2 was measured. Mathematical analysis of the relaxation curves used standard equations for water exchange between cells and plasma. P_d of the whole cell membrane at 25°C is 197 ± 17 μm/sec or 10^{-4} cm^3/sec.cm^2 cell surface area (uncorrected for infoldings) (see also Steward and Garson, 1985). Using the values given above for the apical and basolateral osmotic permeabilities and morphological considerations, $P_{os}/P_d = 23$. P_d decreases in a dose dependent manner with pCMBS to 116 ± 11 μm/sec. Using P_{os}^{cb} values with pCMBS P_{os}/P_d decreases to 3. The E_a for P_d is 5.2 ± 1.0 (control) and 9.1 $\pm$ 2.2 (pCMBS) Kcal/mole. All these experimental results indicate, by analogy with the artificial membranes and red cells, that continuous pathways must be piercing the membrane at a given moment under control conditions. From P_{os}/P_d either the pathways have a radius of 15Å (unreasonable since $\sigma = 1$ for molecules like raffinose, with a molecular radius of about 6Å) or several water molecules file singly through the pathways. The action of pCMBS to change the permeability characteristics of the channel and the reversibility of the action of pCMBS with DTT indicates that sulfhydryl groups (i.e. proteins) are involved in the stability and structure of these water channels, which must be upset by addition of pCMBS so that E_a increases to high values (not different from 10 Kcal/mole). Interestingly, when Hill or Scatchard plots of the degree of inhibition of the osmotic and of the diffusive permeabilities are plotted as a function of [pCMBS], values for N of 4 and 2 are obtained respectively, to be compared with N = 1 for red blood cells (Moura et al., 1984). These values would indicate that in kidney cells more than one, probably 2-4, pCMBS molecules are needed to close each water pathway, while one would suffice in the red cell. Band 3 proteins are being associated with the water channel in the red cell (Berry, 1985; Solomon et al., 1983). It remains an open question whether this is also the case of the cell membrane of the proximal tubule in view of these differing values for N. Alternatively, dimers or tetramers of the protein involved in pore formation in the red cell could be present in the cell membrane of the proximal tubule.

PATHWAYS FOR WATER FLOW ACROSS LEAKY EPITHELIA

There is little doubt that the lateral intercellular spaces facilitate water flow during absorption or secretion in leaky epithelia, because they have actually been seen to dilate during this process (Spring and Hope, 1979). The question is whether the water which flows through them comes from transcellular flow or from transjunctional flow, i.e. from paracellular flow. The ideal experiment would consist in tightly locating one micropipette in front of the cell, and a second one just where cells meet. Comparison of the rates of movement of a physiological solution within the pipettes would indicate their relative importance. Since this experiment is technically difficult at present, two indirect approaches have been sought to look for an answer to this problem. The first studies the solvent drag of large extracellular molecules. The second compares the transepithelial and transcellular water osmotic permeability coefficients.

Paracellular flow in leaky epithelia as explored by studying solvent drag of large extracellular nonelectrolytes

Large extracellular solutes - which are known not to enter the cells and therefore to stay exclusively as extracellular markers - are added to the side of the preparation from which the flow originates and their net flow (J_s) towards the other side is explored as a function of the net volume flow across the preparation (J_v) which is allowed to vary.

J_s is made up of two terms: J_s^d a diffusive, and J_s^c a convective term. $J_s^d = P_s.\Delta C_s$, and $J_s^c = (1-\sigma)\ C_s^*\ J_v$. where P_s is the solute permeability of the membrane, σ the reflection coefficient of the membrane to s, C^* the average solute concentration in the membrane.

$$J_s = P_s.\Delta C_s + (1-\sigma)C_s^* J_v \qquad (1)$$

Therefore a plot of J_s as a function of J_v will have a function of the diffusive term as intercept and the slope $(1-\sigma)C_s^*$ will be different from zero if the convective term is important.

At first sight ΔC_s seems independent of J_v. However, because of unstirred layer effects the volume flow piles up solutes on the side of the membrane from which J_v originates, and washes them away from the other side. Therefore ΔC_s across the membrane proper is indeed a function of J_v and is larger than the ΔC_s originally set up between the two solutions. In consequence, before concluding that J_s is a function of J_v, one must subtract the diffusive term J_s^d taking into account whatever relationship with J_v may exist due to the unstirred layer effects (Eqn. 12 in Andreoli, Schafer and Troutman, 1981; Eqn. 83 in Barry and Diamond, 1984). The observation of a significant relationship between J_s and J_v, after correction for unstirred layer effects, explores J_s^c alone, and therefore it shows the presence of "true" solvent drag which indicates that there exists interaction between solutes and water within the pathway and, therefore, that water and solutes share a common pathway which would be extracellular if the solutes are true extracellular markers.

Such experiments have been performed in the gall bladder (Hill and Hill, 1978; Whittembury et al., 1980). It was found that sucrose and inulin did show a true solvent drag relation with J_v while the small relationship observed between J_s for dextran, haemoglobin and albumin and

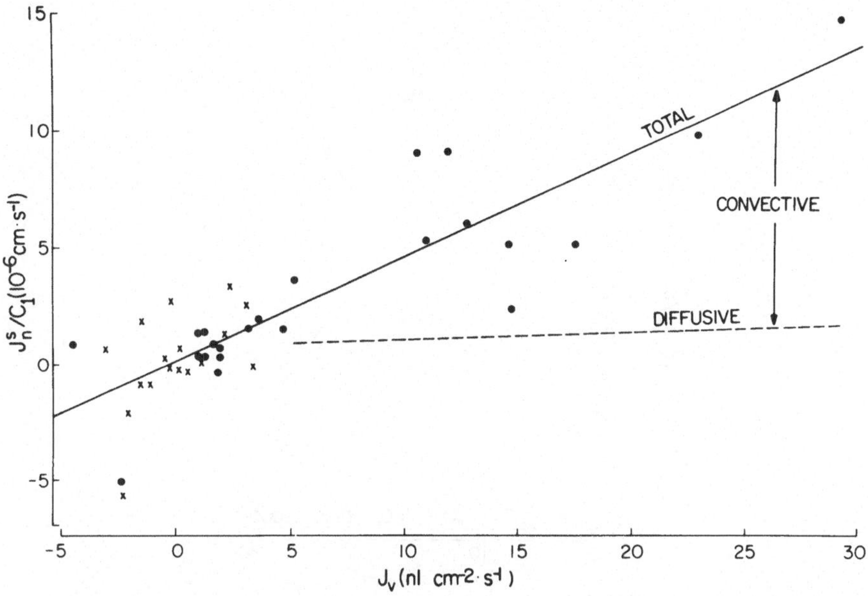

Fig. 1. Net sucrose flux in the absorptive direction, i.e. out of the rat proximal tubule (J_s) per unit average lumen concentration (C_1) examined as a function of the net absorptive volume flow (J_v). (•) represent spontaneous variations in J_v obtained by varying either the length of tubule under observation or the rate of tubular lumen perfusion. (x) corresponds to isosmotic perfusion solutions containing 70 mM raffinose which reduce the rate of net absorption. The full line is the regression line best fitting the experimental points. The dashed line corresponds to the diffusive flow calculated considering unstirred layer effects (Eqn. 12 from Andreoli, Schafer and Troutman, 1981; Eqn. 83 from Barry and Diamond, 1984). It may be noticed that the diffusive flow is slightly dependent on J_v. The difference between the full line (total J_s/C_1) and the line representing the fraction of J_s that is calculated to move exclusively by diffusion corresponds to the flow of sucrose that moves exclusively by convection (from Naftalin and Tripathi, 1986 and G. Whittembury, G. Malnic, M. Mello-Aires and C. Amorena, unpublished).

J_v was due to unstirred layer effects. Similar explorations have been performed in Necturus proximal tubule (Berry and Boulpaep, 1975; Whittembury et al., 1980 and references therein). Flows of mannitol and sucrose have been shown to be truly related to J_v (see Fig. 7 of Whittembury et al., 1980). Fig. 1 shows similar experiments performed in

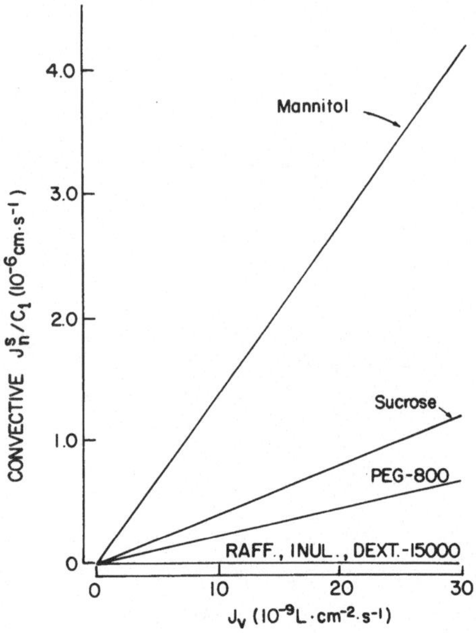

Fig. 2. Net solute flux in the secretory direction, i.e. into the tubular lumen (J_s) per unit bath concentration (C_1), exclusively due to convection, as a function of the net secretory volume flow (J_v) in the isosmotic transporting (distal) segment of the Malpighian tubules of Rhodnius prolixus. The net diffusive flow, which takes into account unstirred layer effects has already been subtracted using Eqn. 12 from Andreoli et al., 1981 and Eqn. 83 from Barry and Diamond, 1984. For the sake of clarity the experimental points have been suppressed (from Whittembury et al., 1986).

the rat proximal tubule using microperfusion (Whittembury, Malnic, Mellow-Aires and Amorena, 1981). Variations in J_v were achieved by various means. Notice that the estimated diffusive flow is only a small fraction of the total flow of sucrose observed. The convective flow is important, leading to the conclusion that sucrose is dragged by solvent during spontaneous absorption in the rat proximal tubular lumen.

Fig. 2 summarizes results of similar experiments performed during secretion in the isosmotic transporting portion (distal segment) of the malpighian tubules from the insect Rhodnius prolixus (Whittembury, Biondi, Pas-Aliaga, Linares, Parthe and Linares, 1986). This preparation was chosen because it secretes (rather than absorbs) an "isosmotic" fluid; the rate of secretion can be varied at will by addition of given doses of serotonin, and the geometrical characteristics of the paracellular pathway are different from the three epithelia studied above. In addition, since the tubules can be extracted from the animal without extracellular material, the spaces of distribution of the non-electrolytes under study can be readily evaluated. Fig. 2 shows only the convective flow of several nonelectrolytes (the diffusive flow has already been subtracted from the total flow). Mannitol shows the highest dependence with J_v. This cannot be judged as due to solute entrainment in the paracellular pathway, since mannitol distributes itself in 90% of the cell water. In the case of sucrose and PEG (polyethylene glycol 800), the slope relating J_s with J_v is clearly significant. These solutes distribute in about 5% of the tubule water, as is the case with raffinose, inulin and dextran MW 15-18000; however the last three substances are not dragged by water during secretion in this preparation. Interestingly sucrose has a molecular radius of 4.5 Å (being quasi-spherical), while PEG is a rod-like molecule with the small diameter of 3.4 Å and a length of 18.5 Å. Raffinose is again quasi-spherical with a diameter 5.9 Å and inulin and dextran are larger. One could therefore deduce that the pathway through which water entrains sucrose and PEG must have dimensions near or smaller than the raffinose diameter, as illustrated in Fig. 3 (Whittembury et al., 1986). Solvent drag of large extracellular solutes has also been observed in the salivary gland (Case, Cook, Hunter, Steward and Young, 1985). Solvent drag of ions has been suggested for the rat proximal tubule (Frömter, 1970,1974).

Pathways for water flow as deduced from P_{os} values

We assume a simple model in which the reciprocals of water osmotic permeabilities are treated as resistances in a circuit. Apical and basolateral cell membrane water osmotic permeabilities (P_{os}^{ca} and P_{os}^{cb}, respectively) would be in series forming $P_{os}^{c} = P_{os}^{ca} . P_{os}^{cb} / (P_{os}^{ca} + P_{os}^{cb})$, the transcellular permeability. The paracellular permeability (P_{os}^{p})would be in parallel with the transcellular one to form the transepithelial permeability, $P_{os}^{te} = P_{os}^{p} + P_{os}^{c}$. Therefore if this simplified model holds, measurements of the transepithelial permeability should equal the transcellular permeability if the paracellular permeability (P_{os}^{p}) is negligible. This would clearly not call for any paracellular water flow. On the other hand, if P_{os}^{te} is larger than P_{os}^{c}, one is bound to conclude that a signficant paracellular water permeability and therefore a paracellular water flow must exist. The value for P_{os}^{cb} of 90 in units of 10^{-4} cm^3/s. Osmolar. / cm^2 of basement membrane (Carpi-Medina, 1986; Carpi-Medina et al., 1984), combined with the value for P_{os}^{ca} of 46 (same units) yields a

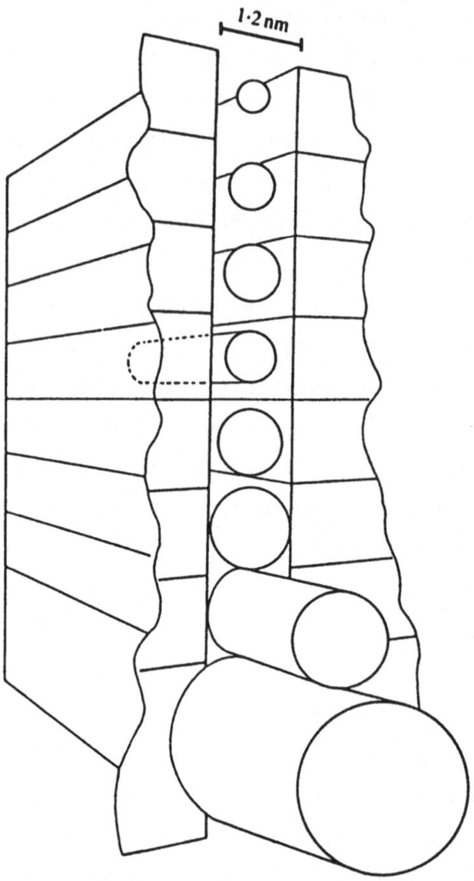

Fig. 3. A 1.2 nm wide slit is compared with the molecular dimensions of the probing molecules. U, E, M, G, PEG, R, I and D stand for urea, erythritol, mannitol, L-glucose, Polyethylene glycol M_r 800, sucrose, raffinose, inulin and dextran M_r 15000-18000. U, E, M, H, S, R, are quasispherical; PEG, I and D are rod-like molecules with major-to-minor axis ratios of about 5, 15-18 and 21, respectively. They are shown with the end of the molecules facing the slit (from Whittembury et al., 1986).

transcellular P^{c}_{os} of 30 $\pm$ 7 (range 16-44) (same units), which is smaller than the transepithelial values P^{te}_{os} of 77 $\pm$ 11 obtained using a fast method to record volume changes (Carpi-Medina, 1986). These values are adjusted to 25°C. In the rabbit PST the measured published value is 94 (also at 25°C) (range 40-138) in the same units (Schafer, Patlak, Troutman and Andreoli, 1978). The conclusion must be drawn that a paracellular P^{p}_{os} of 47 (on average) is required for the transepithelial value of 77.

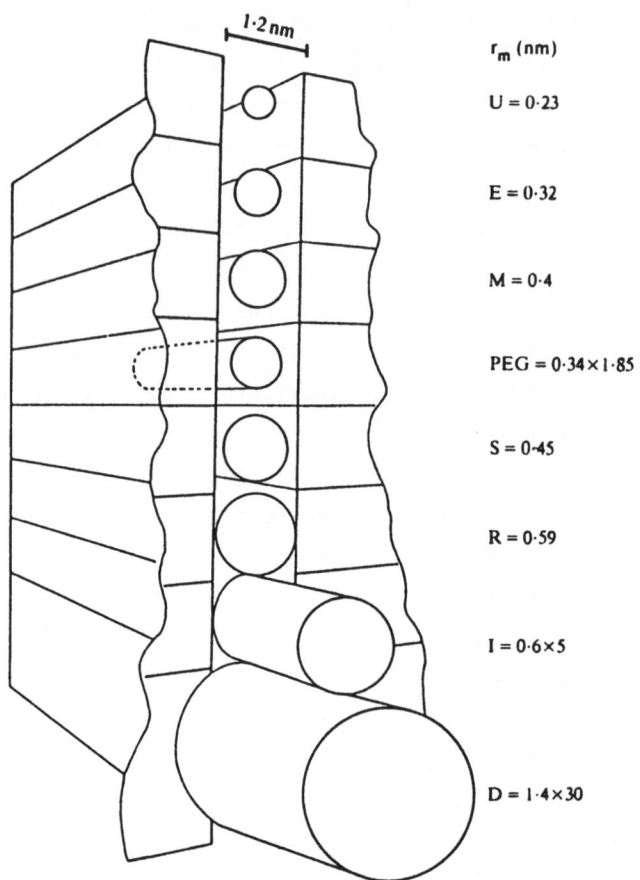

Additional support for the presence of a significant paracellular permeability comes from measurements of P_{os}^{te} under control conditions and in the presence of pCMBS (which inhibits the cellular water osmotic permeability to values not different from zero, see above). These series of experiments yield values for P_{os}^{te} of 59 $\pm$ 11 (control) and 39 $\pm$ 4 (pCMBS) (same units) indicating that the expected total blockage of the transcellular water osmotic permeability described above for the apical and basolateral permeability in the presence of pCMBS is not accompanied by total suppression of the transepithelial one. Opposing views have been presented (Berry, 1985; Berry and Boulpaep, 1975; Preisig and Berry, 1985). Clearly the presence of paracellular water flow bears directly on theories explaining quasi-isosmotic fluid in these epithelia (Berry, 1983; Fromter, 1970,1974; Green and Giebisch, 1984; McLaughlin and Mathias, 1985; Weinstein and Windhager, 1985; Williams, Barfuss and Schafer, 1986; Whittembury, 1985; Whittembury and Hill, 1982; Windhager, 1979).

ACKNOWLEDGEMENTS

Supported in part by grants from Fundacion POLAR, CONICIT, PNUD/UNESCO (RLA 78/024). GW is also a member of Instituto Internacional de Estudios Avanzados, IDEA, Caracas. It is a pleasure to thank Mrs. Mariela Nitta who kindly typed the manuscript. Messrs. José Mora, F. Murillo, A. Cazorla, J. Bigorra and M. Díaz helped in several stages of this research.

REFERENCES

Andreoli, T.E., Schafer, J.A. and Troutman, S.L. (1981) Coupling of solute and solvent flows in porous bilayer membranes. J. Gen. Physiol. 57, 479-493.

Barry, P.H. and Diamond, J.M. (1984) Effects of unstirred layers on membrane phenomena. Physiol. Rev. 64, 763-873.

Berry, C.A. (1983) Water permeability and pathways in the proximal tubule. Am. J. Physiol. 245, F279-F294.

Berry, C.A. (1985) Characteristics of water diffusion in the rabbit proximal convoluted tubule. Am. J. Physiol. 249, F729-F738.

Berry, C.A. and Boulpaep, E.L. (1975) Non electrolyte permeability of the paracellular pathway of *Necturus* proximal tubule. Am. J. Physiol. 228, 581-595.

Carpi-Medina, P. (1986) Estudios de la permeabilidad de los tubulos proximales aislados de riñon de conejo. Tesis Ph. Sc. Instituto

Venezolano de Investigaciones Científicas, IVIC, Caracas.

Carpi-Medina, P., León, V., Espidel, J. and Whittembury, G. (1986) Water diffusive permeability (P_d) of isolated proximal tubule kidney cells measured with proton NMR. 19th Ann. Meeting Am. Soc. Nephrol. 1986.

Carpi-Medina, P., Lindemann, B., González, E. and Whittembury, G. (1984) The continuous measurements of tubular volume changes in response to step changes in contraluminal osmolality. Pflügers Arch. 400, 343-348.

Case, R.M., Cook, D.I., Hunter, M., Steward, M.C. and Young, J.A. (1985) Transepithelial transport of non-electrolytes in the rabbit mandibular salivary gland. J. Membr. Biol. 84, 239-248.

Finkelstein, A. (1984) Water movement through membrane channels. Current Topics in Membranes and Transport 21, 295-308.

Frömter, E. (1972) The route of passive ion movement through the epithelium of Necturus gall bladder. J. Membr. Biol. 8, 259-301.

Frömter, E. (1974) Electrophysiology and isotonic fluid absorption of proximal tubules of mammalian kidney. In: Kidney and Urinary Tract Physiology. Thurau, K., ed., Butterworths: London, 6, 1-38.

Frömter, E. (1970) Elektrophysiologische Untersuchungen am proximalen Tubulus der Rattenniere. Habilitationschrift. Johan Goethe Universität: Frankfurt.

González, E., Carpi-Medina, P., Linares, H. and Whittembury, G. (1984) Water osmotic permeability of the apical membrane of proximal straight tubular (PST) cells. Pflügers Arch. 402, 337-339.

Green, R. and Giebisch, G. (1984) Luminal hypotonicity as a driving force for fluid absorption from the proximal convoluted tubule of the rat. Am. J. Physiol. 246, F167-F174.

Hill, A.E. and Hill, B.S. (1978) Sucrose fluxes and junctional water flow across Necturus gall bladder epithelium. Proc. Roy. Soc. B: 200, 151-162.

Levitt, D.G. (1984) Kinetics of movement in narrow channels. Current Topics in Membranes and Transport 21, 181-197.

McLaughlin, S. and Mathias, R.T. (1985) Electroosmosis and the reabsorption of fluid in the renal proximal tubules. J. Gen. Physiol. 85, 699-728.

Moura, T., Macey, R.I., Chien, D.Y., Karan, D. and Santos, H. (1984) Thermodynamics of all-or-none water channel closure in red cells. J. Membr. Biol. 81, 105-111.

Naftalin, R.J. and Tripathi, S. (1986) The roles of paracellular and transcellular pathways and submucosal space in isotonic water absorption by rabbit ileum. J. Physiol. (Lond.) 370, 409-432.

Pratz, J., Ripoche, P. and Corman, B. (1986) Evidence for proteic water pathways in the luminal membrane of kidney proximal tubules. Biochim. Biophys. Acta. 856, 259-266.

Preisig, P.A. and Berry, C.A. (1985) Evidence for transcellular osmotic water flow in rat proximal tubules. Am. J. Physiol. 249, F124-F131.

Schafer, J.A. (1984) Mechanisms coupling the absorption of solutes and water in the proximal nephron. Kidney Int. 25, 708-716.

Schafer, J.A., Patlak, C.S., Troutman, S.L. and Andreoli, T.E. (1978) Volume regulation in the pars recta. Am. J. Physiol. 234, F340-F348.

Solomon, A.K., Chasan, G., Dix, J.A., Lukacovic, M.F., Toon, M.R. and Verkman, A.S. (1983) Possible relation between anion transport and water flow in red cells. Ann. N.Y. Acad. Sci. 414, 97-104.

Spring, K.R. and Hope, A. (1979) Size and shape of the lateral interspaces in a living epithelium. Science 200, 54-58.

Steward, M.C. and Garson, M.J. (1985) Water permeability of Necturus gall bladder epithelial cell membranes measured by NMR. J. Membr. Biol. 86, 203-210.

Weinstein, A.M. and Windhager, E.E. (1985) Sodium transport along the proximal tubule. In: The Kidney, Physiology and Pathophysiology. Seldin, D.W. and Giebisch, G., eds., Raven Press: New York pp 1033-1062.

Williams, J.C., Barfuss, D.W. and Schafer, J.A. (1986) Transport of solute in proximal tubule is modified by changes in medium osmolality. Am. J. Physiol. 250, F246-F255.

Whittembury, G. (1985) Mechanisms of epithelial solute-solent coupling. In: The Kidney, Physiology and Pathophysiology. Seldin, D.W. and Giebisch, G., eds., Raven Press: New York pp 199-214.

Whittembury, G., Carpi-Medina, P., González, E. and Linares, H. (1984) Effect of para-chloromercuribenzenesulfonic acid and temperature on cell water osmotic permeability of proximal straight tubules. Biochim. Biophys. Acta 775, 365-373.

Whittembury, G. and Hill, B.S. (1982) Fluid reabsorption by necturus proximal tubule perfused with solutions of normal and reduced osmolarity. Proc. Roy. Soc. B: 215, 411-431.

Whittembury, G., Malnic, G., Mello-Aires, M. and Amorena, C. (1981) Flujo paracelular de agua en túbulo renal proximal de rata. Acta Científ. Venezol. 32, suppl 1,40.

Whittembury, G., Biondi, A.C., Paz-Aliaga, A., Linares, H., Parthe, V. and Linares, N. (1986) Transcellular and paracellular flow of water during secretion in the upper segment of the malpighian tubule of Rhodnius prolixus: solvent drag of grades sized molecules. J. Exp. Biol. 123, 71-92.

Whittembury, G. and Rawlins, F.A. (1971) Evidence of a paracellular pathway for ion flow in the kidney proximal tubule: electronmicroscopic demonstration of lanthanum precipitate in the tight junction. Pflügers Arch. 330, 302-309.

Whittembury, G., Verde-Martínez, C., Linares, H. and Paz-Aliaga, A. (1980) Solvent drag of large solutes indicates paracellular water flow in leaky epithelia. Proc. Roy. Soc. B 211, 63-81.

Windhager, E.E. (1979) Sodium chloride transport. In: Membrane Transport in Biology vol 4a, Giebisch, G., Tosteson, D.C. and Ussing, H.H., eds., Springer: Berlin, pp 143-213.

IN SEARCH OF THE PHYSICAL BASIS OF LIFE

Gilbert N. Ling

Department of Molecular Biology
Pennsylvania Hospital
Eighth and Spruce Streets
Philadelphia, PA 19107 U.S.A.

The living cell is the basic unit of life. Toward understanding of the living phenomenon in general and toward the erection of a solid foundation for future biomedical research it is vital to know "What are the fundamental distinctions between a living cell and a dead one?"

A clue may be found in a familiar phenomenon: exposure to dyes like rhodamine B, nigrosin, and trypan blue deeply stains dead cells but not living ones. Indeed different responses of living and dead cells to dyes reflect the fundamental attributes of the living cell to maintain its integrity and hence existence in environments which in chemical composition are different from that within the living cells. Thus the living cell contains a concentration of Na^+ only a fraction of that in the aqueous environment; when killed the cell Na^+ concentration dramatically increases. However, the living cells do not hold all their components at a lower concentration. K^+ is held at a much higher concentration in the cells than is found in the environment. Thus, at the cell level, the transition from life to death is intrinsically related to a dramatic loss of the ability of the cell to maintain its unique composition.

In general principles, there are three ways - and three ways only - whereby the concentrations of a substance can be sustained indefinitely at different levels in two contiguous spaces. Consider the case of a camper at a mosquito-infested site. The camper can keep the mosquitoes away (1) by using a mosquito net, (2) by detecting and catching all mosquitoes that come near and releasing them at a far distance, or (3) by spraying the site with an insect repellent. The molecular equivalent of each of the three basic mechanisms has been invoked to explain the asymmetrical distribution of K^+ and Na^+ between living cells and their environments: (1) in the original membrane theory, a net-like cell membrane permeable to water but not to ions, permanently keeps K^+ inside and Na^+ outside the cells; (2) in the membrane-pump theory, membrane pumps, like the diligent camper, continually capture and transport Na^+ out of the cells and K^+ into the cells; (3) in the protoplasmic theories, distinctive physico-chemical environments within the cells attract K^+ and repel Na^+.

The original membrane theory was disproved in the late 30's and the early 40's when radioactive tracer studies revealed that the cell membrane is permeable to both Na^+ and K^+. After that only two types of theories have survived: the membrane-pump theory is by far the better known and often taught as fact in popular textbooks written by poorly informed authors (see below). There are two versions of the protoplasmic theories: the sorption theory of A.S. Troshin of the Soviet Union (Troshin, 1966) and the association-induction (AI) hypothesis, which I first formally introduced in 1962 and completed in 1965 (Ling, 1962, 1965, 1984). Troshin's sorption theory and the AI hypothesis are in general harmony with each other. However since the 60's, the only protoplasmic theory that has been continually developed and tested is the AI hypothesis.

THE MEMBRANE-PUMP THEORY

The membrane-pump theory was not articulated and defended by one (or more) advocates and in this respect, is quite different from most other serious theories. The scientists whose names have been sometimes cited as the originators of the membrane-pump theory like A. Krogh (1946) and R. Dean (1941), could not be construed as the theory's originators because they only reiterated what had been said long before (see Lillie, 1923) and added little that was new.

By definition, physiology concerns itself with mechanisms of living phenomena. A bona fide theory of the Na-pump must offer a mechanism for the postulated pump. Yet, in the first-of-its-kind review on the subject of the Na-pump, Glynn and Karlish apologized for not being able to compare the "great mass of work" with "a hypothesis accounting for the working of the pumps" because "no such hypothesis exists." (Glynn and Karlish, 1975, p.13)

The Excessive Energy Need of the Na-Pump

In the membrane-pump theory, intracellular K^+ and Na^+ as well as the bulk of cell water are free. The orientation of the Na^+ concentration gradients and the polarity of the resting potential are such that virtually all the Na^+ ion leaving the cell must be by pumping (Ling, 1965). In the late 40's I discovered that frog muscles could maintain their normal K^+ and Na^+ contents at 0°C for hours in the absence of respiration and glycolysis when they were exposed to anoxia which blocks respiration and iodoacetate which blocks glycolysis (Ling, 1952, p.765). Furthermore, the Na^+ efflux rate from frog muscle was also indifferent to the combined action of anoxia and glycolysis (Ling, 1952). The indifference of Na^+ efflux rate of frog muscle to metabolic poison was later confirmed by Keynes and Maisel (1954) and by Conway and his coworkers (Conway, Kernan and Zadunaisky, 1961; for details, see Ling, 1984, pp. 91-92). A possible explanation for this unexpected indifference was the poisoned cells' reliance on their store of ATP and creatine phosphate (CrP), both present in high concentration in frog muscle.

I spent a great deal of time in the early fifties developing suitable methods accurately to assay the ATP and CrP contents of muscle immediately after complete suppression of both respiration and glycolysis and some hours later, being kept at 0°C at all times. In the meantime, the Na^+ pumping rate as well as the resting potential were continually monitored. Assuming that under normal conditions, respiration, glycolysis and the store of ATP and CrP were the only energy sources available to the muscle cells (an assumption fully verified later, see Ling, 1984, p. 125) and that the muscle spent its energy only on the outward pumping of Na^+ (an assumption, obviously untrue) I found, from a total of 28 sets of experiments, that the cell did not have enough energy to operate the Na-pump. Indeed the results of the last three sets of data complete in 1956, showed that the minimal energy need of the Na-pump is from 15 to 30 times higher than the maximally available energy (Ling, 1962; 1984a, p.124)(Table 1).

These data have been part of the literature for nearly a quarter of a century and have never been challenged in print. Three remedial

Table 1. a- The minimum rate of energy delivery required to operate a Na-pump according to the membrane pump theory was calculated from integrated values of the measured rates of Na^+ exchange and the energy needed to pump each mole of Na^+ ion out against the measured electrical and concentration gradients. The maximum energy delivery rate was calculated from the measured hydrolysis of CrP, ATP and ADP, the only effective energy sources available to the muscles, which had been poisoned with IAA and nitrogen.
b- from Ling, 1962.

Energy Balance Sheet for the Na^+ Pump in Frog Sartorius Muscles (0°C)[a,b]

Date	Duration (hr)	Rate of Na^+ exchange integrated average (moles/kg per hr)	$\psi + E_{Na}/\mathcal{F}$ integrated average (mV)	Minimum rate of energy required for Na^+ pump (cal/kg per hr)	Maximum rate of energy delivery (cal/kg per hr)	Minimum required energy / Maximum available energy
9-12-56	10	0.138	111	353	11.57 (highest value, 22.19)	3060%
9-20-56	4	0.121	123	343	22.25 (highest value, 33.71)	1542%
9-26-56	4.5	0.131	122	368	20.47 (highest value, 26.10)	1800%

postulations aimed at keeping the Na-pump concept afloat (Ussing's exchange diffusion, sequestration of cell Na^+ in the sarcoplasmic reticulum, and the nonenergy consuming Na-pump) have all been experimentally disproven (see Ling, Walton and Ling, 1979, p. 274).

However, the energy for the sodium pump described in Table 1 was by no means the only energy need of the living cells for pumping. As years went by it became increasingly clear that pumps are needed for virtually all the solutes found in the living cells as well as for man-made solutes like phenol red which made its first appearance through the past efforts of organic chemists and has been shown also to require pumps (see Troshin, 1966, p.206); so do other pump-requiring solutes yet to be synthesized!

Furthermore, many subcellular particles require their own pumps at their covering membranes because the solute levels in them may differ considerably from those in the cytoplasm. It has been estimated that the liver mitochondrial membrane may be as much as 20 times larger than the liver cell plasma membrane (Lehninger, 1964, p.30) and that the membranes of the sarcoplasmic reticulum of muscle cells may be as much as 50 times that of the muscle plasma membrane (Peachey, 1965). Since energy need for pumping varies directly with the membrane area, the energy needs of the pumps of these subcellular particles may exceed by far those of the plasma membrane pumps.

Other Evidence For and Against the Membrane-Pump Hypothesis

Much change has occured in the last 30 years. As of this date, and to the best of my knowledge, there is no longer any compelling evidence unequivocally in favour of the membrane-pump theory. For a detailed review, the reader must consult my recent monograph, bearing the same title as this article (Ling, 1984a). Some of the earlier evidence once widely regarded as unequivocally supporting the membrane-pump theory (e.g., K^+ mobility in squid axon; K^+ activity in squid axon and other cells measured with K^+ specific intracellular electrodes) have been shown to be erroneous or equivocal (Ling and Negendank, 1980; Ling, 1984, p.245 and p.250; Ling, 1984b). Neither squid axon membrane sacs freed of axoplasm, nor intact erythrocyte ghost membranes freed of haemoglobin achieve net transport of K^+ or Na^+ against concentration gradients (Ling and Tucker, 1983; Ling, 1984, p.129; Ling, Zodda and Seller, 1984), nor do man-made phospholipid membrane vesicles incorporating Na, K-activated ATPase (Ling and Negendank, 1980). In contrast, muscle cells without a functional cell membrane and postulated pumps continue to accumulate K^+ and exclude Na^+ much as intact muscle cells do (Ling, 1978; Ling, 1984a, p.133; see also Kellermayer, Ludany, Jobst, Szucs, Trombitas and Hazlewood, 1986). Still other evidence to be reviewed below demonstrating the adsorbed state of cell K^+ and the bulk of cell water also refute the basic free-K^+ and free-H_2O tenets of the membrane-pump theory.

This then is what I know, as of July 30, 1986, of the true status of the most basic theory of cell physiology and biomedical research. I pause to ask the question, "Why is this disproven theory continued to be taught as unqualified fact in virtually all textbooks and serve as the foundation of the great majority of cell physiological and biomedical research?"

Sociologist Bernard Barber (1961) captioned his article published in Science entitled "Resistance by Scientists to Scientific Discovery" in these words: "This source of resistance has yet to be given the scrutiny accorded religious and ideological sources" (underline mine).

Barber then cited Hans Zinser (1940): "That academies and learned societies - commonly dominated by the older foofoos of any profession - are slow to react to new ideas is in the nature of things. For as Bacon says, scientia inflat, and the dignitaries who hold high honors for past accomplishments do not usually like to see the current of progress rush too rapidly out of their reach." Only in the intervening years since Zinser, has the power of the "dignitaries" been vastly expanded and made virtually unchallengeable through the near universal adoption of the peer review system in evaluating science and scientists. The peer review gives the dignitaries the power to decide who get research grants, who get academic appointments, who receive promotion to tenure, whose work is published in widely-read journals...

As pointed out by Prof. Catchpole, Glynn and Karlish in their review cited above on the "Na-pump", "listed 245 articles in support of the sodium pump and none opposed. Yet Ling's idea had been around for 25 years, so had ours, so had Troshin's..." (Catchpole, 1981). An individual scientist, who deliberately cites only favorable evidence for and ignores negative evidence against a specific theory, commits what 19th century English mathematician Charles Babbage (also from Cambridge University) called "cooking" (Babbage, 1830; see also Honor in Science, 1986). In my view, that Glynn and Karlish should nevertheless choose this course of action (by no means uncommon) attests to the intensity of fear for reprisals that they must have experienced.

Science is a new human experience; only very recently have historians come to agree that past progress in science has been achieved by a dual mechanism: drastic directional changes (called revolutions if eventually proven valid) and fruitful exploration of the new directions thus opened up (Kuhn, 1962; Cohen, 1985, p. 389). Louis Pasteur's work was of the first kind; Jonas Salk's of the second.

The peer review system was adopted before the importance of drastic directional change was recognized and was based on the erroneous idea that science progressed only by the smooth accumulation of knowledge; hence the belief that peers cannot be far wrong. The peer system thus conceived has provided an effective way to promote research that happened to be in the right direction at the start (e.g., genetics). By making it virtually impossible to pursue major directional changes, the peer review system has also become the major road block to further progress in areas of science that are still young and therefore offer the greatest promise for the future (e.g., cell physiology).

It is high time that scientists, concerned about the future of science in general and biomedical science in particular, rise to the challenge and improve the peer review system so that it will foster both modes of progress and not limit its support to currently popular directions. It is not going to be an easy task but one may take heart in remembering that others have already dealt so successfully with even more forbidding and more deeply rooted sexual, racial, religious and ideological bias and abuse of power, in efforts that began not so long ago.

THE ASSOCIATION-INDUCTION (AI) HYPOTHESIS

In the AI hypothesis, the molecules and ions making up a living cell stay together by a mechanism not totally unlike that keeping a school of free-swimming fish together in the open sea. "Near neighbour interaction" among close by fish not only holds the school together; it can, at a signal, also sharply change the direction of swimming of all the fish in an all-or-none manner.

A simple, more cogent model is a chain of loosely tethered nails and iron filings scattered among the nail chain. If a magnet is brought into contact with one terminal nail, the nails will be magnetized one by one. On the magnetized nails, the iron filings are then attracted and organized according to a fixed pattern. Thus interaction with the magnet has brought the whole assembly from a low energy/high entropy state to a high energy/low entropy state. Like the school of fish, it is primarily short-range interactions between immediately neighboring nails and iron filings that provide the coherence and put the nails and iron filings under the controlling influence of the magnet. Of course, in the nail model, the interaction is magnetic polarization. In living cells, the polarization is electronic - or induction. In this case, it is the polypeptide chains of certain cell proteins that function as the equivalents of the nail chain; ions and water molecules then behave like the equivalent of iron filings. The equivalent of the controlling magnet is a class of biologically potent agents, collectively referred to as "cardinal adsorbents". A highly important cardinal adsorbent is ATP.

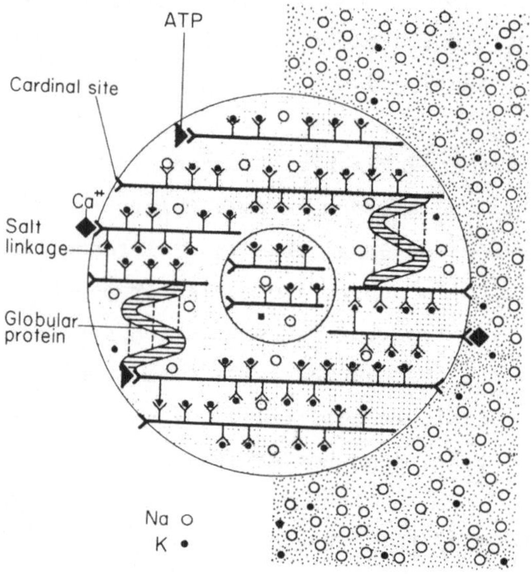

Fig. 1. Diagrammatic illustration of the substance of a living cell. Ordered dots represent water existing in the state of polarized multilayers due to interaction of certain pervasive cell proteins existing in the fully extended conformation. Reduced solubility of polarized multi-layered water for large and complex molecules and hydrated Na^+ (open circles) (and K^+) and inability to compete for β- and γ-carboxyl groups preferentially adsorbing K^+ (solid circles) keep intracellular Na^+ concentration low. Selective adsorption of K^+ and multilayer polarization of cell water depends on the interaction of the involved proteins with ATP and other cardinal adsorbents. Salt linkages among different proteins maintain cell structure.

The high energy/low entropy state of the cell is called the resting living state. It is separated from the active living state (e.g., contracted muscle) by a small energy barrier. Both resting and active living states are separated by a higher energy barrier from the death state of still lower energy and high entropy. A still higher energy barrier separates the death state from the ultimate random state.

Fig. 1 presents a diagrammatic illustration of a living cell. Note that the cell membrane, mitochondria, and other subcellular structures are not shown. Rather the figure illustrates the basic components of all these structures as well as the cytoplasm. Protein-protein interaction, protein-ion interaction, and protein-water interaction link the cell into a coherent unit like the school of fish or the nail-filings model.

In the highly associated state, the cellular proteins as a rule do not assume the same conformation they assume after their isolation. Indeed, in the AI hypothesis, the "native" state of some of the cellular proteins at least, correspond to the dead state. In the resting living state, the conformation of the proteins is different both in terms of steric and electronic conformation.

An expression of the high energy electronic conformation is that the anionic β- and γ-carboxyl groups carried on aspartic and glutamic acid residues of the cell proteins assume such electron density (expressed as the c-value in the AI hypothesis, see Ling, 1984a, p.156) that they adsorb preferentially K^+ over Na^+ by a margin large enough to overcome the concentration bias in favour of Na^+ (100mM) over K^+ (2.5 mM) in the external medium. As a result, these fixed anionic sites predominantly adsorb K^+. Since the cell proteins contain a high concentration of the β- and γ-carboxyl groups the theory offers an explanation why K^+ concentration in the cell is high while K^+ concentration in the external medium is low.

Unable to compete against K^+ for the fixed anionic sites in the cells, a major share of the intracellular Na^+ is free in the cell water. The concentration of Na^+ in the cell water is much lower than that in the surrounding medium, because the cell water exists in a physical state different from that of normal liquid water (i.e., the state of polarized multilayers, see below) and in this state, water has reduced solubility for hydrated ions like Na^+, as well as K^+.

In summary, through its association with other proteins and with cardinal adsorbents, especially ATP, certain proteins in the living cell offer their β- and γ-carboxyl groups for the preferential adsorption of K^+. The same or other proteins, also under the control of ATP, assume the fully extended conformation with their polypeptide NHCO groups directly exposed to the bulk phase water, polarizing the water molecules in multilayers. The exclusion from adsorption onto the β- and γ-carboxyl groups and the reduced solubility of water in this state for Na^+ (and K^+) account for the low concentration of Na^+ in resting living cells. In this model, the physical basis of life is not merely the presence of a suitable concentration of each essential chemical component occupying locations specific to the cell type. It is that but more: the operation of short-range electronic polarization or "induction", mediated through the partially resonating and hence highly polarizable polypeptide chains, enables the closely associated protein-water-ion systems to assume the high energy living state and to behave as a coherent system.

Selective K^+ Accumulation

The idea that K^+ accumulation might be due to its adsorption on cell proteins in a way similar to oxygen accumulation in red blood cells was suggested by Moore and Roaf long ago (Moore and Roaf, 1908). However attempts to demonstrate in vitro similar selective K^+ adsorption by isolated proteins in a scale large enough to explain K^+ association in cells had ended in failure (see Lillie, 1923). In 1952 I suggested that this failure to demonstrate in vitro alkali-metal ion adsorption might reflect the assumption of different conformations of the proteins in vitro and in living cells (Ling, 1952). More specifically, in native proteins the anionic β- and γ-carboxyl groups might be locked in "salt linkages" with cationic ε-amino groups and guanidyl groups and therefore not available for K^+ or Na^+ adsorption. Very recently Ling and Zhang (1984) had taken advantage of the good Na^+-selective electrodes now commercially available and had fully confirmed this "salt-linkage hypothesis". It was found when ε-amino groups, guanidyl groups, etc. of bovine haemoglobin were neutralized with NaOH; for each fixed cationic group discharged, one Na^+ is adsorbed. Furthermore, this stoichiometric binding is ion-specific and exhibits autocooperativity much as the binding of oxygen on the same protein shows autocooperativity (i.e., $n > 1$). These findings on the one hand, removed one of the major objections against selective adsorption as the basic mechanism for the selective accumulation of K^+ in living cells, and on the other hand provided a solid base for the selective adsorption of alkali metal ions as a cooperative phenomenon, a basic concept also introduced in the AI hypothesis (Ling, 1962, 1984a).

The hypothesis that it is the β- and γ-carboxyl groups that specifically adsorb K^+ in living cells leads to two predictions: (1) since in striated muscle over 60% of the β- and γ-carboxyl groups are found in myosin and myosin is found only in the A-band, the first prediction is that most of the K^+ in resting skeletal muscle should be localized in the A-band; (2) since there is evidence that the β- and γ-carboxyl groups are also the protein sites binding uranium (Hodge and Schmidt, 1960) one

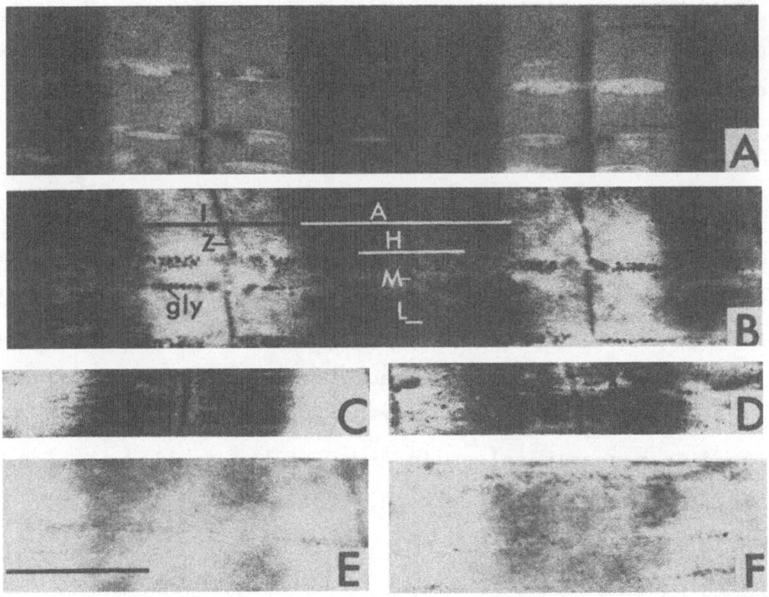

Fig. 2. Electron micrographs of frog sartorius muscle. (A) Muscle fixed in glutaraldehyde only and stained only with uranium by conventional procedure. (B) EM of section of freeze-dried Cs^+-loaded muscle without chemical fixation or staining. (C) Tl^+-loaded muscle without chemical fixation or staining. (D) Same as (C) after exposure of section to moist air, which causes the hitherto even distribution of thallium to form granular deposits in the A-band and Z-line. (E) Section of central portion of (B) after leaching in distilled water. (F) Normal "K^+-loaded" muscle. Scale bar: 1 μm. [(A) from Edelmann, unpublished. (B-F) from Edelmann (1977), by permission of Physiol. Chem. Phys.]

expects to find K^+ in resting cells at subcellular loci stained dark in the EM plates fixed and stained with uranium in the conventional manner.

Both predictions have been confirmed (Fig. 2). When K^+ in frog muscle cells has been stoichiometrically (and reversibly) replaced by the electron dense Cs^+ and Tl^+ and these heavier ions visualized directly in freeze-dried and imbedded, unfixed and unstained muscle cells sections, Cs^+ and Tl^+ are found primarily at the two edges of the A-bands and the Z-line (B,C), looking just like the uranium-stained controls (A).

Other experiments, utilizing the technique of effectively membrane (pump)-less open ended cell (EMOC) preparation, established that the localization demonstrated is due to one site/one ion close contact adsorption (Ling, 1977, 1984a, p.240).

Within the last nine years, extensive evidence has collected demonstrating the adsorbed state of K^+ in muscle cells using no less than four independent methods from three laboratories across the world (for review, see Ling, 1984a; Edelmann, 1984). The conclusion was unanimous except for a single report from one laboratory (Somlyo, Gonzalez-Serratos, Shuman, McClellan and Somlyo, 1981). The cause of this difference was found and the differences resolved in favor of the earlier conclusions of Edelmann, Ling, and Trombitas (Edelmann, 1983, 1984).

Exclusion of Na^+

According to the AI hypothesis, the low concentration of Na^+ maintained in living cells reflects in part the inability of Na^+ to compete successfully against K^+ for the β- and γ-carboxyl groups and in part reflects the physical state of cell water, which in turn is the consequence of the existence of certain intracellular proteins in the fully extended state with their negatively charged CO (or N sites) and positively charged NH (P sites) directly exposed to and polarizing the bulk of cell water. A matrix of such fully extended protein chains provides what has been called an NP-NP-NP system, in which the alternatingly positive P sites and negative sites separated by distances of approximately the distance of one diameter of a water molecule, anchor and polarize rows of water molecules which are further stabilized by the cohesive dipole-dipole interaction between immediately neighbouring water molecules oriented in opposite directions.

For both enthalpic and entropic reasons, water existing in the state of polarized multilayers has reduced solubility for large and complex molecules (e.g., sucrose) and hydrated ions (e.g. Na^+). The enthalpic reason arises from the fact that it takes more energy to excavate a hole in the polarized water to accommodate the Na^+ than the energy gained in filling the hole in the normal water from whence the Na^+ is removed. The entropic reason is due to the greater motional restriction a hydrated Na^+ encounters in the polarized water than in normal liquid water, much as a butterfly caught in a spider web suffers more motional restriction than flying in air. Among the different degrees of motional freedom, both the translational and rotational motion are likely to be reduced. However, since the rotational entropy contributing to the total entropy increases with the size and complexity of the molecules, the reduction of entropy in the polarized water increases with the size and complexity of the molecules involved. Thus both the enthalpic and entropic reasons favour decreased solubility for large and complex molecules.

q-Value and the Size Rule

The true equilibrium distribution coefficient or q-value is the ratio of the concentration of a solute (e.g., Na^+ or sucrose) in the water under study to that in a reference normal liquid water at equilibrium. The apparent equilibrium distribution coefficient, or ρ-value, is at best equal to the q-value but often exceeds it due to the inclusion of unrecognized adsorbed or complexed solute.

In those cases where a low q-value for a solute could be clearly separated from adsorbed solute and studied at more than one temperature, we found that the major cause of the low q-value for ions in frog muscle cells was primarily due to entropic causes (Ling et al., 1979).

Small molecules and molecules of such a chemical structure that can fit into the dynamic water structure, may exhibit a q-value of unity or higher. With increasing size and complexity of the solute, the q-value falls. This relation between molecular sizes and q-values has been called the "size rule".

The ρ-value of Solutions of Native Proteins, Gelatin and Urea-Denatured Proteins

According to the polarized multilayer theory of cell water, only cytoplasmic proteins with their polypeptide chains in the fully extended state polarize water. With their backbone NHCO groups locked in α-helical or β-pleated sheet conformation, native proteins are not expected to polarize water to a level required in living cells. The small amount of hydration water usually in the range of 0.2 to 0.3 gm/gm of proteins is apparently due to polar side chains (Ling, 1972). To test the predictions that native proteins exercise only a minor effect on the solvency of the bulk phase water, we studied the distribution of ^{22}Na-labelled Na^+ in

Table 2. a - Temperature was 25 ± 1°C and test tubes were agitated. The symbols a and b indicate that the media contained initially 1.5 M Na_2SO_4 and 0.5 M sodium citrate, respectively. In (D), PEO (m.w. 600,000) was dissolved as a 10% (w/w) solution, and the viscous solution was vigorously stirred before being introduced into dialysis tubing. Na^+ was labelled with $^{22}Na^+$ and assayed with a gamma counter.

b - from Ling and Ochsenfeld, 1983.

c - temperature was 37°C.

ρ-Values of Na^+ in Water Containing Native Globular Proteins, Gelatin, Polyvinylpyrrolidone, Polyethylene Oxide, and Methylcellulose[a,b]

Group	Polymer	Concentration of medium (M)	Number of assays	Water content (%) (mean ± SE)	ρ-Value (mean ± SE)
A	Albumin				
	Bovine serum	1.5 a	4	81.9 ±0.063	0.973 ± 0.005
	Egg	1.5 a	4	82.1 ±0.058	1.000 ± 0.016
	Chondroitin sulfate	1.5 a	4	84.2 ±0.061	1.009 ± 0.003
	α-Chymotrypsinogen	1.5 a	4	82.7 ±0.089	1.004 ± 0.009
	Fibrinogen	1.5 a	4	82.8 ±0.12	1.004 ± 0.002
	γ-Globulin				
	Bovine	1.5 a	4	82.0 ±0.16	1.004 ± 0.004
	Human	1.5 a	4	83.5 ±0.16	1.016 ± 0.005
	Hemoglobin	1.5 a	4	73.7 ±0.073	0.923 ± 0.006
	β-Lactoglobulin	1.5 a	4	82.6 ±0.029	0.991 ± 0.005
	Lysozyme	1.5 a	4	82.0 ±0.085	1.009 ± 0.005
	Pepsin	1.5 a	4	83.4 ±0.11	1.031 ± 0.006
	Protamine	1.5 a	4	83.9 ±0.10	0.990 ± 0.020
	Ribonuclease	1.5 a	4	79.9 ±0.19	0.984 ± 0.006
B	Gelatin	0.1 b	4	84.0 ±0.78[c]	0.89 ± 0.002
		1.5 a	37	57.0 ±1.1	0.537 ± 0.013
C	Polyvinylpyrrolidone (PVP)	1.5 a	8	61.0 ±0.30	0.239 ± 0.005
D	Polyethylene oxide (PEO)	0.75 a	5	81.1 ±0.34	0.475 ± 0.009
		0.5 a	5	89.2 ±0.06	0.623 ± 0.011
		0.1 a	5	91.1 ±0.162	0.754 ± 0.015

solutions of 12 native proteins and one polysaccharide initially at 20% concentration (w/v). They were placed in dialysis sacs and equilibrated in 1.5 M Na_2SO_4. The ρ-values of Na^+ in the water of these native protein solutions were uniformly close to unity confirming the prediction (Table 2A).

By contrast, a solution of gelatin behaved quite differently (Table 2B). A 16% gelatin solution shows a ρ-value of 0.89±0.002; while a 43% gelatin solution shows a ρ-value of 0.537±0.013. This ability of gelatin to reduce the solvency of water for Na_2SO_4 is not new (see Holleman, de Jong and Modderman, 1934; Troshin, 1966); however our interpretation is. We attributed the different behaviour of gelatin to its possession of repeating units of the triad of amino acid residues glycine, proline and hydroxyproline. Glycine is a helix breaker (Chou and Fasman, 1974), proline and hydroxyproline lack a proton on their backbone N atom to form H-bonds. Together, they prevent the gelatin molecule from existing in the α-helical conformation. Being a denatured collagen, the chain-to-chain H-bonds are also largely broken during the denaturation process even though undoubtedly part of the gelatin molecule does form "collagen-folds" (Veis, 1964) and is not able to react with the bulk phase water. Nevertheless, the difference between gelatin and the native proteins is striking.

Since part of the polypeptide chain of gelatin lacks the positively charged H on its backbone N atom, a solution of gelatin is only partly an NP-NP-NP system and partly what is called an NO-NO-NO system, where the P sites are replaced by vacant O sites. Polyvinylpyrrolidone (PVP)

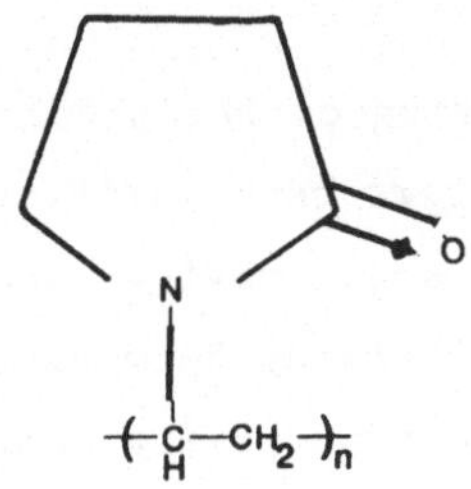

and polyethylene oxide (PEO) $-(CH_2-O-CH_2)_n-$ are examples of pure NO-NO-NO systems. Like gelatin, solutions of these polymers exclude Na_2SO_4 (Table 2C, D). Since these polymers do not form gels, nor do they form "coacervates" (a colloidal phase that remains immiscible with water), clearly the ability of solutions of gelatin, PVP and PEO of excluding solutes like Na_2SO_4 does not directly depend on the assumption of the gel or coacervate state, as is the specific case of gelatin.

Next we studied the effect of four protein denaturants on the effect of proteins on water solvency: urea, guanidine HCl, sodium dodecyl sulfate (SDS) and n-propanol. The first two denaturants are secondary-structure breakers. On the basis of the polarized multilayer theory of cell water, one expects that native proteins, which by themselves have little effect on water solvency, should acquire such an ability after their intramacromolecular H-bonds have been broken and kept in the fully extended state by urea and guanidine HCl. As a result, the water of a solution of these denatured proteins should have reduced ρ-value for a probe molecule like sucrose. This expectation was also confirmed (Ling, 1984a, p. 177).

On the other hand, SDS and n-propanol are tertiary structure breakers with no or even enhancing effects on secondary structures. Therefore they should have little or no effect on the solvency of the water in SDS and n-propanol denatured proteins. None was observed (Ling, 1984a, p. 177).

Of considerable significance was the finding that the ρ-value of urea itself (labelled with ^{14}C urea) in the solution of urea-denatured protein was very close to unity even though that of sucrose in the same solution was much lower. This finding affirms the general principle described as the "size rule". Urea has a ρ-value of unity because it is small and can apparently fit well into the dynamic structure of water in the state of polarized multilayers.

COMPARATIVE STUDIES OF THE MODEL SYSTEMS AND LIVING CELLS

In vitro experimental testing of the predictions of the polarized multilayer theory on model systems have verified some basic predictions of the theory. In this section I shall review results of experimental studies that offer further evidence of harmony between theory and model behaviours and between theory and behaviours of living cells.

Bradley's Multilayer Adsorption Isotherm

In 1936 Bradley derived a (polarized) multilayer adsorption isotherm:

$$\log (p_o/p) = K_1K_3{}^a+K_4, \quad (1)$$

where p_o/p is the reciprocal of the partial vapour pressure, a is the amount of water sorbed. Under specified conditions, K_1, K_3 and K_4 are all constants. Equation 1 can be written in the double-log form:

$$\log (\log (p_o/p) - K_4) = a \log K_3 + \log K_1, \quad (2)$$

which predicts a rectilinear relation between the term on the left and water sorbed, a, obtained in an environment containing water vapour at a specified partial vapour pressure p/p_o.

The theoretical expectations between a and p/p_o were confirmed for both gelatin where the data were those of J.R. Katz published in 1919 (for details, see Ling, 1984a, p.288) and for frog muscle, where 95% of the cell follows Equation 2 rigorously. The remaining 5% is very tightly bound and exists apparently as an adsorbed monolayer (Ling and Negendank, 1970; Ling, 1984a, p.289).

The Size Rule in Solute Distribution

In the model system cross-linked polystyrene sulfonate, the equilibrium distribution of alcohol and sugars were studied. The results showed obedience to the size rule: the ρ-values for methanol (m.w. 32.04), glycerol (m.w. 92.09), D-glucose (m.w. 180.16), and sucrose (m.w. 342.30) were respectively 0.97, 0.464, 0.309, and 0.153. In frog muscle, the equilibrium distribution of methanol (m.w. 32.04), ethylene glycol (m.w. 62.07), glycerol (m.w. 92.09), xylose (m.w. 150.13), D-glucose (m.w. 180.16), α-methyl glucoside (m.w. 194.18), and sucrose (m.w. 342.30) were also studied. The results showed obedience to the size rule. Thus the ρ-values for each of the solutions in the order given above are respectively 1.1, 0.99, 0.71, 0.48, 0.44, 0.36, and 0.18.

ATP AS THE CARDINAL ADSORBENT

Earlier I have described a nail-iron filing model where interaction with a magnet raises the energy and decreases the entropy of the system. I have also mentioned that in the AI hypothesis the equivalent of the loosely tethered nails is the polypeptide chains of cell proteins, the equivalents of iron filings are primarily water and K^+, and the equivalent of the magnet is a cardinal adsorbent which may be other proteins, Ca^{2+}, but above all the ultimate product of cell metabolism, ATP.

Using the one-dimensional Ising method, Yang and Ling derived a cooperative adsorption isotherm for linear protein chains carrying a succession of binding sites for K^+, Na^+ and other solutes. The isotherm differs from the familiar Langmuir type of adsorption isotherm in that the binding sites exhibit nearest neighbour interaction, mediated by, for example, electrical polarization or inductive interaction between pairs of nearest neighbouring sites. When this nearest neighbour interaction, represented by the symbol $-\gamma/2$, is positive, the adsorption isotherm is sigmoid, as is well known in the case of the binding of oxygen on haemoglobin. Indeed it was shown that the empirical Hill coefficient, n, long suspected to involve some sort of cooperativity, in fact equals $\exp(-\gamma/2RT)$, where R is the gas constant and T is the absolute temperature.

Another key parameter of the isotherm is the intrinsic equilibrium constant $K^{oo}_{j \to i}$ between the ith (e.g., K^+) and jth (e.g., Na^+) ions competing for the same binding sites.

The Y-L isotherm accurately describes the binding of oxygen on haemoglobin in vitro (Ling, 1984a, p.216) or in vivo (i.e., in intact erythrocytes) (Ling, 1966, p.967). The Y-L isotherm also describes accurately the uptake of K^+ in frog muscle, canine carotid arteries, rabbit myometrium, guinea pig taenia coli, human lymphocytes, etc. (see Ling, 1984a, Table 11.4 on p.347). In these cases, values of $K^{oo}_{Na \to K}$ ranging from 2.8 to 3.6 Kcal/mole and $-\gamma/2$ from 0.47 to 0.75 Kcal/mole were found. As mentioned above these findings are also in full accord with the cooperative binding of Na^+ on haemoglobin in vitro, where a $-\gamma/2$ of 0.82 Kcal/mole were observed.

The Y-L isotherm in an extended form also describes accurately the changes in the binding of oxygen on haemoglobin in the presence of different concentrations of 2,3-diphosphoglycerate and inositol hexaphosphate (Ling, 1984a, p.223) where the effect on oxygen binding in consequence of the binding of these phosphates (as well as of ATP, see below) (Chanutin and Hermann, 1969; Klinger, Zahn, Brox and Frundes, 1971) was primarily a shift of $K^{oo}_{O_2}$. Of specific interest here is that the oxygen affinity of haemoglobin also reacts allosterically to ATP binding in a similar manner (Chanutin and Curnish, 1967), and it involves no ATP hydrolysis because haemoglobin has no ATPase activity.

We have already shown how K^+ binding on muscle cytoplasmic proteins (probably myosin) and oxygen binding on haemoglobin in vitro and in vivo are both quantitatively described by the Y-L isotherm. The finding just described above shows that like the magnet and nail-filings model, ATP can indeed by binding onto suitable cardinal sites on a cytoplasmic protein, haemoglobin, allosterically control the affinity of a remote solute-binding site.

Given the proper environments (including other essential cardinal adsorbents adsorbed on their respective cardinal sites), we make the following postulates:

Postulate 1 - The binding of ATP exercises an allosteric effect on a "gang" of distant β- and γ-carboxyl groups, causing them to acquire an electron-density (measured by the parameter, c-value) where K^+ is greatly preferred over Na^+.

Postulate 2 - The binding of ATP exercises an allosteric effect on the backbone NHCO groups, causing them to assume a fully extended conformation and polarize cell water in multilayers.

Results of Testing of Postulates

Since ATP has an enormous binding affinity on myosin (10^{10}-$10^{11}M^{-1}$, Goody, Hofmann and Mannherz, 1977, Cardon and Boyer, 1978), one expects virtually all ATP in muscle cells to be adsorbed. Postulate 1 then directly leads to the prediction that at equilibrium, there should be a quantitative relation between the concentrations of ATP and of K^+ in the cells.

This prediction has been verified in two ways. First, Gulati, Ochsenfeld and Ling (1971) showed that by exposing frog muscle to 11 different poisons of widely different pharmacological effects first at 25°C followed by a four-hour incubation at 0°C to permit equilibration, linear relations were shown in all cases between ATP concentrations and K^+ concentrations confirming the prediction. The second confirmation used a different tactic and is described together with the testing of the prediction based on Postulate 2, i.e., that at equilibrium there should be a direct quantitative relationship between the concentration of ATP and the degree of polarization of the cell water as revealed by its q-value for such substances as Na^+ and sucrose.

Results of Testing Postulate 2

The tactic we chose was to expose frog muscle at 0°C to a very low concentration of a metabolic poison (0.2 M Na iodoacetate). Under these conditions, the ATP level slowly declined, reaching its final zero level in about 10 days. This very slow change permits equilibrium to be reached between K^+, as well as Na^+ and labelled sucrose (10 mM) in the external medium and in the cell at any specific ATP concentration reached in the cells.

The results once more confirm the relation between ATP concentration and K^+ levels. While the K^+ and ATP concentrations declined together, the

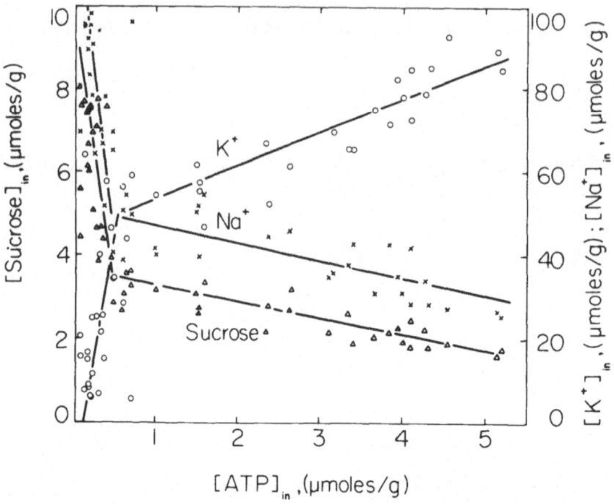

Fig. 3. Equilibrium distribution of K^+, Na^+, and labeled sucrose in frog muscles with different concentrations of ATP, having been exposed to 0.1 mM iodoacetate for up to 9 days (0°C). [from Ling and Ochsenfeld, 1983, by permission of Physiol. Chem. Phys.]

Na^+ concentration rose along a time course that looks like the mirror image of the K^+ time course (Fig. 3). By itself the data could be interpreted in two ways: (1) with ATP depletion, the β- and γ-carboxyl groups decrease their relative preference for K^+ and increase their relative preference for Na^+. As a result a stoichiometric displacement of adsorbed K^+ by adsorbed Na^+ occurs, hence the mirror image profiles; (2) with ATP depletion, the β- and γ-carboxyl groups decrease their preference for K^+ and sharply their preference for fixed cationic ε-amino groups and guanidyl groups. As a result K^+ desorption goes para passu with loss of available K^+ or Na^+ adsorbing sites and the formation of salt linkages. Along with this change, the proteins involved also undergo conformational changes and a concomitant depolarization of the cell water. In this model, there should be an entirely parallel change of Na^+ concentration and sucrose concentration. Since these solutes do not compete for the same sites, their parallel changes can only be due to water solvency changes and its similar effects on the ρ-values of both.

Since with ATP depletion, the change of Na^+ concentration was indeed found to parallel similar changes of sucrose concentration, one reaches the conclusion that Postulate 2 is confirmed: ATP depletion permits cell proteins to return from their high energy/low entropy state in which they preferentially adsorb and polarize water to a low energy/high entropy state seen in native proteins, where the β- and γ-carboxyl groups are locked in salt linkages and the backbone NHCO groups in α-helical and other inter- or intramacromolecular H-bonds. K^+ then leaves the cell; Na^+ (and sucrose) enter. With these changes, the cell dies.

REFERENCES

Babbage, C. (1830) Reflections on the Decline of Science in England and On Some of Its Causes. (Reprinted by Irish University Press, Shannon, 1971)

Barber, B. (1961) Resistance by scientists to scientific discovery. Science 134, 596-602

Bradley, R.S. (1936) Polymolecular adsorbed films. Part I. The adsorption of argon on salt crystals at low temperatures and the determination of surface fields. J. Chem. Soc. 1936, 1467-1474

Cardon, J.W., and Boyer, P.D. (1978) The rate of release of ATP from its complex with myosin. Eur. J. Biochem. 92, 443-448

Catchpole, H.R. (1981) Letter to the Editor. Persps. Biol. Med. 24, 164-165

Chanutin, A., and Curnish, R.R. (1967) Effect of organic and inorganic phosphates on the oxygen equilibrium of human erythrocytes. Arch. Biochem. Biophys. 121, 96-102

Chanutin, A., and Hermann, E. (1969) The interaction of oxygen and inorganic phosphates with hemoglobin. Arch. Biochem. Biophys. 131, 180-184

Chou, P.Y., and Fasman, G.D. (1974) Conformational parameters for amino acids in helical, β-sheet, and random coil regions calculated from proteins. Biochemistry 13, 211-245

Cohen, I.B. (1985) Revolutions in Science, Harvard Univ. Press: Cambridge

Conway, E.J., Kernan, R.P., and Zadunaisky, J.A. (1961) The sodium pump in skeletal muscle in relation to energy barriers. J. Physiol. (Lond.) 155, 263-179

Dean, R.B. (1941) Theories of electrolyte equilibrium in muscle. Biol. Symp. 3, 331-348

Edelmann, L. (1977) Potassium adsorption sites in frog muscle visualized by cesium and thallium under the transmission electron microscope. Physiol. Chem. Phys. 9, 313-317

Edelmann, L. (1983) Electron probe X-ray microanalysis of K, Rb, Cs, and Tl in cryosections of striated muscle. Physiol. Chem. Phys. 15, 337-344

Edelmann, L. (1984) Subcellular distribution of potassium in striated muscles. Scanning Electron Microscopy II, 875-888

Glynn, I.M., and Karlish, S.J.D. (1975) The sodium pump. Ann. Rev. Physiol. 37, 13-55

Goody, R.S., Hofmann, W., and Mannherz, H.G. (1977) The binding constant of ATP to myosin S1 fragment. Eur. J. Biochem. 78, 317-324

Gulati, J., Ochsenfeld, M.M., and Ling, G.N. (1971) Metabolic cooperative control of electrolyte levels by adenosine triphosphate in the frog muscle. Biophys. J. 11, 973

Hodge, A.J., and Schmidt, F.O. (1960) The charge profile of the tropocollagen macromolecule and the packing arrangement in native-type collagen fibrils. Proc. Natl. Acad. Sci. USA 46, 186-197

Holleman, L.W.J., de Jong, H.G.B., and Modderman, R.S.T. (1934) Lyophilic colloids. XXI. Coazervation. 1. Simple coazervation of gelatin sols. Koll.-Beih. 39, 334-420

Honor in Science, (1986) Sigma Xi, The Scientific Research Society: New Haven

Kellermayer, M., Ludany, A., Jobst, K., Szucs, G., Trombitas, K., and Hazlewood, C.F. (1986) Cocompartmentation of proteins and K^+ within the living cell. Proc. Natl. Acad. Sci. USA 83, 1011-1015

Keynes, R.D., and Maisel, G.W. (1954) The energy requirement for sodium extrusion in frog muscle. Proc. Roy. Soc. B: 142, 383-392

Klinger, R.G., Zahn, D.P., Brox, D.H., and Frundes, H.E. (1971) Interaction of hemoglobin with ions. Binding of ATP to human hemoglobin under simulated in vivo conditions. Eur.J. Biochem. 18, 171-177

Kuhn, T.S. (1962) The Structure of Scientific Revolutions, University Press: Chicago

Krogh, A. (1946) The active and passive exchanges of inorganic ions through the surfaces of living cells and through living membranes generally. Proc. Roy. Soc. B: 133, 140-200

Lehninger, A.L. (1964) The Mitochondrion, Benjamin: Menlo Park

Lillie, R.S. (1923) Protoplasmic Action and Nervous Action, University Press: Chicago

Ling, G.N. (1952) in Phosphorous Metabolism (Vol. II), McElroy, W.D. and Glass, B., eds., Johns Hopkins Univ. Press: Baltimore, pp 748-795

Ling, G.N. (1962) A Physical Theory of the Living State: The Association-Induction Hypothesis, Blaisdell: Waltham

Ling, G.N. (1965) Physiology and anatomy of the cell membrane: the physical state of water in the living cell. Fed. Proc. 24, S103-112

Ling, G.N. (1966) All-or-none adsorption by living cells and model protein-water systems: discussion of the problem of "permease-induction" and determination of secondary and tertiary structures of proteins. Fed. Proc. 25, 958-970

Ling, G.N. (1972) Water and Aqueous Solutions, Structure, Thermodynamics and Transport Processes, A. Horne, ed., Wiley: New York, pp 663-699

Ling, G.N. (1977) Thallium and cesium in muscle cells compete for the adsorption sites normally occupied by K^+. Physiol. Chem. Phys. 9, 217-225

Ling, G.N. (1978) Maintenance of low sodium and high potassium levels in resting muscle cells. J. Physiol. (Lond.) 280, 105-123

Ling, G.N. (1984a) In Search of the Physical Basis of Life, Plenum: New York

Ling, G.N. (1984b) Counterarguments against alleged proof of the Na-K pump in studies of Na^+ and K^+ distributions in amphibian eggs. Physiol. Chem. Phys. and Med. NMR 16, 293-305

Ling, G.N., and Negendank, W. (1970) The physical state of water in frog muscles. Physiol. Chem. Phys. 2, 15-33

Ling, G.N., and Ochsenfeld, M.M. (1983) Studies on the physical state of water in living cells and model systems. I. The quantitative relationship between the concentration of gelatin and certain oxygen-containing polymers and their influence upon the solubility of water for Na^+ salts. Physiol. Chem. Phys. and Med. NMR 15, 127-136

Ling, G.N., and Tucker, M. (1983) Only solid red blood cell ghosts transport K^+ and Na^+ against concentration gradients: hollow intact ghosts with K^+-Na^+ activated ATPase do not. Physiol. Chem. Phys. and Med. NMR 15, 311-317

Ling, G.N., and Zhang, Z.L. (1984) A study of selective adsorption of Na^+ and other alkali-metal ions on isolated proteins: a test of the salt-linkage hypothesis. Physiol. Chem. Phys. and Med. NMR 16, 221-235

Ling, G.N., Walton, C.L., and Ling, M.R. (1979) Mg^{++} and K^+ distribution in frog muscle and egg: a disproof of the Donnan theory of membrane applied to the living cells. J. Cell Physiol. 101, 261-278

Ling, G.N., Zodda, D., and Seller, M. (1984) Quantitative relationships between the concentration of proteins and the concentration of K^+ and Na^+ in red cell ghosts. Physiol. Chem. Phys. and Med. NMR 16, 381-392

Moore, B., and Roaf, H.E. (1908) On the equilibrium between the cell and its environment in regard to soluble constituents, with special reference to the osmotic equilibrium of the red blood cells. Biochem. J. 3, 55-81

Peachey, L.D. (1965) The sarcoplasmic reticulum and transverse tubules of the frog's sartorius. J. Cell Biol. 25, 209-231

Somlyo, A.V., Gonzalez-Serratos, H., Shuman, H., McClellan, G., and Somlyo A.P. (1981) Calcium release and ionic changes in the sarcoplasmic reticulum of tetanized muscle: an electron probe study. J. Cell Biol. 90, 577

Troshin, A.S. (1966) Problems of Cell Permeability. (translated by M.G. Hall; Widdas, W.F., ed.), Pergamon: London

Veis, A. (1964) The Macromolecular Chemistry of Gelatin. Academic Press: New York

Zinser, H. (1940) As I Remember Him: The Biography of R.S.. Little Brown: Boston, p. 105

INDEX